WILEY SERIES IN ECOLOGICAL AND APPLIED MICROBIOLOGY

Edited by **Ralph Mitchell**

MICROBIAL LECTINS AND AGGLUTININS: Properties and Biological Activity
David Mirelman   Editor

THERMOPHILES: General, Molecular, and Applied Microbiology
Thomas D. Brock   Editor

INNOVATIVE APPROACHES TO PLANT DISEASE CONTROL
Ilan Chet   Editor

PHAGE ECOLOGY
Sagar M. Goyal, Charles P. Gerba, Gabriel Bitton   Editors

BIOLOGY OF ANAEROBIC MICROORGANISMS
Alexander J. B. Zehnder,   Editor

Phage Ecology

Phage Ecology

edited by

SAGAR M. GOYAL
University of Minnesota
St. Paul, Minnesota

CHARLES P. GERBA
University of Arizona
Tucson, Arizona

GABRIEL BITTON
University of Florida
Gainesville, Florida

A WILEY-INTERSCIENCE PUBLICATION

JOHN WILEY & SONS

New York • Chichester • Brisbane • Toronto • Singapore

Library of Congress Cataloging in Publication Data:

Phage ecology.
   (Wiley series in ecological and applied microbiology)
   "A Wiley-Interscience publication."
   Includes index.
   1. Bacteriophage.   2. Microbial ecology.
I. Goyal, Sagar M., 1944-   . II. Gerba, Charles
P., 1945-   .   III. Bitton, Gabriel.   IV. Series.
[DNLM: 1. Bacteriophages–physiology.   2. Ecology.
QW 161 P532]
QR342.P46   1987   576′.6482   87-10666
ISBN 0-471-82419-4

Printed in the United States of America

10 9 8 7 6 5 4 3 2 1

To Krishna, Peggy, and Nancy

CONTRIBUTORS

GABRIEL BITTON Department of Environmental Engineering Sciences, University of Florida, Gainesville, Florida 32611 USA

ROBERT E. CANNON Department of Biology, University of North Carolina, Greensboro, North Carolina 27412-5001 USA

J. N. COETZEE Department of Medical Microbiology, Institute of Pathology, Box 2034, Pretoria 0001, South Africa

DONNA H. DUCKWORTH Department of Immunology and Medical Microbiology, College of Medicine, University of Florida, Gainesville, Florida 32611 USA

SAMUEL R. FARRAH Department of Microbiology and Cell Science, University of Florida, Gainesville, Florida 32611 USA

K. FURUSE Department of Microbiology, School of Medicine, Tokai University, Bohseidai, Isehara-shi 259-11, Japan

CHARLES P. GERBA Departments of Microbiology and Immunology and Nutrition and Food Science, University of Arizona, Tucson, Arizona 85721 USA

SAGAR M. GOYAL Department of Veterinary Diagnostic Investigation, College of Veterinary Medicine, University of Minnesota, St. Paul, Minnesota 55108 USA

JAMES E. KENNEDY, JR. American Home Food Products, Central Research Laboratory, Milton, Pennsylvania 17847 USA.

LESLEY MANCHESTER Department of Botany, University of Liverpool, P.O. Box 147, Liverpool L69 3BX, England

KARLHEINZ MOEBUS Biologische Anstalt Helgoland, Postfach 850, D–2192, Helgoland, Federal Republic of Germany

A. M. Mortimer   Department of Botany, University of Liverpool, P.O. Box 147, Liverpool L69 3BX, England

Mary Ellen Sanders   Biotechnology Group, Miles Laboratories, Inc., P.O. Box 932, Elkhart, Indiana 46515 USA

S. T. Williams   Department of Botany, University of Liverpool, P.O. Box 147, Liverpool L69 3BX, England

SERIES PREFACE

The Ecological and Applied Microbiology series of monographs and edited
volumes is being produced to facilitate the exchange of information relating
to the microbiology of specific habitats, biochemical processes of importance in microbial ecology, and evolutionary microbiology. The series will
also publish texts in applied microbiology, including biotechnology, medicine, and engineering, and will include such diverse subjects as the biology
of anaerobes and thermophiles, paleomicrobiology, and the importance of
biofilms in process engineering.

During the past decade we have seen dramatic advances in the study of
microbial ecology. It is gratifying that today's microbial ecologists not only
cooperate with colleagues in other disciplines but also study the comparative
biology of different habitats. Modern microbial ecologists, investigating ecosystems, gain insights into previously unknown biochemical processes, comparative ecology, and evolutionary theory. They also isolate new microorganisms with application to medicine, industry, and agriculture.

Applied microbiology has also undergone a revolution in the past decade.
The field of industrial microbiology has been transformed by new techniques
in molecular genetics. Because of these advances, we now have the potential
to utilize microorganisms for industrial processes in ways microbiologists
could not have imagined 20 years ago. At the same time, we face the challenge of determining the consequences of releasing genetically engineered
microorganisms into the natural environment.

New concepts and methods to study this extraordinary range of exciting
problems in microbiology are now available. Young microbiologists are increasingly being trained in ecological theory, mathematics, biochemistry,
and genetics. Barriers between the disciplines essential to the study of modern microbiology are disappearing. It is my hope that this series in Ecological and Applied Microbiology will facilitate the reintegration of microbiology
and stimulate research in the tradition of Louis Pasteur.

Phage plays a central role in the ecology of environments as dissimilar as
the oceans, the food industry, and fermentation processes using modern

techniques in molecular genetics. Yet we know surprisingly little about the distribution and survival strategies of phage in the natural environment. This volume should provide the reader with an understanding of the many dimensions of phage ecology, including its importance in microbial processes in soil and aquatic habitats. New insights are provided into the role of phage as an indicator of public health hazards.

RALPH MITCHELL

Cambridge, Massachusetts
April 1987

PREFACE

Soon after the discovery of phage by F. W. Twort and F. d'Herelle some seventy years ago, there was great hope of using "phage therapy" to treat bacterial diseases. The discovery of antibiotics in the 1940s brought this type of research to a halt. Since then, molecular biologists and geneticists have focused on the use of phage as a convenient tool for examining the processes of nucleic acid replication, genetic recombination, and gene structure. Phage are also commonly used as convenient diagnostic tools for phage-typing of bacteria. Meanwhile, the ecology of phage was left somewhat unexplored. Studies on phage distribution in the environment have demonstrated their occurrence in feces from humans and animals, domestic sewage, seawater, freshwater, soils and food.

Phage Ecology is the first concise compilation of the scientific literature pertaining to the distribution and behavior of bacterial phage in the environment.

After some general considerations pertaining to the distribution of DNA and RNA phage in the environment, the book provides a detailed coverage of phage occurrence and behavior in specific environments, namely seawater, freshwater, soil, water and wastewater treatment plants, and food. The book also examines the problems resulting from phage infection of industrial processes in the dairy, sausage, and wine industries. An aspect of importance to public health microbiologists is the potential use of phage as indicators of fecal contamination. Their potential use as indicators of viral pollution is also discussed in the book. A discussion of the ecology of cyanophage and their possible significance in regulating the growth of cyanobacteria in lakes is also provided.

We hope that *Phage Ecology* will serve as a valuable starting point for those concerned with some of the environmental and practical aspects of phage biology.

This book should serve as a supplementary text in environmental microbiology and environmental virology courses. It should also serve as a useful reference book for microbial ecologists, food microbiologists, and public health microbiologists.

We wish to thank all the authors who have been so willing to contribute their time in the preparation of this volume. Their dedication is greatly appreciated. The secretarial assistance of Mary Thurn, Laurene Rehnblom, and Lezlie Larson is gratefully acknowledged.

SAGAR M. GOYAL
CHARLES P. GERBA
GABRIEL BITTON

St. Paul, Minnesota
Tucson, Arizona
Gainesville, Florida
August 1987

CONTENTS

HISTORY AND BASIC PROPERTIES OF BACTERIAL VIRUSES

DONNA H. DUCKWORTH

Department of Immunology and Medical Microbiology, College of Medicine, University of Florida, Gainesville, Florida 32610

1.1.   THE DISCOVERY OF BACTERIOPHAGE

Bacteriophages were the last of the three major classes of viruses to be discovered. Plant viruses were discovered in 1892 by Ivanowski and in 1897 by Beijerinck, and animal viruses were first recognized in 1902 by Loeffler and Froesch. It was more than another decade before the first publication appeared on what we now know to be bacterial viruses or bacteriophages. Again, two names are associated with this discovery. In the case of animal and plant viruses, however, there is general agreement that the two discoverers should be awarded equal credit, whereas in the case of bacterial viruses there is no agreement as to who first discovered them or even the date of their discovery! The observed phenomenon, spots of lysis on a lawn of bacteria or a vitreous ("glassy") transformation of bacterial colonies, was known for a number of years as the Twort–d'Herelle phenomenon. This harmonious nomenclature belies the depth of the conflict that existed—and still does exist—regarding the discovery of bacteriophage (29,82). The facts go something like this:

F. W. Twort, a pathologist and superintendent of the Brown Institute in London, founded "for the care and treatment of Quadrupeds or Birds useful to man," in attempting to grow vaccinia virus on artificial medium, observed that colonies of a contaminating micrococcus often appeared to be afflicted with some disease (87). "Inoculated agar tubes often showed watery-looking areas, and in cultures that grew micrococci it was found that some of these colonies could not be subcultured, but if kept they became glassy and transparent." Twort found that the agent causing the "glassy transformation" could be filtered without loss of its biological activity and concluded that the cause was an infectious, filterable agent that killed bacteria and multiplied itself in the process. He refrained from concluding that he had discovered bacterial viruses, although he did consider this. He thought it equally possible that he had found a fluid form of life or "an enzyme with the power of growth" (86). Although Twort's results were published in the *Lancet* of December 4, 1915 (86), his paper was virtually unrecognized for more than 5 years.

In the meantime a young bacteriologist had been extremely busy. Felix d'Herelle's story has been reprinted a number of times (23, 29, 82). It is a good story and the ingenuous enthusiasm for science that it exhibits—missing so often in today's scientific papers—makes its further retelling worthwhile. In addition, it has implications for the ecology of bacteriophage that are, even now, not fully understood.

In 1910, I was in Mexico, in the state of Yucatan, when an invasion of locusts occurred; the Indians reported to me that in a certain place the ground was strewn with the corpses of these insects. I went there and collected sick locusts, easily picked out since their principal symptom was an abundant blackish diarrhoea. This malady had not as yet been described, so I studied it. It was a septicemia with intestinal symptoms. It was caused by bacteria, the locust coccobacilli, which were present almost in the pure state in the diarrhoeal liquid. I could start epidemics in columns of healthy insects by dusting cultures of the coccobacillus on plants in front of the advancing columns: the insects infected themselves as they devoured the soiled plants.

During the years which followed, I went from the Argentine to North Africa to spread this illness. In the course of these researches, at various times I noticed an anomaly shown by some cultures of the coccobacillus which intrigued me greatly, although in fact the observation was ordinary enough, so banal indeed that many bacteriologists had certainly made it before on a variety of cultures.

The anomaly consisted of clear spots, quite circular, two or three millimetres in diameter, speckling the cultures grown on agar. I scratched the surface of the agar in these transparent patches, and made slides for the microscope; there was nothing to be seen. I concluded from this and other experiments that the something which caused the formation of the clear spots must be so small as to be filtrable, that is to say able to pass a porcelain filter of the Chamberland type, which will hold back all bacteria.

In March, 1915, during the first World War, a large invasion of locusts appeared in Tunisia, threatening to destroy the harvests which were then so vital; I was given the job of starting an epidemic amongst them. As the result of the infection there was a considerable mortality, and, even more interesting, when the following year all the rest of North Africa was again invaded, Tunisia remained free: the illness had continued to rage amongst those swarms of locusts which had survived long enough to move South at the end of the season, and had brought about their destruction during the winter.

In the course of this campaign, I again observed my clear spots, and before returning to France I stayed for a time at the Institut Pasteur in Tunis, to investigate their significance.

I showed them to Charles Nicolle, then director of the Institute, and he said to me: "That may be the sign of a filtrable virus carried by your coccobacilli, a filtrable virus which is the true pathogenic agent, while the coccobacillus is only a contaminant." So I filtered emulsions of cultures grown on agar and showing the clear spots, and tried to infect healthy locusts with the filtrate, but without result.

On my return to Paris in August 1916, I was asked by Dr. Roux to investigate an epidemic of dysentery which was raging in a cavalry squadron, then resting at

Maisons-Laffitte. I thought that the hypothesis put forward for the locusts' illness might be helpful in understanding human dysentery. I therefore filtered emulsions of the faeces of the sick men, let the filtrates act on cultures of dysentery bacilli and spread them after incubation on nutritive agar in petri dishes: on various occasions I again found my clear spots, but the feeding of these cultures to guinea pigs and rabbits produced no disease.

At this time we often got cases of bacillary dysentery in the hospital of the Institut Pasteur in Paris. I resolved to follow one of these patients through from the moment of admission to the end of convalescence, to see at what time the principle causing appearance of the clear patches first appeared. This is what I did with the first case which was available.

The first day I isolated from the bloody stools a Shiga dysentery bacillus, but the spreading on agar of a broth culture, to which had been added a filtrate from the faeces of the same sick man, gave a normal growth.

The same experiment, repeated on the second and third days, was equally negative. The fourth day, as on the preceding days, I made an emulsion with a few drops of the still bloody stools, and filtered it through a Chamberland candle; and to a broth culture of the dysentery bacillus isolated the first day, I added a drop of the filtrate; then I spread a drop of this mixture on agar. I placed the tube of broth culture and the agar plate in an incubator at 37°. It was the end of the afternoon, in what was then the mortuary, where I had my laboratory.

The next morning, on opening the incubator, I experienced one of those rare moments of intense emotion which reward the research worker for all his pains: at the first glance I saw that the broth culture, which the night before had been very turbid, was perfectly clear: all the bacteria had vanished, they had dissolved away like sugar in water. As for the agar spread, it was devoid of all growth and what caused my emotion was that in a flash I had understood: what causes my clear spots was in fact an invisible microbe, a filtrable virus, but a virus parasitic on bacteria.

Another thought came to me also: "If this is true, the same thing has probably occurred during the night in the sick man, who yesterday was in a serious condition. In his intestine, as in my test-tube, the dysentery bacilli will have dissolved away under the action of their parasite. He should now be cured." I dashed to the hospital. In fact, during the night, his general condition had greatly improved and convalescence was beginning.*

In this report, written by d'Herelle in 1947 (23), one still feels the enthusiasm he must have felt 30 years previously even though his scientific papers from that time are more staid.

d'Herelle's first communication on bacteriophage was in 1917, 2 years after Twort's first paper had been published (86). In this short note in the Comptes Rendus of the Societé de Biologie (20) he described the filterable

---

* From Felix d'Herelle, 1949. The bacteriophage. *In* J. L. Crammer (ed.), Science News *14*:44–47. Reprinted by permission of Penguin Books, Ltd.

agent that killed the dysentery bacillus. It was noted that the agent occurred in the intestines of persons recovering from dysentery but was not present in the acute stage of the disease. Furthermore, because this lytic agent could protect rabbits from a lethal dose of the dysentery bacillus it was suggested that it might operate in the production of immunity to disease.

The effect of this announcement on a world rife with bacterial disease and death therefrom can hardly be overestimated. Humankind's only defense against these diseases had been to produce immunity to them, and this was successful in only a very few cases. Finding the agent of this immunity and a way to produce it outside of the body was surely as exciting then as a cure for cancer would be now. And, in fact, d'Herelle was greatly honored for a number of years for his discovery.

1.1.1. The Dispute Over the Discovery of Phage

In 1921, however, 4 years after d'Herelle's first publication, it was pointed out by Jules Bordet, a Nobel Prize-winning immunologist, that d'Herelle's discovery had been preceded by the "remarkable work," published in 1915, of F. W. Twort (12). Rather than graciously acceding to the possibility that Twort had first discovered phage, d'Herelle campaigned vigorously against this idea, insisting that what Twort had observed had nothing to do with phage. There arose, then, a terrible controversy over the nature of phage and over its first discoverer (29). Bordet gained a strong ally in the young Belgian microbiologist, Andre Gratia, and as time passed they became increasingly adamant in their views—views that went so far as to invoke spontaneous generation to explain the phenomenon of bacteriophage (42)—and in their insistence that Twort and not d'Herelle had discovered phage. d'Herelle's steadfast adherence to his views, including his "remembrance" of the fact that he had actually first observed phage in 1910, seemed to enrage his opponents, with the end result being the accusation that d'Herelle *stole* the idea of phage from Twort (29). One version of the story, favored by Twort's son (A. Twort, personal communication), has d'Herelle hearing Twort lecture in Egypt on the lytic phenomenon he had discovered and going back to Paris to publish his own paper about phage.

There has, however, been no evidence presented in favor of this idea. It seems more likely that d'Herelle in 1910, like the many bacteriologists he mentions, had observed the effects of phage, the small circular clear areas on a lawn of bacteria that are now known as plaques, but had not, at that early date, understood the cause.

1.1.2. Immunology and the Factors that Caused Bordet to Vilify d'Herelle

Why, then, was Bordet, a Nobel-Prize-winning immunologist and pupil of the highly regarded Elie Metchnikoff, so insistent in his support of Twort and so vicious in his abasement of d'Herelle? What is the fairest interpretation of this plexus of petty intrigue?

To fully understand Bordet's adamancy, a brief look at the history of immunology is necessary. It is a field that, because of its complexity, has suffered more than its share of confusion. Jan Klein has said, "The progress of the science [of immunology] has been slow and tortuous. There have been periods during which all the carefully accumulated data seemingly contradicted each other; all was confusion and nobody seemed to know what to do next. There have been blind alleys, irreproducible results, erroneously interpreted data, and misleading experiments. There have been controversies, skirmishes—yes, even battles in which great salvos were fired by the opposing sides" (53).

Great battles were fought over the question of whether immunity was humoral or cellular; other controversies concerned the nature of antibodies and whether these were actually discrete substances. The names associated with these battles—the great generals—include Metchnikoff, Ehrlich, Buchner, von Behring, and yes, . . . Bordet.

Many of the very early studies on the immune system, studies done during the latter part of the nineteenth century, concerned the fate of bacteria in different animals or after contact with various cells, cellular extracts, or exudates. Many of the effects observed were very similar to the effects that can be produced by bacteriophage. Bacteria would, for instance, disappear when injected into an animal's bloodstream, just as they would lyse and disappear when treated with phage. It was also shown that bacteria injected into the peritoneal cavity of a guinea pig "dissolved," and this was termed bacteriolysis (15). It was assumed that this lysis was caused by an immune reaction, but d'Herelle later utilized the peritoneal cavity as a prime source of what he called bacteriophage.

In vitro, the colony-forming ability of some bacteria was reduced by incubation in blood, but this property was lost with time or by heating the blood at 56°C. Metchnikoff, the great Russian scientist who discovered phagocytosis, showed that the heated serum could be reactivated even *in vitro* by adding peritoneal exudate; he, of course, believed it was the cells in the exudate that performed the feat of reactivation. Bordet then showed that normal serum could reactivate heated serum but that heated normal serum could not (15). Thus we have the discovery of what is now called complement, although it is safe to say that Bordet had no idea of what caused its activity then or even 20 years later when he began to attack d'Herelle.

In 1920, Bordet and Ciuca showed that a leukocytic exudate could cause normal bacteria to produce an agent of transmissible lysis (11). This transmissible lysis is what had been for several years known as the phenomenon of d'Herelle, first published in connection with its appearance in recovering dysentery patients, and considered by d'Herelle to be a bacterial virus. But the immunologist, Bordet, had other ideas. Says he (11),

In view of the fact that the stools of patients with dysentery are rich in leucocytes, and that the lysinogenic power is only observed toward the period of convalescence, we have asked ourselves if the phenomenon of d'Herelle is not the result of *a defensive activity of the organism,* and particularly of an activity of the leucocytic exudate. This produces in the bacterium an hereditary nutritive vitiation, consisting in the production by the bacterium of a sort of lytic ferment, which is capable, moreover, of diffusing in the ambient fluid and as a result, reacting in the same fashion on normal bacteria of the same species.

For Bordet the idea of a lytic enzyme that could reproduce itself was much more sensible than a bacterial virus. Although we now think the idea of an enzyme "with the power of growth" is absurd, it should be remembered that at that time—and even for some 30 years after—the best guess as to the molecule that carried genetic information was that it was a protein (52, 69). So a little protein "gene" that also had some normal enzymatic activity was really quite an ingenious, rather than a ridiculous, idea. Bordet apparently thought that because Twort had first suggested it, Twort should receive the credit for the discovery of this "remarkable" idea (12).

Bordet was, in 1921, what could be considered the elder statesman of immunology. Between 1915 and 1917, Ehrlich, von Behring, and Metchnikoff had died. In addition, Bordet had just won the Nobel Prize for his work on complement, peritoneal exudates, and leukocytic exudates, so he was probably feeling very noble (!) when he suggested that Twort deserved the credit for first discovering transmissible lysis of bacteria. And unquestionably Bordet thought the discovery was very "hot."

Scientists had tried for years to rationalize the observations supporting the concept of cellular immunity with the concept of humoral immunity. Metchnikoff had been the most ardent and articulate supporter of cellular immunity, yet his own pupil Bordet discovered at least one humoral component of immunity (15,53). Would it not be satisfying if Bordet could find the link between these seemingly opposing theories? By finding that leukocytes could simulate what would be a humoral bacteriolytic activity, it seems quite possible that Bordet had indeed thought that he had found that "missing link." Had it been true it would have been one of the greatest discoveries of immunology up to that time. And Bordet surely did not want an upstart like d'Herelle to get credit for it. Hence he was probably very relieved to find that Twort had published, 2 years before d'Herelle, the observation of transmissible lysis and an interpretation that fit in very well with his "nutritive vitiation–lytic enzyme" theory.

1.1.3. The Dual Nature of Phage and Its Contribution to the Dispute

There is another factor, however, in the d'Herelle vs. Bordet-Twort fight that must be taken into account: Bacterial viruses definitely have a dual nature—either they can be virulent and replicate with destruction of the host or they can be temperate and exist for generations as part of the host's genetic complement. d'Herelle very accurately delineated the steps in the reproduction of virulent bacteriophage (21). But he did not contemplate the ability of some phages to become a part of their host's genetic complement. When this happens, the bacterium becomes "lysogenic" and may, under certain environmental influences, begin to produce phage in the apparent absence of any infection. Thus in this case, the phage does appear to be a cellular product; this was, in fact, Bordet's view.

1.1.4. Early Views on Lysogenic Bacteria

It had been recognized for some time that some cultures of bacteria could cause the lysis of other cultures. These were termed lysogenic, although the term had a very different meaning then. These lysogenic cultures were often simply mixtures of bacteria and phage in a more or less stable equilibrium. These would be what we now call carrier or pseudolysogenic strains. But there were also undoubtedly true lysogenic strains in which the bacterial genome contains the entire genetic complement of the phage and each isolated colony has the power to produce phage, given the correct environmental stimuli (61,83).

d'Herelle had observed phage-producing cultures but thought they were a symbiotic mixture of virus and host. Others, among whom were a number of Bordet's disciples, thought the ability of a bacterial culture to produce phage meant that phage was not a virus but a harmful pathogenic enzyme produced by the bacteria. In 1925 it was shown that single colony isolates of some strains, subcultured as many as six times serially, could still produce phage. This would not, of course, be possible with carrier or pseudolysogenic strains. McKinley, in Bordet's laboratory, then made antiphage antiserum and finally showed conclusively that lysogeny could be maintained in the absence of free phage. Bail and Bordet both then showed that lysogenic strains could sometimes by made by adding the appropriate "phage." Hence all the particulars of the lysogenic state had been discovered by 1925, but it was far from being understood (61). Had DNA been known then as the carrier of genetic information Bordet might have been able to figure out what was happening, for he wrote, rather prophetically, that the faculty to produce bacteriophage is "inserted into the heredity of the bacterium." By this, though, he apparently only meant that bacteria possess the property to autolyse given the right stimulus, and used this fact to maintain his tiresome denial of the existence of bacterial viruses. As late as 1931 he was still saying, "the invisible virus of d'Herelle does not exist. The intense lytic action, to which the name bacteriophage is given, represents the pathological exaggeration of a function belonging to the physiology of the bacteria (61)."

It was not, in fact, until after the structure of DNA had been published that a true understanding of the ability of some bacteria spontaneously to produce phage was attained. Andre Lwoff, in a series of painstakingly tedious experiments, provided a number of crucial facts that, when the role of DNA was understood, made an accurate understanding of this phenomenon possible.

Lwoff worked with a lysogenic strain of *Bacillus megaterium* and set out to answer the following questions (83): (*a*) Can the ability of lysogenic bacteria to produce phage be perpetuated without intervention of exogenous phage? (*b*) How do lysogenic bacteria liberate the phage they produce? and (*c*) What factors induce the production of phage in a population of lysogenic bacteria?

To Lwoff it seemed evident that the study of mass cultures of lysogenic bacteria, as had been the practice of his predecessors, could furnish only partial and not definitive answers to these questions, and he thought that only the observation of individual bacteria could lead to unambiguous conclusions. He therefore proceeded to cultivate an individual cell of *B. megaterium* in a microdrop of growth medium and watch its division under the microscope. Immediately after the division, one of the two daughter cells was withdrawn from the microdrop and plated on agar to see whether it would give rise to a colony of lysogenic descendant bacteria. As soon as the cell that remained in the culture fluid had once more divided, one of the two new daughter cells was withdrawn and plated on agar, this process being continued for a total of 19 divisions. In the same experiment, samples of the culture medium were also withdrawn from the microdrop and assayed for the presence of any free phage. The results of this experiment were that every colony derived from the cells in the microdrop was lysogenic, and none of the samples of culture fluid contained any infective phage. Lysogeny was thus shown to persist for at least 19 successive divisions in the absence of any exogenous phage.

Lwoff repeated his microscopic study a number of times. On occasion, he observed the spontaneous lysis of the cell in the microdrop. Whenever the culture fluid was assayed for the presence of free phage after such lysis, hundreds of phage were found to be present in the drop. He could conclude, therefore, that lysogenic bacteria liberate their phage by lysis, each lysing cell yielding a burst of many phage particles.

Lwoff's experiments allowed him to describe lysogeny in the terms in which the phenomenon is now understood. We now know that each bacterium of a lysogenic strain harbors a noninfective structure, the prophage, which gives the cell the ability to produce infective phage *without the intervention of exogenous phage particles* (83). We also know that in a small fraction of a population of growing lysogenic bacteria the prophage will become "induced" to produce a burst of infective phage. This prophage induction leads to death and ultimate lysis of the cell. Later it was shown that ultraviolet light could induce phage production in lysogenic cells and that phage produced in this way could convert a nonlysogenic bacterium into one that could later produce phage. Additionally, it was shown that a temperate (having the ability to make cells lysogenic) phage could be mutated so that it became virulent. It was later recognized that the prophage was a piece of DNA; Campbell (49), in his prophage integration model in 1962, explained how this could be. This model, of course, ended for all time the argument about the nature of "phage," if not the argument about who discovered phage.

1.1.5. The Search for the Curative Powers of Phage

Whereas Bordet may have thought that he had found the "missing link" between cellular and humoral immunity, d'Herelle surely thought he had found a way to control bacterial disease. He was uniquely attuned to the possibilities of biological control and was perhaps the first proponent of this method of pest control (23). He was apparently successful in controlling locusts with a locust bacterial disease, so very naturally he thought of using bacteriophage for the control of bacterial disease. In the preantibiotic era, it was certainly a welcome idea. And, according to d'Herelle, it was an idea that was successful. His book, *The Bacteriophage and its Behavior,* published in 1925 (22), contains almost 200 pages detailing the use of bacteriophage in both the prevention and treatment of disease. Here he describes the behavior of bacteriophage in dysentery, streptococcal and staphylococcal infections, typhoid fever, and other "colon bacillus infections." He talks about how he stopped epidemics by using phage—epidemics of avian typhosis, hemorrhagic septicemia of the buffalo, bubonic plague, dysentery, and flacherie of the silk worm. He then describes how "immunization" against a number of diseases may be accomplished with phage, and how a formidable list of bacterial diseases—dysentery, typhoid, plague, wound infections, and streptococcal and staphylococcal infections—can be cured with phage. Typical of his observations are the following dysentery cases (22):

Robert K. . . . (eleven years). This was a case of bacillary dysentery of moderate severity with 5 to 7 bloody stools a day. August 1. The stool examination showed: *B. dysenteriae Shiga* present. The intestinal bacteriophage with virulences as follows: *B. coli* ++, Shiga 0, Flexner 0, Hiss 0. August 2. At 10 o'clock in the morning the patient ingested 2 cc. of a Shiga-bacteriophage. This filtrate had been held for thirty-five days. During the afternoon of this day there were 3 bloody stools, in the evening there was one stool and that was free of blood. August 3. During this day there was only the one formed stool. Examination showed: *B. dysenteriae Shiga* absent. The intestinal bacteriophage with virulences as follows: *B. coli* ++++, Shiga ++++, Flexner +++, His +++. August 8. The intestinal bacteriophage was active as follows: *B. coli* +++, Shiga +, Flexner 0, Hiss +. August 9. The patient was discharged from the hospital.

Robert D. . . . (twelve years). This patient had a very severe dysentery, with vomiting, cold sweats, chilling of the extremities, and involuntary and uncountable stools. September 8. The stools could not be counted. They were fetid, purulent, and streaked with blood. Examination showed: *B. dysenteriae Shiga* present; about 1 out of every 10 colonies on the plates was the dysentery bacillus. The intestinal bacteriophage showed no virulence for *B. coli,* or for the Shiga, Flexner, or Hiss organisms. September 9. Two cubic centimeters of a suspension of Shiga-bacteriophage were ingested at 11 o'clock. This suspension was three and one-half months old. During the afternoon and the night the stools became less numerous but continued bloody. September 10. There were 6 fluid stools, without blood. Examination showed: *B. dysenteriae Shiga,* not present. Intestinal bacteriophage active as follows: *B. coli* + + + +, Shiga + + + +, Flexner + + + +, Hiss + + +. September 11. There were 2 normal, formed stools. September 20. The patient was discharged from the hospital.

No controls are mentioned, although it could be argued that it would not be ethically appropriate to withold treatment for this purpose. But he also discussed a number of experiments with chickens, and again control experiments were not performed. The following experiments (22) are typical:

Experiment III. This experiment was conducted at Provins, with the aid of M. Sorriau, D.V.M., in an important poultry-yard where typhosis was present in endemic form. For several months the daily mortality had been 2 or 3 fowls. On January 25 the 225 survivors were immunized [author's note: this was with bacteriophage]. The epizootic immediately and permanently disappeared from the date of the immunization.

Experiment IV. Performed at Rouillac, Charente, with the assistance of M. Chollet, D.V.M. On December 15, 100 fowls were immunized [author's note: this again refers to use of bacteriophage] in a poultry-yard where typhosis had appeared about ten days previously. The daily mortality had been from 4 to 6 animals. With the immunization there was an immediate and permanent cessation of the epizootic. Typhosis continued to prevail on all the neighboring farms. Among the 100 chickens inoculated, about 12 were already affected. Of these only 2 died, 2 and 3 hours after the injection.

He does mention that the disease continued on neighboring farms. Presumably, this is evidence that his "immunization" using phage was working. But injecting only half the chickens on a single farm would have been more appropriate.

But controls or no, d'Herelle's work seems to have excited a number of people, resulting in trials of phage therapy at a number of places throughout the world. Even the Department of Bacteriology and Experimental Pathology at Stanford University, under the direction of Professor Edwin Schultz, set up a laboratory to determine whether or not these bacterial predators might be used therapeutically. They "advertised to physicians that the laboratory would try to adapt a bacteriophage—of which there was a large collection—to any bacterium isolated from a patient. This was done at a cost to the laboratory, then about $2.00. The physician was expected to report briefly on the use of the phage and subsequent fate of the patient. After several years, the project was abandoned for lack of adequate reporting and because of the advent of sulfonamides'' (71). Better results were reported from Brazil, however, where phage were employed to treat bacillary dysentery. One laboratory reported (22):

> Up to the present, we have prepared about 10,000 ampoules of the bacteriophage which have been distributed to several hospitals and to a number of physicians in the different States of Brazil.

> It is possible to find only a few rare cases in which the administration of the bacteriophage has not been followed by benefit to the patient. Of these cases, the first was a patient infected with the Hiss strain of the bacillus, perfectly susceptible *in vitro* to the bacteriophage, but in whom the bacteriophage failed, for reasons not yet discovered, to exert its action in the intestinal tract. The second was an infant who died of a very severe infection caused by the Shiga bacillus, in spite of the fact that both serum and bacteriophage treatment were used.

> The therapeutic effect—the sudden change in the condition of the patient—consequent to the administration of the bacteriophage has astonished all of the physicians who have worked with it. A few hours after the administration of the first dose improvement has been noted, and the rapidity of the recovery has been a cause of amazement. In examining the collected data pertaining to the patients treated it appears that the majority of them had first received all known treatments, and had been given the bacteriophage as a last resort. We are, therefore, absolutely convinced that it has saved the lives of a great many patients.

Independently it was observed that (22)

> . . . the anti-dysentery bacteriophage constitutes a specific treatment for the bacillary dysenteries, whether they be caused by the Shiga bacillus or by the Paradysentery organisms.

> Without causing general reactions, the ingestion of this principle modifies completely, within a very short space of time, the evolution of the dysenteric syndrome. Recovery, in a clinical and in a bacteriological sense, is complete.

> By the use of the bacteriophage the development of carriers is avoided, for dysentery bacilli are never found in the excreta of patients who have been treated by the ingestion of the bacteriophage.

Despite these glowing reports and the absence of any definitive studies showing that phage were *not* therapeutic, the discovery of antibiotics in the 1930s led to an almost complete cessation of attempts to use phage in the prevention or treatment of disease. This rather hasty withdrawal from the medical arena would lead one to believe that perhaps phage had not been so wonderful after all—at least for curing disease.

1.1.6. The Finding that Phage Can Cause Disease

The dreams and subsequent disillusionment of those hoping to cure disease with phage have been fictionalized by Sinclair Lewis in his *Arrowsmith*. Dr. Martin Arrowsmith, disappointed because he was unable to stop an epidemic of bubonic plague with an antiplague bacteriophage, charted the course for bacteriophage research that it ultimately took. Arrowsmith resolved to study the fundamental nature of bacteriophage, instead of its practical applications (82). The intense study of the basic properties of phage—studies that led to the establishment of the science of molecular biology—began after World War II. But it would be unfair to say that no one was interested in the basic properties of phage before that time. In fact, one fairly "cruel" fact for those believing that phage had medicinal value came out of a basic study utilizing phage that can infect the diphtheria bacillus.

It is now known that the virulence factors—toxins, adherence factors, and anti-immunity factors—for many diseases are coded for by extrachromosomal genetic elements, either plasmids or temperate phage. This very unexpected fact was first illustrated in the case of diphtheria by the discovery that the genetic information for the diphtheria toxin was carried by a phage. This fact, now understood in terms of very sophisticated molecular biology, actually had its roots in an observation made by d'Herelle, who isolated "two strains of bacteriophage, *active for only atoxic strains of Bact. diphtheriae* [italics mine]." (21). This observation actually indicates what was concluded some 30 years later—that the toxin gene is a phage gene—because lysogenized bacteria are "immune" to the phage they carry (see Section 1.2.9). Therefore, had the phage d'Herelle isolated been ones that carried the toxin gene and converted non-toxin-producing strains into toxin producers, the toxin producers would have been immune to the phage. d'Herelle did not further study the problem but did note that it "certainly offers much of interest." Other people thought so, too, with workers at Brown, Yale, and Columbia Universities all studying bacteriophage of *Corynebacterium diphtheriae* (10,76,84). Many observations pointed to the possibility of pathogenic or virulent strains of the bacteria being lysogenic, but this was never directly shown.

Then in 1951, Victor Freeman reported that avirulent strains of *C. diphtheriae*, when infected with bacteriophage, yielded virulent *C. diphtheriae* "mutants" (36). Just exactly what made Freeman perform the experiments he did is not entirely clear; it had been suggested that phage might be useful in understanding some of the differences between avirulent and virulent strains of *C. diphtheria*, and Freeman was likely just trying to do this. But what he found was "exposure of the phage-susceptible avirulent cultures to phage B resulted in the production of virulent (pathogenic) lysogenic strains of *C. diphtheria*." He initially thought "the most likely hypothesis to explain the phenomenon of virulence conversion is the spontaneous development of toxigenic mutants, with selection by phage lysis." Later, however, he concluded (37):

> It would seem logical that the simultaneous acquisition of lysogenicity and virulence in the same bacterial cell is a related, rather than coincidental phenomenon. The change to toxin production might well be interpreted as being due directly to the acquired lysogenicity. Conceivably, the bacteriophage may make possible the toxin production through some as yet undetermined association with the metabolic processes of the bacterial cell.

Hypotheses for the association of toxin production and virulence with the lysogenic state included activation of a bacterial toxin gene or transduction of a toxin gene from one cell to another. It was conclusively shown in 1971, however, that the toxin gene was an integral part of the phage. The final proof was provided by Uchida et al. (88), who isolated a "CRM"-producing mutant of the toxin-inducing phage. The mutant did not cause cells to produce any active toxin, but *C. diphtheriae* lysogenized with the mutant phage did produce a protein that could cross-react immunologically with the toxin protein in that it could be precipitated by either horse or rabbit antitoxic sera. This production of a mutant protein, or cross-reacting material (CRM), by a mutant phage established without a doubt that the structural gene for the toxin was carried by the phage.

This was a fairly cruel twist for those who had desired to show that phage could cure disease. Rather than having therapeutic properties, the phage was, in the case of diphtheria, shown actually to cause the disease.

1.1.7. Basic Phage Research

It would be very difficult to say with certainty who first considered using phage to study the basic properties of living organisms. The thought could hardly have escaped d'Herelle, although for him the practical uses of phage held forth more promise.

Muller, the famous fruit fly geneticist, probably earliest and best articulated in 1921 what some 20 years later came to be an obsession among a fairly large group of scientists—Muller noted that both genes and "d'Herelle bodies" are heritable and mutable and suggested that although "it would be very rash to call these bodies genes, we must confess that there is no distinction known between the genes, and them. Hence we cannot categorically deny that perhaps we may be able to grind genes in a mortar and cook them in a beaker after all" (52).

F. M. Burnet, working in the 1930s, was one of the first individuals actually to research the basic biochemical properties of phage, and he was followed closely by M. Schlesinger. Burnet showed that there were many different types of phage, not just one, and also began experiments on how the phage grow, showing that phage first accumulate inside the cell and are released in one "burst." Schlesinger measured the approximate dimensions of phage particles and their mass. He also showed the chemical composition to be about 50% protein and 50% DNA—a composition very close to that of chromosomes. Emory Ellis was another of the early pioneers and he also studied the kinetics of phage reproduction (82), confirming and extending the results of Burnet.

Max Delbruck and Salvador Luria must, however, be credited with laying the foundations of modern phage research. Although they were not actually interested in "cooking" genes in a beaker, they were the major originators and proponents of the idea that phage could be used to understand the basic coding properties of the gene, and it was the vision and enthusiasm of these two men that raised the desire to do basic research with phage to a fever pitch.

Delbruck was an immigrant in two senses: he came from an intellectually aristocratic family in Germany to this country in 1937. He had been trained as a quantum physicist but his interest in the physical properties of the gene was evidenced quite early and he eventually "migrated" from physics to biology, hoping to apply his physical knowledge to biological problems. He did his first work on phage at California Institute of Technology with Emory Ellis, and it is they who definitively established the "one-step" pattern of phage growth. In 1940 Delbruck moved to Vanderbilt University in Tennessee (52).

In 1940 Luria and Delbruck met. Luria, a physician, had done some phage research in his native Italy and was captivated by Delbruck's idea of using phage to understand the nature of the gene. These two men, then, along with Alfred Hershey and Thomas Anderson, formed the nucleus of what came to be called the "phage group"—a spiritual rather than a physical group of scientists, scattered from New York to California, whose one abiding notion was to figure out how life "worked," using phage as their prototype (52).

New disciples were garnered, beginning in 1945, by Delbruck's course on bateriophage given every summer at the Cold Spring Harbor Laboratories on Long Island. It was here, too, that the very important decision to concentrate on only seven different *E. coli* phages took place. The phage group and all those who joined them were characterized by openness and candor, and they lacked a concern of many scientists today—that of individual priority. This, in combination with the obvious import and interest of their mission, is surely the secret of their success (52).

And they *were* successful. In the course of just a very few years, the basics of phage composition and structure had been determined, as had essential facts regarding their replication. It was shown, in the one-step growth experiment of Ellis and Delbruck, that phage growth is characterized by a latent period (during which time the number of plaque-forming units in a culture is constant), a rise period, and a sudden burst. It was also shown, by breaking open infected cells and counting intracellular viruses, that phage undergo what is called an eclipse, a period of time during which not even the infecting phage can be found. This was shown by A. H. Doerman, who also showed that during the latent period, when the number of plaque-forming units in a culture is not changing (because the infected cells are not multiplying), the number of phage inside each bacterium is increasing rapidly. The basis for the unusual eclipse (unknown in any other "life" form) was elucidated by the famous experiment of Hershey and Chase.

The historical background to this experiment is important. It had been thought for nearly half a century that the biochemical substance that carried genetic information would be proved to be protein. Proteins, being made of 20 possible subunits, were the only structure thought to be able to generate enough variability to encode genetic information. Nucleic acids, made up of only four possible subunits, even though known to be found everywhere genes were, were thought more likely to provide some structural, scaffolding role. Ellis and Delbruck, in their report on the one-step growth experiment (33) stated (in a way highly reminiscent of Twort) that "certain large protein molecules (viruses) possess the property of multiplying within living organisms." No data existed to say that genes were proteins, and in fact, very good data had been published that strongly implicated DNA as the genetic substance. These were the experiments of Avery, McCarthy, and McCloud, who showed that bacteria could be "transformed" with a substance that was almost pure DNA from a rough form—a form that does *not* produce a polysaccharide capsule—to a smooth, polysaccharide-producing form. Even when the transforming substance was purified many times over and shown to contain virtually no protein, people were still not struck by the importance of DNA. Even Luria, as late as 1951, suggested that protein might be the genetic substance of the phage (52). But the Hershey–Chase experiment was to change all that.

Alfred Hershey and Martha Chase are credited with being the first to demonstrate that the nucleic acid of a virus contains the genetic information necessary for virus replication. They showed that for phage T2, at least, the protein coat was dispensable, confirming what Avery, McCarthy, and Mc-Cloud had shown almost a decade previously, that nucleic acid functioned as a carrier of genetic information. The Hershey–Chase experiment was very simple in design and utilized a common kitchen aid for its implementation. The protein coat of the phage was labeled with radioactive ^{35}S, an element found only in proteins, and the nucleic acid with ^{32}P, an element unique to nucleic acid. The differently labeled phage were then allowed to adsorb to their host, *E. coli*. At various times after this, the phage–bacterial complexes were subjected to a high shearing force in a Waring Blender—this has the effect of stripping off anything protruding from the cell surface—and centrifuged. The amount of each kind of radioactive label removed by the shearing force of the blender was then determined. From the data, which Hershey and Chase interpreted as indicating that most of the ^{35}S or protein could be removed from the cells without inhibiting their ability to replicate the virus, they concluded that the protein coat of the virus "probably has no function in the growth of the intracellular phage" and "DNA has *some* function" (50). From this rather modest statement the scientific world concluded that they meant that DNA was the carrier of genetic information and the only component necessary for virus replication (98). The data themselves were not actually terribly convincing, showing, in fact, that only 80% of the protein could be removed and that as much as 20% of the nucleic acid could be removed (50), but it was an experiment the world had been waiting for. Watson's and Crick's newly proposed structure for DNA (92), which allowed an explanation of how DNA could replicate and also carry genetic information, made possible a belief in the popular interpretation of the data of Hershey and Chase, that only the nucleic acid was necessary for viral replication.

We don't know exactly what made Hershey design the Hershey–Chase experiment. Hershey cites experiments on isotope transfer from parental to offspring phage (48). Also instrumental were the findings of Anderson and of Roger Herriott that phage could be osmotically shocked to separate the DNA from the protein and that the protein coats of T2 actually had a different biological activity, distinguishable from the activity of the DNA. In fact, it may have been a communication from Herriott in which he suggested that T2 was a syringe filled with "transforming principles" that led Hershey to his experiment. Whether this idea was to be taken seriously or not is not known. Anderson remembers himself, Hershey, and "possibly Herriott . . . discussing the wildly comical possibility that only the viral DNA finds its way into the host cell, acting there like a transforming principle in altering the synthetic processes of the cell" (2). But the "joke" turned out to be not so ridiculous after all, and for this finding Hershey was awarded, along with Delbruck and Luria, the Nobel Prize in 1968.

1.1.8. The Enzymology of Virus-Infected Bacteria

Absent from Stockholm in that year, though, was an individual who had surely done as much to unravel the basic facts of phage replication as had the three prize winners. Seymour Cohen was a biochemist who had studied for his Ph.D. at Columbia University. His mentor was Erwin Chargaff, who had first observed that, in DNA, the amount of adenine (A) was equal to the amount of thymine (T), and the amount of guanine (G) was equal to the amount of cytosine (C). As a postdoctoral fellow Cohen worked with Wendell Stanley on the structure of tobacco mosaic virus. He began to study nucleic acids before their importance in replication was known—a mountain to be climbed because it was there—and chose phage as a convenient vehicle. From the beginning, his orientation was different from that of the phage group; they were interested in phage almost exclusively as an agent of biological information, whereas Cohen looked at the virus-infected cell as a metabolic unit that could be characterized (52).

Cohen started his phage work measuring the most basic metabolic characteristics of infected cells. Studying initially the respiration and energy metabolism of phage-infected cells, Cohen showed that the rate of respiration that prevailed immediately before infection was maintained without change during infection and did not increase exponentially, as did the rate of respiration in growing cells. He concluded that although maintaining their energy metabolism intact (as common sense would dictate) infected cells had lost the ability to synthesize the cellular enzymes involved in respiration. He also observed that the amount of RNA in the infected cell did not change dramatically but that the rate of DNA synthesis was increased some 5- to 10-fold. This DNA synthesis was shown to precede the appearance of active intracellular phage. Cohen notes, "This result suggested that phage synthesis involved a stepwise synthesis and assembly of virus components to form virus from materials such as DNA, itself inactive as free phage" (16). Phage multiplication was, then, totally unlike cell multiplication, in which a fully assembled cell never disappears.

Cohen also showed that the rate of protein synthesis in infected cells stayed approximately at its preinfection level. In 1947 he showed that inhibition of protein synthesis (using 5-methyltryptophan) prevented the synthesis of phage DNA. With the newly developed tracer techniques using radioisotopes, Cohen also was one of the first to detect the synthesis of what we now know to be messenger RNA (mRNA), although, as he ruefully notes, he was unable to prove that the ^{32}P seen in the RNA fraction really represented RNA; instead, he suggested that it was only a contaminant, thereby missing the existence of mRNA (16).

In 1952 at a famous phage meeting at the Abbaye de Royaumont near Paris, Cohen presented perhaps his greatest discovery—that the DNA of the T-even phages contained a pyrimidine that was not present in uninfected cells. Hydroxymethylcytosine (H) replaced cytosine, so that rather than having a DNA composed of A, T, G, C, as all other previously characterized DNA was, T-even phage DNA was composed of A, T, G, and H (16). This led to the discovery that the T-even phage DNA was highly glucosylated and, in addition, led Cohen to the discovery of virus-induced enzymes.

In looking to see if hydroxymethylcytosine would replace the requirement of a thymine auxotroph for thymine (thymine being a methylated pyrimidine and hence closely related in structure to hydroxymethylcytosine), Cohen also made the chance discovery that after phage infection, cells previously unable to synthesize thymine could now do so. This result suggested to him the idea of virus-induced acquisition of function, a totally novel idea in 1953.

Cohen then searched for and discovered the enzyme responsible for the synthesis of hydroxymethylcytosine and showed, in an extremely rigorous fashion, that the enzyme was synthesized *de novo* after infection: it was not injected into cells by the phage, nor was it present in an inactive state before infection. He also showed that the synthesis of thymine after phage infection was accomplished by a completely new protein not present in the cell before infection (16).

This work was the catalyst for a virtual explosion of work on the enzymology of infected cells and led to the discovery of dozens of induced enzymes and to a search for the genes controlling their synthesis. Enzymes such as dCTPase, glucosylase, dCMP deaminase, deoxynucleotide kinase, dihydrofolate reductase, DNA polymerase, and DNA ligase were only a few of the enzymes discovered to be induced by the T-even phages. Efforts to prove that the phage actually carried the genetic information for the synthesis of these activities led to the discovery of the amber and temperature-sensitive mutants, mutants whose conditional lethality allows them to be grown under certain conditions and then studied under other conditions for their adverse effects on the metabolism of the infected cell. It also allowed for the concept of early versus late protein synthesis and the establishment of the steps in the very orderly process of virus development.

Hence it is fair to say that although Delbruck, Hershey, and Luria provided the system, breakthroughs made by Cohen greatly accelerated advancements in the field and were instrumental in making the study of phage so productive.

And the study of phage, no one could deny, was extraordinarily productive. Besides important discoveries relating to virology and biochemistry, some of the most basic discoveries of modern molecular biology were made using phage. These include, in addition to the fact that DNA is the carrier of genetic information, the finding of the intermediate between DNA and protein, messenger RNA (44), as well as some of the basic properties of genes and the genetic code. In experiments with rII mutants of T2, Benzer defined various properties of a "gene" with the terms muton, cistron, and recon (5). Through a set of ingenious experiments, Crick and his co-workers showed that the genetic code is read from a fixed starting point in triplets and that it is degenerate; that is, one amino acid can be coded for by more than one triplet (18). The mechanism of action of many mutagens was shown with phage (38,83), and the colinearity of DNA and protein was confirmed by experiments utilizing phage (72). Also elucidated by experiments with phage was the fact that DNA is synthesized discontinuously, in "Okazaki" fragments, and that these are primed by RNA (14,68). The power of using conditional lethal mutants in deciphering gene–protein relationships was also first shown with phage (34,85,97). The fact that specific subunits of RNA polymerase controlled transcription initiation was discovered with phage T4, and the observation that some *B. subtilis* phage cannot infect their host if the bacterium has begun to sporulate led to the discovery of the role of sigma factors in sporulation (57). This in turn led to the realization that these subunits of RNA polymerase often control the sequence of events in developing systems. The finding that Mu phage inserts itself at any sequence in the DNA, causing mutations at virtually any site (51), previewed the discovery of transposons and insertion sequences.

No discovery within this century, however, with the exception of the discovery of atomic energy, has wider implications than another discovery made with bacteriophage. This is the discovery of restriction endonucleases (3; see also Section 1.2.9.). Resulting from a study of a seemingly obscure phenomenon—that phage sometimes grew better in one host than another— these enzymes, with their ability to recognize and cut specific nucleotide sequences, form one cornerstone of the whole new field of genetic engineering (66,77), a discipline that is dramatically changing the face of science and medicine, if not the face of the world. Things such as cloned human genes for the detection and correction of genetic abnormalities are no longer a dream but a reality, as are recombinant viral vaccines, whereby a whole host of diseases may be eradicated with one wave of the inoculating needle. In addition, genetic engineering has provided dramatic new ways of studying the way life "works," so that the results of the original discovery will exponentially increase for some time to come—and d'Herelle's dream of curing diseases with phage may come to pass after all.

1.2. THE REPLICATION OF BACTERIOPHAGE

Although viral reproduction is frequently called "growth," it is fundamentally different from cell growth, its principles being, in fact, more closely analogous to the principles of automobile manufacture than to the principles of cellular reproduction. Doerman was one of the first to remark on this fundamental difference when he observed that the infecting phage particles became "eclipsed" shortly after infection. Said he, "One must therefore conclude that the phages (of the T-even series) do not multiply by binary fission and further, that the first completed phage particles are not themselves the parents of those viruses which appear later within the same cell" (25). Instead, the virus can be thought of as carrying within it a blueprint that turns the infected cell into a little "factory" for the production of more viruses. All the parts are made assembly line fashion, then put together to form the new virus particles. The process is stepwise and carefully controlled and, in general, follows the plan presented in Table 1.1.

The biggest difference in the replication of bacteriophage and animal viruses, aside from differences in the way prokaryotes and eukaryotes synthesize proteins, seems to be at step 2. With animal viruses, the whole virus is often taken into the cell and the nucleic acid "uncoated" therein (60). With bacterial viruses, this separation of nucleic acid and protein always takes place at the cell surface and is commonly referred to as nucleic acid injection, although as indicated later, this may not be the most appropriate term.

An overview of each of the steps in bacteriophage replication follows.

1.2.1. Adsorption

The adsorption process alone is quite complex, being composed of three stages: (*a*) initial contact, (*b*) reversible binding, and (*c*) irreversible binding.

TABLE 1.1. Steps in the Process of Viral Replication

1. Adsorption
2. Separation of the nucleic acid from the protein coat
3. Expression and replication of the nucleic acid
   a. Pre-early mRNA and protein synthesis
   b. Early mRNA and protein synthesis
   c. Nucleic acid replication
   d. Late mRNA and protein synthesis
4. Assembly of new virus particles
5. Release of new virus

The first stage is accomplished, for freely moving cells in a liquid medium, by the mobility of the cells and the diffusion and Brownian or random movement of the virus. The reversible binding step may include an association of any phage structure with the cell surface, with electrostatic forces acting to implement a nonspecific attachment. This nonspecific association is followed by a specific, reversible interaction with a defined structure on the cell surface. This defined structure may or may not be identical to the receptor for irreversible binding.

The number and type of receptors on the bacterial surface vary greatly but often the molecules recognized by the phage are essential components of the cell surface. Thus although it is possible for bacteria to mutate and become phage resistant by virtue of losing their receptors, it is often a disadvantage for them to do so. Lambda phage, for instance, attaches to the maltose receptor, an essential part of the maltose transport system (49). Another phage, BF-23, attaches to the vitamin B_{12} receptor, a protein that functions in transporting this essential vitamin into the cell (24). Phage T5 uses a part of the very critical iron-scavenging system of the bacteria as its adsorption site (13), binding to the *FhuA* (formerly *TonB*) protein, a receptor for an iron-scavenging protein.

Essential cell surface components are not the only virus attachment sites, however. In fact, virtually any component on the surface is probably used by one virus or another as its specific attachment site (56). Two types of phage, for instance, attach to the pili that male (F^+) or genetic-donor bacteria use to transfer genetic information to other bacteria. One type of these phage, the MS-2-like phage which contain RNA as their genetic information, attach to the sides of these pili on which there are many adsorption sites. The other type of male-specific phage is a filamentous DNA-containing phage that went undetected for years because its morphology was so similar to the pili themselves. These phage attach to the distal end of the pilus, on which there is a site for one phage only. A different group of phage attach to the flagella of motile bacteria and "crawl" down to the cell surface (56,60).

Carbohydrates can also serve as receptors, with T5 coliphage binding reversibly to polymannose 0-antigens (46) before binding irreversibly to the *FhuA* protein (13), and the T-even phage binding to the *E. coli* LPS (41). Other phage attach to the capsule surrounding the bacteria and must hydrolyze it before their nucleic acid can enter the cell (99).

For those phage with long tail fibers, it appears that reversible attachment involves an interaction between the distal end of the tail fiber and the cell surface. In both T4 and T5, which attach to cells reversibly by means of long tail fibers, this reversible attachment allows the phage to probe the surface for a site of irreversible attachment (41,47). With T5 the site on the phage that irreversibly binds the cell is not on the tail fibers or even at the tip of the tail but at a site slightly above the conical end of the tail. This is thought to allow the actual end of the phage tail to be placed in a favorable spot for the DNA to enter the cell (47).

In T4, after the tail fibers reversibly attach and the site for irreversible attachment is "found," the six short tail pins that are part of the phage base plate then make contact with the host surface and firmly anchor the phage tail to it. This attachment of the tail pins to the surface renders the adsorption irreversible. The conformation of the hexagonal tail or base plate (an extremely complex structure consisting of at least 16 different proteins) then changes from a hexagon to a "star" form, making an opening through which the tail tube extrudes. Following this, the tail sheath contracts, driving the tail tube or core through the cell wall and positioning the end of the core (through which the DNA will pass) in juxtaposition to the cell membrane (41). Interestingly, the tail core apparently does not penetrate the cell membrane. This was shown by the very beautiful electron microscopic study of T4 phage infection done by Simon and Anderson (75). The DNA is thus actually deposited not into the cytoplasm but right outside of it.

Ever since Roger Herriott first suggested that perhaps the phage coat was acting like a hypodermic syringe to inject transforming principles into the cell (2), people have referred to the process by which phage DNA enters bacterial cells as injection. It has been assumed that this injection is caused by contraction of the phage tails. How phage with no tails or those with noncontractile tails operate was never explained. For this reason, and because the term injection implies an external force to reduce the head volume, something for which there is no evidence at all, many people no longer like to use the term injection. However, inaccurate or not, it is a convenient term and will likely creep in—inadvertently—in the following pages.

1.2.2. Separation of Nucleic Acid and Protein

Rather than a forcible injection of nucleic acid into the cytoplasm, it appears that irreversible attachment of the phage to the bacterium triggers a change of conformation in the phage coat that allows the DNA to be released at a site where it may be easily taken into the cell. The evidence that the DNA is deposited outside of the cytoplasm, in T4 at least, is quite strong. The

morphological evidence, mentioned above, is supported by two other experimental findings: (*a*) naked T4 DNA will not normally "infect" spheroplasts
but can be made to enter spheroplasts of *E. coli* by urea-treated "helper"
phage (see below); and (*b*) DNA from a superinfecting homologous phage
(i.e., a T4 phage that attaches to a cell that is already infected by T4) is
ejected from the phage, but cannot enter the cell. It is found degraded in the
medium (1). This "superinfection exclusion" depends on the synthesis of a
new phage-coded protein (90), but how the protein acts to make the membrane impermeable to the phage DNA is not known.

Some of the earliest experiments examining the process of phage DNA
entry into bacteria were those using urea-treated phage. If a suspension of
T4 is treated with 8 *M* urea, its capacity to infect normal *E. coli* is lost and its
DNA, unlike that of whole phage, becomes susceptible to degradation by
DNAase. The phage's proteins can likewise by degraded by trypsin after
urea treatment. These urea-treated phage or pi particles, however, retain an
ability to cause an infection of *E. coli* if the outer layers of the bacterial cell
wall are removed prior to the addition of the pi particles. Normal T4 phage
result from such an infection. Phage not treated with urea are not able to
infect spheroplasts because the phage receptors are lost during the preparation of the spheroplasts. It can also be shown that isolated DNA, freed from
all phage structural proteins, can enter *E. coli* spheroplasts if the pi particles
are added at the same time (91). This has been interpreted to mean that some
component of the phage coat that is exposed either by the contraction and
change of conformation of the tail that occurs during the course of a normal
infection, or by the urea treatment, causes changes in the cell's membrane,
allowing the DNA to enter. Soon after infection, however, a new protein is
made that renders any additional phage DNA unable to go through the cell
membrane.

Very little is known about how the phage DNA actually crosses the cytoplasmic membrane. During bacterial transformation, in which isolated DNA
is taken up by certain species of bacteria, the cell membranes must be
modified in some way before the cell is "competent" to take up the DNA.
This is apparently done by the introduction of new proteins into the cell's
membranes (100). Although they have never been identified, phage coats
may contain such membrane-altering proteins. One group of phage, the T5
and BF-23 group, causes the infected cell to synthesize some new proteins
before the whole DNA molecule can enter the cell (64). These new proteins
do appear to affect the cell membrane (8,30,80), and it may be that a study of
these proteins will help to elucidate the problem of DNA entry into cells.

Based on work with ϕX174 phage, Arthur Kornberg hypothesized the
existence of what he called pilot proteins. These were phage-adsorption
proteins that were also attached to the viral nucleic acid, and were thought
to guide the nucleic acid to its appropriate site for replication within the cell.
The hypothesis fit very nicely with what was known about the infection
process of ϕX-174, the filamentous phage, and the MS-2-like phage. The H

protein of ϕX-174, for instance, is the capsid protein of the phage which attaches irreversibly to the bacterium. In addition, it is bound to ϕX's single strand of DNA and is required for the DNA to replicate. Evidence to confirm the existence of pilot proteins in all bacteriophages has not been forthcoming, however (54).

It has been suggested that in T4 the product of gene 2 is a pilot protein because of its association with the nucleic acid and also the capsid. When phage mutants are produced that have a defective gene 2 protein, they cannot replicate because their DNA is broken down to acid-soluble fragments as soon as it enters the cytoplasm. But if these mutants infect bacteria that have no exonuclease V (recBC$^-$), they can replicate (41). So whether gene 2 is a true pilot protein or merely serves to protect the DNA from exonucleolytic digestion is not clear.

The process of DNA uptake, in some phage at least, is an energy-requiring one. In T4, for instance, energy poisons such as cyanide, azide, 2,4-dinitrophenol, and carbonyl cyanide-m-chlorophenylhydrazone prevent phage DNA entry. All these compounds have the effect of reducing the chemiosmotic gradient. Experiments with valinomycin in the presence of a high potassium concentration also showed no DNA uptake, indicating that it is the membrane potential and not the pH gradient that is important. This led to the hypothesis that the voltage across the membrane controls the conformation of a membrane pore or channel through which the T4 DNA must pass (41). With T5 phage, transient depolarization of the membrane occurs during the two phases of this phage's DNA entry into cells (40,55). In spite of this, metabolic energy does not seem to be required for the T5 DNA entry (62).

Knowledge of the structure and function of cell membranes has lagged somewhat behind our knowledge of systems of genetic information. It is conceivable that in solving the problem of phage DNA transport, new principles regarding cellular membranes will be elucidated, just as so many of the principles of molecular genetics were garnered from phage studies.

1.2.3. Expression and Replication of the Nucleic Acid: General Considerations

After the phage nucleic acid enters a bacterium there are two types of productive responses possible, the two very different responses that so confounded the earliest phage workers:

1. The lytic response, caused by virulent phage, results in the production of progeny phage and the ultimate lysis of the cell. The lytic response is often described in terms of the parameters of the one-step growth curve (60,83). These parameters include the lengths of the eclipse, latent, and rise periods as well as the total time needed for the number of phage in a culture to reach a plateau. Also measured is the burst size, or the number of phage that are produced per infected cell. Many of the T phages produce bursts of

upwards of 200 phage per cell, and 20 min is not an uncommonly short time for this to happen in. The smaller phage can be even more prolific, producing bursts of thousands of phage per cell.

2. Phage can also produce a lysogenic response, in which the phage DNA is made a part of the host's chromosome and replicated along with it without any adverse effect on the cell (54,60,83). In some very rare cases of lysogeny, the phage DNA stays in the cell as a plasmid instead of being inserted into the host chromosome. Phage that are able to produce a lysogenic response are known as temperate, because the infected cell remains alive for as long as the virus stays in its latent, or phophage, state. The latency occurs as a result of a protein repressor that is synthesized very soon after the phage DNA enters the cell. This repressor binds to certain sites on the phage DNA and prevents the normal functioning of the DNA. If the cell is subjected to certain stresses, however, the repressor may become inactivated and the prophage excised from the chromosome. This initiates a lytic response which, for all intents and purposes, is indistinguishable from the lytic response of a virulent phage. A temperate phage, then, may initiate either a lytic or a lysogenic response (49,54,60,83).

Only one group of phage does not fall into this classification of either virulent or temperate. These are the filamentous phage which attach to the pili of F^+ or male bacteria. Cells infected by filamentous phage continue to grow and produce viruses that are secreted through the cell wall without the cell undergoing lysis (19). Because latency does not occur the phage is neither temperate nor, because it does not kill the cell, virulent.

In certain hosts, some phage undergo a third type of response, but one that is not productive (31). These abortive responses are considered in Section 1.2.9.

The biochemical events that occur in an infected cell are now known to occur in a highly controlled, orderly progression somewhat analogous to the events that occur in differentiating cell systems. The stages in the process are generally referred to as pre-early (or immediate early), early and late. The reactions that occur during the pre-early stage are often concerned with "taking over" the host's protein synthesizing machinery and preparing the cell for the production of viral macromolecules. Other pre-early functions ensure that only the correct viral nucleic acid is replicated and that no foreign virus interferes in the process. The early phase occurs before nucleic acid replication begins and often involves the synthesis of enzymes and substrates needed for the replication of nucleic acid. Phage that contain unique bases, for instance, must synthesize new enzymes for the production of these new components of DNA. The late phase of infection involves the synthesis of a large quantity of viral structural proteins and any enzymes needed for morphogenesis of the coat and for packaging the nucleic acid.

The orderly progression of reactions during phage infection results from a sequential expression of the phage genome. A great deal of control of gene expression, in DNA-containing phage as well as bacteria, has been found to be at the transcriptional "level." In *E. coli,* transcription is catalyzed by a single enzyme, the DNA-dependent RNA polymerase. This enzyme consists of a core of relatively tightly associated proteins forming a complex composed of a beta subunit, beta prime subunit, and two alpha subunits. This core enzyme associates with a number of other proteins, the most prominent being the sigma protein, which allows the enzyme to select only certain DNA sequences for transcription into RNA (57,70). Temporal changes in sigma and the sigma-like proteins allow certain genes to be expressed only at certain times. If a gene that is expressed at an early time produces a product that initiates the transcription of other genes, a very orderly progression of gene expression ensues. This "positive" control has been found to operate in the sequential gene expression found in many differentiating systems, as well as in phage-infected cells. It is perhaps best illustrated by T7 phage, which, during the pre-early phase of infection, induces an entirely new RNA polymerase which then transcribes the rest of the genome.

In elucidating the life cycles of both virulent and temperate phage, phage that infect *E. coli* (coliphage) have figured most prominently. Much of what we know about virulent phage comes from the study of the very large and complex T-even phage (T2, T4, and T6) (63). T3 and T7, smaller than the T-evens but still highly complex, have also added much to what we know about virulent infections (64,85). ϕX-174, among the smallest of phage, has also been extensively studied (19). In what follows I will use these phage as examples of large, medium-sized, and small virulent phage, respectively. Of the temperate bacteriophage, the coliphage λ has been studied from every point of view imaginable and it is probably fair to say that we know as much about λ phage as any biological entity (49).

1.2.4. The Pre-early Phase of Phage Replication and the "Take-over" of the Host's Synthetic Machinery

The pre-early, or immediate early, genes are those genes of the phage that are expressed immediately after infection. If the phage genetic material is DNA these genes must be transcribed by the unmodified host RNA polymerase—except in the very rare cases where new polymerases or modifying proteins are injected with the DNA—and the mRNA translated by unmodified ribosomes and tRNA. They are genes whose structure and transcripts must resemble fairly closely those of the host bacterium. In the case of the RNA phage it is more difficult to categorize their genes into particular stages because, in the MS-2 group of coliphage at least, there are only three genes (60). But their RNA must have a structure that will readily bind host ribosomes.

The major purpose of the pre-early genes appears to be that of taking over the hosts' synthetic machinery and preparing the cell for new phage production. One of the most fascinating, but least well understood, aspects of phage replication is how a tiny piece of phage nucleic acid can enter a cell and completely subvert all its synthetic capabilities toward the production of new phage. Arthur Kornberg has stated (54) that this is one of the major unsolved problems of molecular biology.

In T-even phage-infected cells the process is fairly rapid, with all host DNA, RNA, and protein synthesis terminating within the first several minutes after infection (79). Inducible enzymes, such as β-galactosidase, can no longer be turned on even when the necessary mRNA is present in the cell (27). Hence although the phage is known to cause changes in the host RNA polymerase (70), all shutoff of host syntheses cannot be explained thusly.

Roger Herriott first observed that the empty protein coats or ghosts of the T-even phage, produced by osmotic shock, can also stop all macromolecular synthesis in cells they attach to. It was at one time thought that understanding this phenomenon of ghost killing would give some insight into the way the whole phage takes over the cell (26). The two processes are now known to be quite different, however, with the killing by ghosts happening very rapidly and with no new protein synthesis, and the take-over by phage being somewhat slower and requiring phage protein synthesis. It has even been found that phage-infected cells synthesize a new protein, the *imm* protein, that causes them to be immune to killing by superinfecting ghosts (28,90). Rather than the simple attachment of the phage coat, the inhibition of host macromolecular syntheses after T-even phage infection requires a number of simultaneously occurring reactions. These reactions include modifications of the host nucleoid, RNA polymerase, and ribosomes (79).

There are three major modifications of the host nucleoid that have been described after T4 infection: nuclear disruption, during which the host DNA moves from its central location to the periphery of the cell in juxtaposition with the cell membrane; this can be observed electron microscopically. In addition, there is an unfolding process that also serves to block transcription from cytosine-containing DNA, and degradation of the host DNA with the resulting nucleotides reutilized for the synthesis of phage DNA. Each of these modifications is effected by different phage genes, and the three processes seem to be largely independent of one another. Details of these reactions, and their presumed roles, may be found in Ref. 79.

T4, besides inhibiting the translation of preexisting messenger RNA, can inhibit translation of the mRNA of the RNA-containing phage MS-2 when the two phage simultaneously infect a cell. Much evidence indicates that this is because the ribosomes are changed after T4 infection so that T4 messages are preferentially translated (94).

As indicated previously, much of the regulation of gene expression in DNA-containing *E. coli* phage occurs at the transcriptional level, that is, at the point of synthesis of RNA. Hence the RNA polymerases are critical in the regulation of gene expression and changes in RNA polymerase operate in the "take-over" of the cell by the phage. An unusual phage, N4, injects its own RNA polymerase for the specific transcription of its genes, and one of

T7's pre-early genes codes for an entirely new phage RNA polymerase. Another of the pre-early genes of T7 produces an inhibitor of the *E. coli* polymerase that serves to prevent further transcription of the host DNA (70). In the large T-even and T5 phages, the host RNA polymerase is necessary throughout infection, but the enzyme is modified several times during phage development. In T4, during the pre-early phase of development there are apparently two changes. The first change is catalyzed by an enzyme that is injected into the cell along with the DNA and that modifies one of the alpha subunits of the RNA polymerase by adding an ADP-ribose group to it. At a slightly later time a different ADP ribosylase is synthesized that modifies both alpha subunits of the polymerase. These changes are concomitant with a lowered affinity of the polymerase for the *E. coli* sigma factor and may contribute to the turnoff of host mRNA synthesis (70).

Besides coding for functions that help to take over the host's metabolic machinery, some pre-early functions contribute to protecting the phage DNA from nucleolytic enzymes. T3, T5, and T7 all code for an anti-endonucleolytic function (64), and T4 codes for an activity that inactivates exonuclease V (41).

Several of the most unusual pre-early functions of any phage are coded for by T5. When T5 (and its close relative BF23) attaches to cells its DNA enters the cell in two stages. Initially, only 8% of the DNA is injected. Genes on this 8% piece, or first-step transfer (FST) DNA, must be expressed before the remainder of the DNA can enter the cell (64). Hence one of the pre-early functions of T5 is to get its whole DNA molecule into the cell! The two proteins necessary for this second-step DNA transfer, the A1 (30) and A2 (80) proteins have both been reported to be membrane associated; in addition, the A2 protein is bound to the phage DNA (81) and the A1 protein to RNA polymerase (65). How all these components operate to get the DNA inside the cell is not known, however. A third protein on the FST DNA which is expressed during the pre-early phase causes the infection of T5 to be abortive if there is a colicin Ib plasmid in the cell (31). The purpose of this abortive infection gene of T5 is not easy to understand, for it is very difficult to conceive of how this could ever be of use to the phage. Hence the gene may have another, undiscovered, function in the absence of the plasmid. But there do appear to be times when phage find conditions in the cell that make it very important *not* to replicate. Temperate phage, under these conditions, can lysogenize their hosts and wait for a more favorable time to replicate. Other phage may have to cause their infections to abort. In either event, it appears to be the function of the pre-early DNA to "decide" whether or not to replicate.

1.2.5. Early Gene Expression

The early genes of the DNA-containing bacteriophage are those genes that must be expressed before phage DNA synthesis can commence. They are generally transcribed by an RNA polymerase that has either been modified during the pre-early phase of infection or by an entirely new, phage-induced RNA polymerase. The T-even phage are among the largest and most complex of phage so their need for early proteins or early enzymes is fairly easy to understand, for they have DNA that is fundamentally different from bacterial DNA. In these phage, the normal pyrimidine, cytosine, is replaced by a new base, hydroxymethylcytosine (H). The free hydroxyl group on this base then serves as a site for the glucosylation of the DNA, thus protecting it from most restriction enzymes. Before T-even phage DNA can be synthesized, there must be a pool of hydroxymethylcytosine in the cell and also enzymes to phosphorylate it and to glucosylate the DNA. Besides the hydroxymethylase to produce the new base there are deoxycytidine tri- and diphosphatases to produce dCMP, substrate for the hydroxymethylase, and a kinase to produce dHTP. New enzymes involved in folate biosynthesis are also produced as increased amounts of methyl group donors are needed for both dTMP and dHMP synthesis. To ensure than no cytosine is incorporated into the phage DNA there is, besides the dCTP phosphatase, a dCMP deaminase that converts the cytidine to uridine. The uridine is in turn converted to thymidine by a phage-coded thymidylate synthetase. There is also an enzyme to increase the supply of deoxyribose at the expense of ribose and nucleases that hydrolyze the bacterial DNA. In addition to these enzymes that ensure a pool of nucleotides there are a number of enzymes produced that are instrumental in the mechanics of DNA synthesis, including a new DNA polymerase. Several enzymes are also made in the early phase that act on the phage DNA after it is synthesized. These include two glucosylating enzymes, a methylase, and a DNA kinase (54,60).

Some phage that infect *B. subtilis* have hydroxymethyluracil as a component of their DNA (54) and these phage have problems similar to the T-even phage in obtaining a pool of nucleotides for DNA synthesis. These phage synthesize a number of enzymes analogous to the T-even induced enzymes.

T7 uses all the normal bases for its DNA, yet before its DNA can be replicated several new early enzymes must be synthesized. Two of these, an endonuclease and an exonuclease, contribute to the breakdown of the host DNA to supply the nucleotides needed for T7 DNA synthesis. A new DNA polymerase is also synthesized. This new polymerase is unusual in that it consists of two subunits, one supplied by the host and one by the phage. The former is *E. coli* thioredoxin; why it should be needed for DNA polymerization is a complete mystery (54).

The phage ϕX-174 contains a single-stranded circle of DNA as its genome and requires a number of phage-coded enzymes, described later, for its synthesis. Studies with ϕX-174 showed that viruses have a previously unimagined capacity for efficiently coding for the functions they need to replicate. In ϕX several stretches of DNA were found that could simultaneously code for more than one protein (4). The gene A region, for instance, is now known to contain all of gene B and part of gene K. In addition, a small gene

A protein (A*) is formed by initiation of translation in the middle of gene A (54,78). Thus four different proteins are read from one "gene," something that, before the study of φX, was thought to be impossible.

Several principles emerge from a study of "early" functions of bacteriophage. One of these is fairly obvious: if a phage DNA is different from host DNA either because of a difference in composition or a structural difference, new enzymes have to be synthesized to cope with these differences. But even in those cases where the DNA of the phage is not drastically different from host DNA new enzymes are often synthesized to make the process of phage DNA replication more efficient. Finally, phage have a remarkable ability to maximize new synthetic capabilities in a cell while minimizing the amount of DNA "baggage" needed. This they do by overlapping genes and also by utilizing host factors wherever possible, either alone or in combination with newly synthesized phage proteins.

1.2.6. Nucleic Acid Synthesis

The discovery in 1953 of the complementary, double-stranded (duplex) structure of DNA (92) led to a recognition of how DNA could be replicated. The complementary base-pairing, semiconservative mechanism that the structure suggested is now known to be the way all DNA replicates (54).

The process has four basic requirements: (*a*) a template strand to be copied; (*b*) enzymes that synthesize the template's complement; (*c*) a primer from which the complement can be elongated; and (*d*) deoxynucleoside triphosphate precursors. The necessary proteins include not only a polymerizing enzyme but a variety of other proteins including those needed for unwinding the superhelical strucutre, untwisting or relaxing the double helix, supplying energy, and removing errors and repairing nicks (54).

Since Kornberg's isolation of the first DNA polymerase in 1956, a number of DNA polymerizing enzymes have been purified from both prokaryotic and eukaryotic cells. All of these appear to work in much the same way— they all join the 5′ α-phosphate of an incoming nucleoside triphosphate onto the 3′ OH end of a growing DNA strand; that is, they all work in just one direction. Other evidence indicates, however, that DNA synthesis proceeds in both directions simultaneously. How this could happen was a problem for a number of years. A study of T4 phage replication indicated that both chains of a double helix are replicated simultaneously in a discontinuous fashion, by small bursts of DNA synthesis (68). Synthesis of each strand proceeds locally in the 5′–3′ direction, producing overall synthesis of one strand in the 5′–3′ direction and of the other in the 3′–5′ direction. The small fragments that are made are called Okazaki fragments, after their discoverer. Each small fragment is primed by an even shorter segment of RNA that is later removed and the gaps filled with DNA. Even the 3′–5′ strand, which could be copied continuously from one end to the other, seems to be copied by the synthesis of the small Okazaki pieces. The reason for this is unknown but it has been hypothesized that there is less chance of copying error this way. The small fragments are eventually "ligated" to form a continuous strand.

A second problem regarding the synthesis of DNA that was unsolved for quite some time is the problem of replicating the ends of the DNA molecules; again, a great deal of light was shed on the problem by phage research. Studies of phage DNA molecules showed that many were terminally redundant. These terminal redundancies or terminal repetitions (TRs), which are simply repeats of the sequence at one end of a DNA molecule and at the other end, can be explained by the need for a mechanism to prime DNA synthesis at the end of a molecule. The TRs can serve to circularize the molecules, allow concatemer formation, or provide the primer for one end of the molecule by replication of a short piece of DNA on the other end, which then moves and becomes a primer.

Genetic studies of T4 phage have indicated that more than 20 genes are essential for DNA replication (67). Some of these are the genes for the early enzymes and some are genes that provide proteins that actually participate in the polymerization of new DNA. *In vitro* studies of T4 DNA synthesis have revealed which enzymes are actually necessary for the process of polymerization of this phage DNA. Seven phage proteins have been purified and these, together with the necessary deoxynucleotide and ribonucleotide triphosphates and magnesium ions, synthesize DNA at close to the *in vivo* rate. The seven proteins include the actual polymerase and a single-stranded binding protein that serves to keep the two old strands separated while they are being copied. Two other proteins are apparently responsible for synthesizing the RNA primer, while three proteins combine to unwind the DNA and break down ATP to provide energy for the reaction (67).

Although the T4 genetic map is circular the DNA molecules are linear, with the circular map occurring because a population of phage contains a population of "circularly permuted" linear molecules—that is, a population of molecules, each the same size, that begin and end at a different place. The generation of the circular permutations occurs shortly after DNA replication has begun, when the newly synthesized molecules recombine with one another through their terminal redundancies. This recombination produces concatamers that are equivalent to many genomes in length. The DNA that will go into the newly synthesized heads is cut from these long molecules. Each new phage receives about 105% of a genome, thus generating the circularly permuted set of molecules. The ends of the long concatemers, which cannot be replicated, are discarded.

The genomes of the T7 group of phage (T3 and T7) are also linear with TRs, but rather than being circulary permuted, the molecules have a "unique" structure; that is, all are identical. Electron microscope studies indicate that T7 DNA replication starts at a point 17% from the genetic left end of the molecule (to the right of gene 1) and proceeds from there in both directions. A "bubble" is thus formed that enlarges to an "eye." When replication has reached the left (nearest) end, a Y-shaped structure is produced. Eventually two new duplex molecules are produced, each having a single-stranded (unreplicated) 3′ end on the template. The unreplicated ends can recombine with other molecules also having unreplicated ends to form concatemers. The gaps in the concatemers are filled and then the concatemers are cut at a precise sequence of the T7 molecule (54).

The phage with single-stranded genomes (either DNA or RNA) presented a particular problem. At the time it was discovered that ϕX-174 contains single-stranded DNA, it was thought that perhaps an exception to the semiconservative model for DNA replication had been found. Instead, it was found that the first step in the synthesis of the DNA was its conversion to a duplex or RF (replicating form). Thus there are three stages in the synthesis of a single-stranded (SS) DNA molecule: SS to RF, RF to RF, and RF to SS (19,54).

The SS to RF step in ϕX is completed totally by cellular DNA-synthesizing enzymes. For the replication of the RF, four proteins are necessary: the phage-coded gene A protein, which nicks the supercoiled RF form, the *E. coli* rep product, which (in combination with A) unwinds the duplex, *E. coli* DNA polymerase III, and a single-stranded binding protein. The RF to SS step depends on seven virus-coded proteins. The coat protein genes of the phage are all necessary, and it is thought that these proteins act to complex and encapsidate the progeny single strands so that they will be unavailable to be made back into duplex RFs. Consequently, accumulation of viral strands is coupled to morphogenesis of the phage particle (54).

The RNA of the single-stranded RNA phage is likewise replicated through a duplex RF form. A phage-coded RNA polymerase is the enzyme that synthesizes both the RF form and the progeny single strands.

1.2.7. Synthesis of Late Proteins and Morphogenesis

Late proteins are generally capsid structural proteins in addition to any proteins required for the morphogenesis of the new phage particles. Phage make as few as 2 capsid proteins (MS-2) or more than 30, with dozens of nonstructural proteins required to assemble them (T4). The medium-size phage (T7) have a coat consisting of 10 proteins, with three more needed to assemble the capsid and package the DNA.

A very early observation about T-even phage was that the synthesis of the late proteins absolutely dependeds on phage DNA synthesis; mutants missing any of the enzymes required for synthesis of the phage DNA could also not synthesize any capsid proteins. The reason for this appears to be that only DNA with some single-strand breaks can act as a template for late transcription. The single-strand breaks that are in the replicating T4 DNA apparently expose the late promoters which are transcribed by a modified RNA polymerase. Starting at about 6 min after infection the polymerase associates with the products of genes 33 and 55. Both these products are required for late transcription (70). Also required is the product of gene 45, a protein also necessary for DNA synthesis, where it has a role in unwinding the DNA.

The components of the T-even phage coat are synthesized apparently simultaneously along three separate pathways. The heads, tails, and tail fibers are each synthesized before the three assemble into mature phage. The regulation of the assembly process is controlled at the level of protein interactions rather than at the level of genome expression. The pathways of synthesis of components and their assembly were determined both by genetic (*in vivo*) studies and by the *in vitro* study of assembly of phage precursors (96). Details of the process fill many pages (6,9,95), and the studies providing the details resulted in the development of a number of fascinating concepts.

In 1955, Fraenkel-Conrat and Williams made the dramatic discovery that the isolated RNA and protein components of tobacco mosaic virus (TMV) could reconstitute themselves into active virus particles (35). This report was the root of the concept of "self-assembly" wherein all the information needed for the viral assembly process is built into the components of the virus particle. Although this is true for many of the smaller phage, experiments with other viruses have shown that, for the more complicated viruses especially, the process is not so simple. This was first indicated by work with T4 which showed that a very large number of genes were involved in morphogenesis (96). In T4, some two dozen gene products are required for the assembly of the head alone, and fewer than half of these are found in the finished viral particle. The other gene products are morphogenetic "factors," either enzymes that catalyze noncovalent association of structural proteins or scaffolding proteins required for the correct assembly of the phage head. These scaffolding or core proteins, which are eliminated from the final structure, control the size and shape of the head (9).

It has been found that proteolysis is one of the prime controlling factors in the assembly process. Study of the morphogenesis of the T4 tail has shown that critical regulatory interactions in assembly are between the growing structure and soluble subunits, rather than between soluble subunits themselves. Subunits with a nonreactive precursor conformation are switched into a reactive conformation upon incorporation into the growing structure (6). Often these changes occur as a result of proteolysis.

One of the big surprises to come from the study of bacteriophage morphogenesis was the mechanism of packaging DNA into an icosahedral head. Three models for the packaging of DNA can be imagined: (*a*) the DNA condenses into a small ball, such as would be found in the virus, and the protein capsid forms around it; (*b*) the capsids are formed first and the DNA is then brought inside to fill the empty head; (*c*) the DNA and protein coat condense simultaneously. Of these models, the least likely seemed to be the second; it was thought to be unlikely because of the difficulty in imagining the insertion of a very long flexible thread through a tiny opening into a very small space. However, this has been shown experimentally, in both animal virus and bacteriophage sysems, to be a mechanism of DNA packaging.

First an empty head or prohead is made. In T4 the prohead assembly is initiated at the cell membrane with the membrane acting to concentrate the needed subunits. A primer is formed utilizing one of the minor coat proteins, after which the coat is formed around a protein core or scaffold which helps to control the head assembly. The finished empty prohead has the shape of the final head although it is slightly smaller and somewhat more rounded than the finished head. It contains the protein core and, for T4 at least, is bound to the inner cell membrane (9).

The conversion of a prohead into a mature head filled with DNA is almost always accompanied by proteolytic cleavage of some of the proteins. In T4, the major coat protein as well as some of the internal "scaffolding" proteins are cleaved. The cleavage of the major coat protein apparently causes a change of shape of the head and the breakdown of the internal components allows the DNA to enter (9). Exactly how this happens remains a mystery, however. One suggestion is that when the prohead changes its shape and expands, the pressure in the head is lowered and the DNA is sucked in. Another suggestion is that the proteolytic cleavage of the internal proteins releases small acidic peptides that cause the DNA to collapse inside the head.

Phage apparently have two mechanisms to determine how much DNA to put into a head. In one (T4), a "headful" of DNA is inserted. In others (T7 and λ), the precursor DNA is cut at specific sequences, in lambda called *cos* sequences, and only DNA bounded by those sequences is packaged. One consequence of the latter mode of packaging is that variations such as additions, deletions, or substitutions that alter the DNA length result in packaging of genomes of other than wild-type length (54).

Even though MS-2 has one of the simplest capsids imaginable, its synthesis is not without interest. The capsid is composed of 180 subunits of coat protein plus one A (adsorption) protein. Its genome acts as messenger RNA for the three genes carried by the phage, the two genes for the capsid proteins and an RNA replicase. In DNA-containing viruses the amount of protein made can be controlled by the amount of message transcribed but MS-2 is faced with needing 180 times the amount of coat protein as A protein and has no transcriptional controls available. The affinities of the ribosome binding sites on each gene are vastly different, however, so that by virtue of greater binding to the ribosome, about 10 times as much coat protein is made as A protein. The replicase is also not needed in very large amounts and its synthesis is blocked by the binding of coat protein subunits to the RNA, preventing efficient translation of that gene. Thus a capsid protein does double duty by acting also as a repressor of translation (54).

1.2.8. Lysis of Cells

Liberation of phage and the end of the lytic cycle requires, generally, two types of events: (*a*) the rigid peptidoglycan or murein layer of the bacterial wall must be hydrolyzed, and (b) the cytoplasmic membrane must be damaged. In T4, lysis is preceded by a sudden cessation of respiration, perhaps as a result of the product of the t gene, whose protein product is unknown. The peptidoglycan is then cleaved by T4 lysozyme, an enzyme similar to egg-white lysozyme, which cleaves the bond between the N-acetylglucosamine and the muramic acid in the murein (63). In λ phage, the "S" protein appears to damage the cytoplasmic membrane, allowing the λ endolysin to attack the murein. Lambda's endolysin is not a true lysozyme but rather a peptidase that hydrolyzes cross bridges in the murein (54).

The lysis gene of φX-174 provided another example of this phage's ability to encode several proteins in one gene. Translation of the E protein, needed for lysis, starts from a signal within gene D. The start site for E is frameshifted by one nucleotide from the translation frame of D, producing two entirely unrelated proteins from one DNA sequence (4).

1.2.9. Inhibition of Phage Replication

Extremely important in understanding the ecology of bacteriophage is an understanding of how phage interact with each other and with other extrachromosomal elements. Both temperate and virulent phage have the ability to inhibit either homologous or heterologous superinfecting phage, resulting in what is often called mutual exclusion between related or unrelated phages (45,93). Plasmids likewise can inhibit the replication of a variety of phage. These interactions among bacterial extrachromosomal elements, more than any other single factor, dictate whether or not a particular bacterium will be infected by a certain phage. Mechanisms used by extrachromosomal elements (either plasmids or phage) to prevent the infection of their hosts by phage include the induction of immunity, the alteration of the cell surface to produce resistance (an inability to adsorb), "exclusion" of superinfecting DNA, endonucleotytic restriction of the DNA, and abortive infection or growth interference (31).

Temperate phage can inhibit homologous phage by the induction of immunity in their hosts. In lysogenic cells, none of the viral genes needed for productive, virulent infection are transcribed. Instead, a repressor gene produces a product that binds to the prophage DNA and prevents transcription of most of the viral genes. If a lysogenized cell is infected by another homologous phage, the entering DNA is repressed and cannot initiate a productive cycle of development. The lysogenized cells are thus immune (49,54,60,83). The replication of phage that are not totally homologous can also be affected if they contain an homologous immunity region.

That lysogenized phage could change the cell surfaces of their hosts and cause resistance was first indicated when it was found that the antigenic type of Salmonella was changed after lysogenization by certain phage. If *Salmonella anatum,* classified antigenically as O type 3,10, is lysogenized by the phage ε-15, it becomes O type 3,15, or *Salmonella newington* (89).

The O 3,10 antigen is composed of repeating units of a trisaccharide joined in an α-galactosyl linkage. The trisaccharide contains mannose, rhamnose, and O-acetylgalactose. In contrast, the subunits of the O 3,15 antigen are joined via a β-galactosyl link and the galactose is not acetylated. In mutants of Salmonella that cannot polymerize the trisaccharide into O-antigen, ε-15 phage restores polymerizing activity with the resulting polymers linked in the beta conformation. This suggests that a new beta polymerase is produced upon lysogenization (58). The same change also occurs after infection with a virulent mutant of ε-15, strongly implicating a phage-coded enzyme (89). The receptors for phage C-341, which is adsorbed only by cells with antigen 10, disappear following infection by ε-15. However, another phage, ε-34, can adsorb only to O type 15, so it can infect only after the change in the lipopolysaccharide (LPS) is induced by ε-15. After ε-34 infects a cell, it modifies the LPS again, so that no more ε-34 can adsorb. Other similar "lysogenic conversions" are not uncommon and can either produce new phage receptors (73)—making a cell susceptible to a phage that it was previously resistant to—or by destroying an existent receptor, making the cell resistant (58,89).

When *E. coli* is infected by the virulent phage T5, its surface receptors are changed so that they no longer can bind T5 (32). This is an example of a virulent phage inhibiting infection by homologous phage. The T-even phage can likewise prevent infection of superinfecting homologous phage, but in this case the adsorption of the superinfecting phage is normal, as is the ejection of its DNA. The DNA is unable to enter the cell, however, and is found degraded in the medium (1). The *imm* protein that controls the superinfection exclusion also induces in the cells the ability to resist the effects of the T-even phage ghosts (28,90) and so may alter the conformation of an essential membrane structure.

Restriction is that process in which "invading" DNA is broken down by nucleases in the cell (3). The phenomenon usually refers to cases where unmodified or unmethylated DNA is hydrolyzed by site-specific restriction endonucleases (66), but this is not necessarily the case. The phenomenon was first observed by Luria and Mary Human using phage T2, a phage that we now know does not undergo the most common type of restriction and modification. Luria had noticed that certain mutants of *E. coli* would lyse when infected by T2, but that the lysate seemed to contain no new virus—at least none that would grow on *E. coli*. Later, Luria and Human wanted to add the lysate to streptomycin-resistant *E. coli,* but having none at hand, used streptomycin-resistant Shigella, because T2 had been observed to replicate in this genus of bacteria as well as it did in *E. coli.* By virtue of their having no streptomycin-resistant *E. coli* they serendipitously found that the phage produced in their mutant *E. coli,* even though it would not reinfect *E. coli,* did grow in Shigella. After one cycle of growth in Shigella, moreover, the phage could again grow in *E. coli* (59). It is now known that the mutant *E. coli* was unable to glucosylate the T2 DNA and that unglucosylated DNA was broken down when injected into other *E. coli*. Shigella is apparently missing the enzymes needed to hydrolyze the DNA but has the glucosylating enzymes and so can "modify" the DNA for growth in *E. coli*. This host-controlled variation of phage was soon thereafter found to occur with the coliphage λ as well as with phage that infect Salmonella and Staphylococcus, with these phage undergoing the more common type of restriction and modification.

It was found that if λ phage was grown on *E. coli* K it could not grow on *E. coli* B strains or on *E. coli* K lysogenized by phage P1. If lambda was grown on *E. coli* K lysogenized by P1 it could be grown on *E. coli* K and *E. coli* K (P1). If λ was grown on *E. coli* B it could not be grown on either lysogenized or unlysogenized K strains. Bertani and Weigle excluded nonadsorption of phage as a possible reason for restriction. They also showed that adsorption of restricted λ neither kills nor even delays the growth of the host cells (7). They later showed that radio-labeled DNA was broken down inside the bacterium, in contrast to superinfection exclusion, in which the DNA never enters the cytoplasm (3).

It is now known that many genetic elements—chromosomes, lysogenized phage, or plasmids—code for restriction and modification enzymes. The restriction enzymes recognize specific sequences in the DNA (sequences of four to six base pairs) and hydrolyze the DNA if these sequences are not protected or modified. The modification usually consists of methyl groups added to the bases in the specific recognition sequence. Restriction is not limited to phage DNA but rather involves any entering or invading DNA, be it from transformation, conjugation, or phage infection.

Abortive infections, on the other hand, appear to be a way by which plasmids or lysogenized phage can specifically inhibit the replication of virulent phage. Many instances of apparent abortive infections by extrachromosomal elements, either plasmids or lysogenized phage, have been reported (31). Anderson, for instance, indicated that the reaction of certain strains of Salmonella to a large number of phage is changed by the introduction of antibiotic resistance plasmids into the cells. Similar findings exist for *E. coli,* several species of Streptococcus, and *Bacillus subtilis*. In many of these cases the results suggest that adsorption and injection are normal and that no breakdown of the phage DNA occurs, eliminating other mechanisms of inhibition. However, only a few cases have been studied in detail.

The earliest case that was studied was that of the inhibition or "exclusion" of T-even rII mutant phage by λ lysogens (5,31,39). In this case the infection by the T-even phage proceeds normally for about 10 min, at which time an event that appears to alter the normal function of the cell membrane occurs (74). The gene of λ that causes the infection to abort, the *rex* gene, is the only gene, other than the repressor gene, expressed by the prophage (43). It does not appear to have any function other than inhibiting these phage mutants, but its mechanism of action is not known.

Two cases of phage inhibition by plasmids have been extensively studied.

These are the inhibition of T7 phage by the F factor, the plasmid that confers genetic donor ability or maleness on *E. coli,* and the inhibition of T5 by the colicin Ib (Col Ib) plasmid. In both these cases, the infection is normal for the first 3–5 min, after which time abnormal cell membrane permeability is seen (31). In the case of T7's inhibition, a controversy exists as to whether the abnormal membrane function is the cause of, or an effect of, the abortive infection (17). In the Col Ib inhibition of T5, cell membrane depolarization has been observed at the time the normal infectious process stops, and it is thought that this actually causes the abortive infection (40).

Although many of the mechanisms are unclear, one thing is certain; the interactions between virulent phage and other extrachromosomal elements are very common and "an important part of the life cycle of these subcellular entities" (31). Just how important these interactions are will likely become ever clearer as the ecological aspects of phage are thoroughly investigated.

1.3. CONCLUDING REMARKS

When bacteriophages were first discovered, during the time of World War I, it was hoped that their ability to kill bacteria could be used for the prevention and treatment of bacterial disease. Although some work indicated that phage had a positive effect on a variety of bacterial diseases, the discovery of antibiotics caused a virtual cessation of research in this area. But studies of the basic genetic properties of phage led to the development of an entirely new science—that of molecular biology—which is allowing unprecedented advancements in all the biological and medical sciences. Experiments with bacteriophage confirmed that DNA is the carrier of genetic information and unveiled the concepts of virus-induced acquisition of metabolic function and virus-induced enzymes. In addition, the way all viruses reproduce was first indicated by work with phage. These fundamental discoveries were followed in short order by the discovery of messenger RNA, the elucidation of the triplet and degenerate nature of the genetic code, proof the the colinearity of the gene and its protein product, discovery of the RNA priming and discontinuity of DNA synthesis, discovery of the existence and value of conditional lethal mutants in genetic studies, and the role of sigma factors in controlling RNA transcription. These findings and more, all of which read like chapter headings in a textbook of molecular biology, were the unexpected results of studies of the fundamental nature of bacteriophage.

Nothing in the history of science, however, so well illustrates the unforeseen consequences of basic research as does the discovery of restriction enzymes. This discovery occurred as a result of the study of a fairly obscure phenomenon concerning the behavior of bacteriophage grown on two different strains of bacteria. It had been observed that phage grown on one strain of bacteria grew very poorly (i.e., were restricted) on a second strain, and vice versa. Research into the cause of this "restriction" showed that if unmarked or inappropriately marked DNA entered a bacterium, it could be broken down by restriction endonucleases. These enzymes, because of their nucleotide sequence specificity, form one cornerstone of the whole new discipline of genetic engineering, a discipline that is dramatically changing the face of both science and medicine. Almost daily, new identification or therapeutic innovations occur as a result of techniques made possible by the discovery of restriction enzymes. In addition, genetic engineering has provided dramatic new ways of studying the way life "works," so that the results of the original discovery will exponentially increase for some time to come.

In spite of these dramatic advancements, many biological mysteries remain unsolved and many microbes remain unconquered. Research on bacteriophage is now at a lower ebb than at any time since the early 1950s. It is quite likely, however, that phage will again prove to be useful, either as a practical means of detecting or eliminating bacteria, or simply as biological "tools" to help to uncover the still-buried secrets of nature.

ACKNOWLEDGMENTS

I am very grateful to Drs. Ed Siden, Tom Pinkerton, and Conie Young for critically reading this manuscript and providing many helpful suggestions. I also thank Antony Twort for traveling to Florida to talk about the discovery of bacteriophage.

REFERENCES

1. Anderson, C. W., and J. Eigner. 1971. Breakdown and exclusion of superinfecting T-even bacteriophage in *Escherichia coli*. J. Virol. *8*:869–886.
2. Anderson, T. F. 1966. Electron microscopy of phages, pp. 63–78. *In* J. Cairns, G. S. Stent, and J. D. Watson (eds.), Phage and the Origins of Molecular Biology. Cold Spring Harbor Laboratory, Cold Spring Harbor, New York.
3. Arber, W. 1965. Host-controlled modification of bacteriophage. Ann. Rev. Microbiol. *19*:365–378.
4. Barrell, B. G., G. M. Air, and C. A. Hutchinson, III. 1976. Overlapping genes in bacteriophage PhiX174. Nature (London) *264*:34–41.
5. Benzer, S. 1966. Adventures in the rII region, pp. 157–165. *In* J. Cairns, G. S. Stent, and J. D. Watson (eds.), Phage and the Origins of Molecular Biology. Cold Spring Harbor Laboratory, Cold Spring Habor, New York.
6. Berget, P. B., and J. King. 1983. T4 tail morphogenesis, pp. 246–258. *In* C. K. Mathews, E. M. Kutter, G. Mosig, and P. B. Berget, (eds.), Bacteriophage T4. American Society for Microbiology, Washington, DC.
7. Bertani, G., and J. Weigle. 1953. Host controlled variation in bacterial viruses. J. Bacteriol. *65*:113–121.
8. Billmire, E. W., and D. H. Duckworth. 1976. Membrane protein biosynthesis in bacteriophage-infected *Escherichia coli*. J. Virol. *19*:475–489.

9. Black, L. W., and M. K. Showe. 1983. Morphogenesis of the T4 head, pp. 219–245. *In* C. K. Mathews, E. M. Kutter, G. Mosig, and P. B. Berget, (eds.), Bacteriophage T4. American Society for Microbiology, Washington, DC.

10. Blair, J. E. 1924. A lytic principle (bacteriophage) for *Corynebacterium diphtheriae*. J. Infect. Dis. *35*:401–406.

11. Bordet, J., and M. Ciuca. 1920. Le bacteriophage de d'Herelle, sa production et son interpretation. C.R. Soc. Biol. Paris *83*:1296–1298.

12. Bordet, J., and M. Ciuca. 1921. Remarques sur l'historique des recherches concernant la lyse microbienne transmissible. C.R. Soc. Biol. Paris *84*:745–747.

13. Braun, V., K. Schaller, and H. Wolff. 1973. A common receptor protein for phage T5 and colicin M in the outer membrane of *E. coli* B. Biochim. Biophys. Acta *323*:87–97.

14. Brutlag, D., R. Schekman, and A. Kornberg. 1971. A possible role for RNA polymerase in the initiation of M13 DNA synthesis. Proc. Natl. Acad. Sci. USA *68*:2826–2830.

15. Bulloch, W. 1938. The history of bacteriology. Oxford University Press, London.

16. Cohen, S. S. 1968. Virus-induced enzymes. Columbia University Press, New York.

17. Condit, R. C. 1975. Who killed T7? Nature (London) *260*:287–288.

18. Crick, F. H. C., L. Barnett, S. Brenner, and R. J. Watts-Tobin. 1961. The general nature of the genetic code. Nature (London) *192*:1227–1232.

19. Denhardt, D. D. 1975. The single-stranded DNA bacteriophages. CRC Crit. Rev. Microbiol. *4*:161–224.

20. d'Herelle, F. 1917. Sur un microbe invisible antagonistic des bacilles dysenteriques. C. R. Acad. Sci. Paris *165*:373–375.

21. d'Herelle, F. 1922. The bacteriophage; its role in immunity (English translation). Williams and Wilkins, Baltimore, Maryland.

22. d'Herelle, F. 1926. The bacteriophage and its behavior. (English translation). Williams and Wilkins, Baltimore, Maryland.

23. d'Herelle, F. 1949. The bacteriophage. Sci. News *14*:44–59.

24. Di Masi, D. R., J. C. White, C. A. Schnaitman, and C. Bradbeer. 1973. Transport of vitamin B12 in *Escherichia coli*: Common receptor sites for vitamin B12 and the E. colicins on the outer membrane of the cell envelope. J. Bacteriol. *115*:506–513.

25. Doermann, A. H. 1951. The intracellular growth of bacteriophages. I. Liberation of intracellular phage T4 by premature lysis with another phage or with cyanide. J. Gen. Physiol. *35*:645–656.

26. Duckworth, D. H. 1970. Biological activity of bacteriophage ghosts and "take-over" of host functions of bacteriophage. Bacteriol. Rev. *34*:344–363.

27. Duckworth, D. H. 1970. Inhibition of translation of preformed *lac* messenger RNA by T4 bacteriophage. J. Virol. *5*:653–655.

28. Duckworth, D. H. 1971. Inhibition of T4 bacteriophage multiplication by superinfecting ghosts and the development of tolerance after bacteriophage infection. J. Virol. *7*:8–14.

29. Duckworth, D. H. 1976. "Who discovered bacteriophage?" Bacteriol. Rev. *40*:793–802.

30. Duckworth, D. H., G. B. Dunn, and D. J. McCorquodale. 1976. Identification of the gene controlling the synthesis of the major bacteriophage T5 membrane protein. J. Virol. *18*:542–549.

31. Duckworth, D. H., J. Glenn, and D. J. McCorquodale. 1981. Inhibition of bacteriophage replication by extrachromosomal genetic elements. Microbiol. Rev. *45*:52–71.

32. Dunn, G. B., and D. H. Duckworth. 1977. Inactivation of receptors for bacteriophage T5 during infection of *Escherichia coli* B. J. Virol. *24*:419–421.

33. Ellis, E. L., and M. Delbruck. 1939. The growth of bacteriophage. J. Gen. Physiol. *22*:365–384.

34. Epstein, R. H., A. Bolle, D. M. Steinberg, E. Kellenberger, E. Boy de la Tour, R. Chevalley, R. S. Edgar, M. Susman, G. H. Denhardt, and I. Lielausis. 1963. Physiological studies of conditional lethal mutants of bacteriophage T4D. Cold Spring Harbor Symp. Quant. Biol. *28*:375–392.

35. Fraenkel-Conrat, H., and R. C. Williams. 1955. Reconstitution of active tobacco mosaic virus from its inactive protein and nucleic acid subunits. Proc. Natl. Acad. Sci. USA *41*:690–698.

36. Freeman, V. 1951. Studies of the virulence of bacteriophage-infected strains of *Corynebacterium diphtheriae*. J. Bacteriol. *61*:675–688.

37. Freeman, V. 1952. Further observations on the change to virulence of bacteriophage-infected avirulent strains of *Corynebacterium diphtheriae*. J. Bacteriol. *63*:407–414.

38. Freese, E. 1959. The specific mutagenic effect of base analogs on T4. J. Mol. Biol. *1*:87–98.

39. Garen, A. 1961. Physiological effects of rII mutations in bacteriophage T4. Virology. *14*:151–163.

40. Glenn, J., and D. H. Duckworth. 1980. Fluorescence changes of a membrane-bound dye during bacteriophage T5 infection of *Escherichia coli*. J. Virol. *33*:553–556.

41. Goldberg, E. 1983. Recognition, attachment, and injection, pp. 32–39. *In* C. K. Mathews, E. M. Kutter, G. Mosig, and P. B. Berget, (eds.), Bacteriophage T4. American Society for Microbiology, Washington, DC.

42. Gratia, A. 1932. Antagonisme microbien et "bacteriophage." Ann. Inst. Pasteur *48*:413–437.

43. Gussin, G. N., and V. Peterson. 1972. Isolation and properties of *rex* mutants of bacteriophage lambda. J. Virol. *10*:760–765.

44. Hall, B. D., and S. Spiegelman. 1961. Sequence complementarity of T2 DNA and T2 specific RNA. Proc. Natl. Acad. Sci. USA *47*:137–146.

45. Hausman, R. 1973. The genetics of T-odd phages. Ann. Rev. Microbiol. *27*:51–67.

46. Heller, K. J., and V. Braun. 1982. Polymannose O-antigens of *Escherichia coli,* the binding sites for the reversible adsorption of bacteriophage T5 via the L-shaped tail fibers. J. Virol. *41*:222–227.

47. Heller, K. J., and H. Schwarz. 1985. Irreversible binding to the receptor of bacteriophages T5 and BF23 does not occur with the tip of the tail. J. Bacteriol. *162*:621–625.

48. Hershey, A. D. 1953. Functional differentiation within particles of bacteriophage T2. Cold Spring Harbor Symp. Quant. Biol. *18*:135–139.

49. Hershey, A. D. (ed.). 1971. The Bacteriophage Lambda. Cold Spring Harbor Laboratory, Cold Spring Harbor, New York.

50. Hershey, A. D., and M. Chase. 1952. Independent functions of viral protein and nucleic acid in growth of bacteriophage. J. Gen. Physiol. *36*:39–56.

51. Howe, M. M., and E. G. Bade. 1975. Molecular biology of bacteriophage Mu. Science. *190*:624–632.

52. Judson, H. F. 1979. The Eighth Day of Creation. Simon and Schuster, New York.

53. Klein, J. 1982. Immunology: The Science of Self–Nonself Discrimination. Wiley, New York.

54. Kornberg, A. 1980. DNA Replication. Freeman, San Francisco.

55. Labedan, B., and L. Letellier. 1981. Membrane potential change during the first steps of coliphage infection. Proc. Natl. Acad. Sci. USA. *78*:215–219.

56. Lindberg, A. A. 1973. Bacteriophage receptors. Ann. Rev. Microbiol. *27*:205–241.

57. Losick, R., and J. Pero. 1976. Regulatory subunits of RNA polymerase, pp. 227–246. *In*

R. Losick and M. Chamberlin (eds.), RNA Polymerase. Cold Spring Harbor Laboratory, Cold Spring Harbor, New York.

58. Losick, R., and P. W. Robbins. 1967. Mechanism of epsilon 15 conversion studied with a bacterial mutant. J. Mol. Biol. *30*:445–455.

59. Luria, S. E. 1955. The T2 mystery. Sci. Am. *192*:92–98.

60. Luria, S. E., J. E. Darnell, D. Baltimore, and A. Campbell. 1978. General Virology. Wiley, New York.

61. Lwoff, A. 1953. Lysogeny. Bacteriol. Rev. *17*:269–337.

62. Maltouf, A. F., and B. Labedan. 1983. Host cell metabolic energy is not required for injection of bacteriophage T5 DNA. J. Bacteriol. *153*:124–133.

63. Mathews, D. K., E. M. Kutter, G. Mosig, and P. B. Berget (eds.). 1983. Bacteriophage T4. American Society for Microbiology, Washington, DC.

64. McCorquodale, D. J. 1975. The T-odd bacteriophages. CRC Crit. Rev. Microbiol. *4*:101–159.

65. McCorquodale, D. J., C. W. Chen, M. K. Joseph, and R. Waychik. 1981. Modification of RNA polymerase from *Escherichi coli* by pre-early gene products of bacteriophage T5. J. Virol. *40*:958–962.

66. Nathans, D., and H. O. Smith. 1975. Restriction endonuclease in the analysis and restructuring of DNA molecules. Ann. Rev. Biochem. *44*:273–293.

67. Nossal, N. G., and B. M. Alberts. 1983. Mechanism of DNA replication catalyzed by purified T4 replication proteins, pp. 71–81. *In* C. K. Mathews, E. M. Kutter, G. Mosig, and P. B. Berget (eds.), Bacteriophage T4. American Society for Microbiology, Washington, DC.

68. Okazaki, R. T., K. Okazaki, K. Sakabe, K. Sugimoto, and A. Sugino. 1968. Mechanism of DNA chain growth. I. Possible discontinuity and unusual secondary structure of newly synthesized chains. Proc. Natl. Acad. Sci. USA *59*:598–603.

69. Olby, R. 1974. The path to the double helix. University of Washington Press, Seattle, Washington.

70. Rabussay, D. 1983. Phage-evoked changes in RNA polymerase, pp. 167–173. *In* C. K. Mathews, E. M. Kutter, G. Mosig, and P. B. Berget, (eds.) Bacteriophage T4. American Society for Microbiology, Washington, DC.

71. Raffel, S. 1982. Fifty years of immunology. Ann. Rev. Microbiol. *36*:1–26.

72. Sarabhai, A. S., A. O. W. Stretton, and S. Brenner. 1964. Co-linearity of the gene with the polypeptide chain. Nature (London) *201*:13–17.

73. Schnaitman, C. A., D. Smith, and M. F. de Salsas. 1975. Temperate bacteriophage which causes the production of a new major outer membrane protein by *E. coli*. J. Virol. *15*:1121–1130.

74. Sekiguchi, M. 1966. Studies on the physiological defect in rII mutants of bacteriophage T4. J. Mol. Biol. *16*:503–522.

75. Simon, L. D., and T. F. Anderson. 1967. The infection of *Escherichi coli* by T2 and T4 bacteriophage as seen in the electron microscope. I. Attachment and penetration. Virology *32*:279–297.

76. Smith, G. H., and E. F. Jordan. 1930. *Bacillus diphtheriae* and its relationship to bacteriophage. J. Bacteriol. *21*:75–88.

77. Smith, H. E. 1979. Nucleotide sequence specificity of restriction endonucleases. Science *205*:455–462.

78. Smith, M., N. L. Brown, G. M. Air, B. G. Barrell, A. R. Coulson, C. A. Hutchinson III, and F. Sanger. 1977. DNA sequence at the C termini of the overlapping genes A and B in bacteriophage PhiX174. Nature (London) *265*:702–705.

79. Snustad, D. P., L. Snyder, and E. Kutter. 1983. Effects on host genome structure and expression, pp. 40–55. *In* C. K. Mathews, E. M. Kutter, G. Mosig, and P. B. Berget, (eds.), Bacteriophage T4. American Society for Microbiology, Washington, DC.

80. Snyder, C. E., Jr. 1984. Bacteriophage T5 gene A2 protein alters the outer membrane of *Escherichia coli.* J. Bacteriol. *160*:1191–1195.

81. Snyder, C. E., and R. Benzinger. 1981. Second-step transfer of bacteriophage T5 DNA: purification and characterization of the T5 gene A2 protein. J. Virol. *40*:248–257.

82. Stent, G. S. 1963. The molecular biology of bacterial viruses. Freeman, San Francisco.

83. Stent, G. S., and R. Calendar. 1978. Molecular Genetics. Freeman, San Francisco.

84. Stone, F. M., and G. L. Hobby. 1933. A coccoid form of *C. Diphtheriae* susceptible to bacteriophage. J. Bacteriol. *27*:403–417.

85. Studier, F. W. 1972. Bacteriophage T7: genetic and biochemical analysis of this simple phage gives information about basic genetic processes. Science, 176:367–373.

86. Twort, F. W. 1915. An investigation on the nature of ultra-microscopic viruses. Lancet *2*:1241–1243.

87. Twort, F. W. 1949. The discovery of the bacteriophage. Sci. News *14*:33–34.

88. Uchida, T., D. M. Gill, and A. M. Pappenheimer. 1971. Mutation in the structural gene for diphtheria toxin carried by temperate phage beta. Nature New Biol. *233*:8–11.

89. Uetake, H., S. E. Luria, and J. W. Burrows. 1958. Conversion of somatic antigens in Salmonella by phage infection leading to lysis or lysogeny. Virology *5*:68–91.

90. Vallee, M., and J. B. Cornett. 1972. A new gene of bacteriophage T4 determining immunity against superinfecting ghosts and phage in T4-infected *Escherichia coli.* Virology *48*:777–784.

91. Wais, A. C., and E. B. Goldberg. 1969. Growth and transformation of phage T4 in *E. coli* B/4, Salmonella, Aerobacter, Proteus, and Serratia. Virology. *39*:153–161.

92. Watson, J. D., and F. H. C. Crick. 1953. A structure for deoxyribose nucleic acid. Nature (London) *171*:737–738.

93. Weigle, J. J. and M. Delbruck. 1951. Mutual exclusion between an infecting phage and a carried phage. J. Bacteriol. *62*:301–318.

94. Wiberg, J. S., and J. D. Karam. 1983. Translational regulation in T4 phage development, pp. 193–201. *In* C. K. Mathews, E. M. Kutter, G. Mosig, and P. B. Berget (eds.), Bacteriophage T4. American Society for Microbiology, Washington, DC.

95. Wood, W. B., and R. A. Crowther. 1983. Long tail fibers; Genes, proteins, assembly, and structure, pp. 259–269. *In* C. K. Mathews, E. M. Kutter, G. Mosig, and P. B. Berget (eds.). Bacteriophage T4. American Society for Microbiology, Washington, DC.

96. Wood, W. B., and R. S. Edgar. 1967. Building a bacterial virus. Sci. Am. *317*:3–13.

97. Wood, W. B., and H. R. Revel. 1976. The genome of bacteriophage T4. Bacteriol. Rev. *40*:847–868.

98. Wyatt, H. V. 1974. How history has blended. Nature (London) *249*:803–806.

99. Yurewicz, E. C., M. A. Ghalambor, D. H. Duckworth, and E. C. Heath. 1971. Catalytic and molecular properties of a phage-induced capsular polysaccharide depolymerase. J. Biol. Chem. *246*:5607–5616.

100. Zoon, K. C., M. Habersat, and J. J. Scocca. 1976. Synthesis of envelope polypeptides by *Haemophilus influenzae* during development of competence for genetic transformation. J. Bacteriol. *127*:545–554.

2

BACTERIOPHAGE TAXONOMY

J. N. COETZEE

Department of Medical Microbiology, Institute of Pathology, Box 2034, Pretoria 0001, South Africa and the Bacterial Genetics Research Unit of the South African Medical Research Council

2.1. INTRODUCTION

Microbial taxonomic literature abounds with phrases such as "All classifications are artificial but some are more artificial than others" (114), "All classifications are subjective and, like religious and political opinions, have a large element of aesthetic unreason about them" (69), and "Taxonomy is as much an art as a science and its interpreters are artists" (73). There are also religious overtones: Linnaeus, initially at least, believed that each species was created separate and distinct. Bringing together in an arrangement those plants or animals that were alike was to follow the plan of the Creator. For this reason taxonomy was regarded as a highly religious duty (56). To this day an element of religious fervor can be detected in the way taxonomists tend to expound (or defend) their particular taxonomic persuasions (158). Nevertheless, the underlying assumption of scientific classification is that there may be a natural order, a system of similarities, and a general biological affinity present among microorganisms, which may be discovered. It is the purpose of this presentation to place the taxonomy of bacteriophages (phage) into this perspective.

Phage are considered to be bacterial viruses and, according to rule 3 of the International Committee on Taxonomy of Viruses (ICTV) (157), conclusions reached should also be applicable to other viruses. This possibility is not considered here (but see Sections 2.3.5, 2.4, and 2.5).

In January 1981 there were about 2100 listed bacteriophages, making them the largest viral group described (5). This number continues to grow by yearly increments of about 100 (5,107,157). To be able to communicate quickly and intelligently about phage, to make generalizations, to venture predictions, and to satisfy the intellectually tidy, it is desirable that phage be arranged in named groups sharing some properties.

2.2. SYSTEMS OF TAXONOMY AVAILABLE FOR BACTERIOPHAGES

There is confusion about the derivation of the word *taxonomy*: In Greek *taxis* means arrangement. The latter part of the word could be from *nomia* (distribution); *nomos* (law, in which case it should have been *taxinomy*); or a Greek-Latin hybrid, *nomen* (name) (73). Be that as it may, taxonomy is generally taken to be the equivalent of biological systematics (71) and entails the orderly arrangement of units into groups and the naming of these entities. Taxonomy may also embody a third activity, namely identification of unknown specimens with the above units.

2.2.1. Arrangement

There are a number of ways in which living objects may be classified: First, if the intention is to pigeonhole them, each is given a distinctive name or number, which is then arranged in order. This is merely the enumeration or cataloging, and is not the grouping of objects together that is the essence of classification. Mere cataloging is the fate of many, especially tailed phage, at present (4). For aesthetic reasons and also for the practical reason that it is difficult to remember the contents of catalogs, classification is resorted to.

To make the information about strains manageable in practice, this information must be compressed. This is why hierarchical systems of classification are used with ranks of taxa to indicate degrees of similarity. A few of these systems are discussed here, the artificial and some of the natural groups of classifications (147,161).

The description *artificial* is applicable to any classification based on one or a few criteria as discriminants that happen to provide a ready means of subdivision (114,212). Such classificiations, also called *monothetic* (205), are noted not only for the small number of criteria on which they are based, but also for the small number of purposes they serve. Most diagnostic keys fall into this category and are characterized by the fact that each class at every level is uniquely definable by attributes shared in common by its component elements and no organism is tolerated in any class that does not conform to *all* the diagnostics of the class.

The word *natural* is notoriously imprecise (55,161) as applied to arrangements of biological entities and may have a minimum of three meanings. First, it may be taken to imply the natural classifications of some preevolutionary systematists including Linnaeus during the eighteenth century. These were based on Aristotelian notions of the real *natures* or *essences* of the objects involved. Thus a plane triangle is defined as a plane figure bounded by three straight lines. This definition states its *essence*. Attributes that arise as inevitable consequences of the definition are called *properties*: the sum of the internal angles must be 180° and the sum of two sides is greater than the third. Once the essence of a plane triangle is defined it is possible to divide the *genus* triangle, on the basis of the proportions of boundary lines, into the three possible *species*. Other properties that these forms may have, such as size or color, are merely *accidental,* for they do not influence the essential nature of the triangle. The terms *genus* and *species* had definite usages in logic (55,70) and were taken over by the systematists mentioned above. Aristotelian logic, though suitable for classifying the subject matter of geometry, does not lend itself to biological arrangement (57). The reason is that although we can descriptively define an organism we do not know its real essence (nature) because *unanalyzed entities* (55) are being dealt with and we thus have difficulty in excluding mere accidental characters: only deductive reasoning can be applied to situations that call for inductive treatment (57,210).

Second, a classification may be natural in the sense that it is based on a maximum correlation of attributes. Such a classification differs merely in degree from the artificial classifications mentioned above. This degree difference is enough, however, to ensure that such a natural assemblage of organisms need not owe its unity to the possession by all its members of one or a few characteristics in common: any pair of the assemblage share more characteristics than would any member with one of another group. These are the natural phenetic classes of organisms not based on ancestry but produced by grouping according to overall likeness. This may be accomplished intuitively or, more objectively, by following the method of Adanson as developed by Sneath (205,207). In this latter system at least 50, and preferably many more, independent unit characters of strains are needed for estimating similarity coefficients between the operational taxonomic units (strains) used (163). The characters must be truly independent of one another because if two tests measure the same property, that property is being weighted (4,5,205).

Through this process the principle of a hierarchy of characters is reintroduced, but it would no longer be a hierarchy based on preconceived opinions of the importance of characters. The hierarchy would be based *a posteriori* on the observed distribution of characters in different taxa. Designations of taxa are then made on the basis of the ranges of degrees of similarity.

Third, the word "natural" is also equated with phylogenetic classifications, which aim to express evolutionary relationships among organisms. This post-Darwinian development has been the source of much confusion: A tendency has arisen to regard only those arrangements that make phylogenetic sense, which portray the order or design in nature that we think exists (or should exist), as *the* or "good" classification (182,187). It is quite true, of course, that the greater the number of independent characteristics shared by members of a natural group of organisms, the more probable it is that they belong to a single lineage. There may thus indeed be phenetic taxonomies that represent phylogeny more or less closely (100,204), but in the absence of a reliable fossil record this can only be speculative (205). Bisset (36; and see also 182) has defended a phylogenetic concept in bacterial taxonomy, in the absence of a paleontological basis, on the grounds that it is wrong to assume that paleontology has been the major source of information regarding the study of the evolution of animals and plants. The converse is true in that modern concepts of evolution arose from the classification of contemporary forms and paleontology served mainly to test hypotheses derived from those sources: The foundation of theories of evolution lies in the comparative anatomy of existing forms.

At present bacteria are classified in *Bergey's Manual,* ninth edition (33), on a largely nonphylogenetic basis into an hierarchical system of species, genera, and families up to a kingdom *Procaryotae.* However, the definition of a bacterial species is still very subjective, there is no general agreement on the definition of a genus, and "classificatory relationships at the familial and ordinal levels are even less certain than those at the genus and species levels" (212). Phylogenetic information regarding bacteria, obtained largely by the use of methods mentioned in Section 2.3.3 (and see 91,125,211), is still fragmentary but some preliminary rearrangements of bacterial taxa, based on this information, have already been included in the above edition (29,148). The International Committee on Taxonomy of Viruses (ICTV) has set its sails to follow a similar course for viruses (157) although explicitly denying evolutionary or phylogenetic implications (Section 2.4).

When a classification deals with living units it is usual, but by no means essential, to cast them in terms of a hierarchical system often portrayed as a dendrogram and use a terminology based on a genealogical tree with names such as family, genus, and species (or phenons and operational taxonomic units) (135,205). The prestige of phylogeny is so high and ingrained that a common and totally unwarranted assumption is often made that the tree-like arrangement of the groups now represents the cladogeny of a phylogenetic system (70,206).

2.2.2. Nomenclature

"Nature makes individuals agreeing in certain characters and disagreeing in others" (225). Only the individual strain ("variety" in the Linnaean sense) exists (181), but as mentioned (Section 2.2.1) we need to compress information to make it intellectually manageable. We group similar varieties. By a process called logical division (55) Linnaeus distinguished *species* among a particular *genus* of interbreeding plants (56,57)—taxa that he borrowed from Aristotelian logic (Section 2.2.1; 70,210). In this usage a species is thus a category occupying a rank between a variety or strain and genus (156), the latter being the cornerstone of Linnaean taxonomy (56). Those addicted to the Linnaean system would do the same even for individuals constituting asexual groups. To accommodate the latter within the system, a rash of different definitions of species have been devised (114,159). "Species" has even been defined in terms of "genus" (70). It has been said that the fact that criteria for species designation are not necessarily comparable does not make the category any less useful (156). Some low-level taxon is essential: To cope with the number of "different" phage isolated (Section 2.1) it is essential to pool some data from isolates in a taxon of the above-mentioned rank.

Pheneticists (209) have objected to "distortions" of the meaning of the word "species" by phylogeneticists. It is maintained that the latter gave the word an evolutionary or biological definition, though their evidence for calling a population a species was based in most cases on purely phenetic data. If the evidence concerning the limits of a species is phenetic, why not define the species phenetically, that is, on the basis of the number of shared characters? For phylogeneticists (201) such a definition would make taxonomy "superficial" and incomplete. The reason given is that it is at the species level that self-reproduction, the trait that distinguishes organisms from nonorganisms, is incorporated: To use a phenetic definition of "species" is to treat organisms as if they were not fundamentally different from nonorganisms.

Arbitrary rules have been constructed to validate the species concept for viruses (see Section 2.4 for new rules 11, 19, 21 of the ICTV; 157). Excellent judgment is required in such ventures: If viruses are subdivided into too many "species," the purpose is defeated (see above). Should too few be created then there are many exceptions to the rule. A nice balance should be maintained between extremes (34). Apart from the ICTV definition (Section 2.4) other current definitions of a bacteriophage species include (*a*) a group of phage with "significant" genomic relatedness (107); and (*b*) any phage or group of phage that cannot be confused with any other (9). The frequent use of the word species in quotation marks (7,186) is an indication of the embarrassing uncertainty that prevails. A respected bacterial taxonomist has labeled the species concept for microbes (free-living but, like phage, largely asexual) as a macromyth (70).

It is true that Linnaeus used the words "species" and "genus" during the pre-Darwinian era (114) before there could have been intrusion of evolutionary ideas (70,164,182,204). However, these taxa have subsequently become closely associated with phylogeny and sexuality through their successful use in the taxonomy of animals and plants (Section 2.5). In order to avoid any implied relationship with phylogeny, Sneath (205) has adopted the term "phenon" to name clusters of strains (operational taxonomic units) in the system of numerical taxonomy mentioned (Section 2.2.1).

Many plant virologists (92) anticipate two major obstacles to arranging plant viruses according to the Linnaean system: first, the emphasis that is placed on phylogeny as an integral part of classification, and second, the ability of two organisms to be capable of interbreeding as the ultimate test of conspecificity.

Matthews (156) has attempted to allay fears that the use of "species" could have unwarranted evolutionary implications for viruses. Nevertheless a system (here referred to as the "group" system of nomenclature) employing the taxonomic categories "virus group," "virus subgroup," and "member," plus English vernacular names, remains more acceptable to most plant virologists and now applies to many plant viruses (157; Section 2.5). The above ranks correspond roughly to those of family, genus, and species but carry no phylogenetic or sexual implications. These taxa are reminiscent of the process of logical division mentioned above but carried through several more hierarchical stages—Genus summum, Genus intermedium, Genus proximum, Species, and Individuum, which were used by several of Linnaeus's predecessors as their arrangement of ranks (55). The system of "groups" was recommended by the Virus Subcommittee of the International Committee on Bacteriological Nomenclature (ICBN) in 1953 (15,16; Section 2.4). It is a more flexible and less committed series of taxa, and additional terms that could possibly be used if this nomenclature comes into widespread use are "virus supergroup," "node" (69), "cluster," or "clone." This is not a mere juggling with words like "families," "genera," or "species," as Matthews (156) contends. It can also better accommodate variations in viruses and should not prejudice other taxonomies of the future (164).

Ideally, nomenclature should be independent of classification (188) so that any change in the arrangement of an organism need not occasion a change in name. This is not so for the system developed by Linnaeus: Closely associated with Linnaean classification is the latinized binomial nomenclature he used—a generic name followed by the specific epithet. The latter expresses the distinctness of the organism and the former name indicates a relationship to some other cluster(s). The use of Latin has been amply defended (159), and by rules 4, 5, and new rule 13 in its current report (157; Section 2.4), the ICTV recognizes and supports the association between the species concept and latinized binomials.

The vernacular system of naming possesses independence but it provides little information of importance about the organism. It lacks discipline and encourages the proliferation of synonyms. Various compromises have been suggested varying from uninomials (56) to binomials consisting of the vernacular name (or its siglum) and a cryptogram in code for indicating some salient features (51,99,100). A coded identification complex (for bacteria) consisting of numbers and letters has also been proposed (71,203). Many of the names suggested are unwieldy and some convey too little taxonomic information. These systems of nomenclature have not been generally accepted and the ICTV has discouraged the use of cryptograms (156) in favor of a latinized (preferably) binomial nomenclature (157). The "group" system of nomenclature (mentioned above), which is also independent of classification (and phylogeny and interbreeding abilities), is presented in Sections 2.4 and 2.5.

2.3. PROPERTIES OF BACTERIOPHAGES THAT MAY BE USED FOR ARRANGEMENT

Bacteriophages are associated with almost all bacterial genera, the blue-green algae or cyanobacteria (60), and perhaps green algae (8). Their very diversity calls for an arrangement.

A phage may readily be grouped by the use of a few gross characteristics. It requires a much more detailed knowledge of the phage and its habits to detect the subtle distinguishing elements necessary for its arrangement in one of the systems mentioned (Section 2.2.1). Ackermann and colleagues (4,5,8) have appealed for a more rational and thorough investigation of phage for classification purposes. Here some possible discriminants are dealt with merely as an indication of the diversity which exists.

2.3.1. Host Range

Phage usually respect host genus boundaries. This specificity applies to adsorption to receptors (18,35,108,138,183,215,235) that are usually present in only a very limited number of closely related bacterial strains (123). This distribution is affirmed by the large number of papers that have appeared relating bacterial genera to particular sets of phage, and a proposal has been made (5) that the propagating host be identified in the name of a phage. There are some exceptions to this rule of host specificity, notably for phage of bacteria belonging to the family *Enterobacteriaceae,* where polyvalency is fairly common (42). This is also true for phage adsorbing to bacterial appendages such as flagella (17, 120) or plasmid-coded sex pili (44,65,66). In the case of these appendages, phage adsorption is a manifestly reversible step during which the phage is transferred (by various mechanisms) from appendages to secondary somatic receptors (35,101,178,179,196,234).

Adsorption proceeds by steps, the first being recognition by a phage appendage of a specific site on the bacterial surface. This receptor may be part of a protein, a lipopolysaccharide, a teichoic acid, the peptidoglycan, or an exopolysaccharide. The specificity of the receptor depends on both its composition and spatial configuration. Initial attachment following recognition is usually reversible, allowing the phage to position itself. This is followed by irreversible steps involving triggering of conformational changes in the particle. These eventually result in nucleic acid ejection and possible uptake of the ejaculate by the cell. The successive stages follow very rapidly on one another but may be separated with the use of appropriate mutants of the host or phage (35,101,194,228). There are very many variations on the general theme, often partly determined by the morphology of the phage attachment appartus (35).

A phage-sensitive strain of a bacterium may become resistant to a phage by a single mutational step. The strain may also, in some cases, acquire resistance by becoming lysogenic. It is therefore not surprising to find that phage that attack only strains of a single species often do not attack all strains of that species.

These are minor drawbacks that should not be allowed to influence the importance of host range as a distinguishing feature of phage. The relationship is so delicate and yet so lasting in most cases that the words of Lwoff and Tournier (145), ''A virus transcends the host and the disease,'' should not be applied to bacterial viruses.

2.3.2. Morphology of Virions

The primacy of morphological criteria of virions in phage taxonomy was established by Bradley (42) with the introduction of six basic morphological types of bacteriophage (Figure 2.1.). The original types have been increased to 18, of which tailed morphotypes constitute half (8,187). All the present names of bacteriophage families and genera established by the International Committee on Virus Taxonomy (ICTV) are derived from morphological features of phage (157) and most phage catalogs rely heavily on morphology (e.g., 7,9).

Owing to the large number of genes involved in determining the morphol-

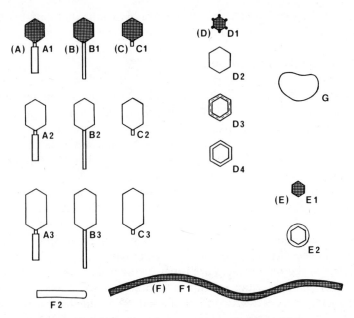

Figure 2.1. Morphological types of bacteriophages. Shaded virions (A–F) represent types A–F of Bradley (42). Partially overlapping these are the morphologies A1–G listed by Reanney and Ackermann (187). These include virions depicted by Ackermann and Eisenstark (8).

ogy of a phage, the latter property is bound to remain a most important distinguishing feature, often allowing "instant categorization" (5). Members of the PRD1 group of closely related phage can be distinguished from one another by their structural protein patterns in SDS gel electrophoresis (27) and filamentous phage have been divided into two classes on the basis of X-ray fiber diffraction patterns of capsid proteins (74,153,154). More attention should be given, however, to the coat proteins of tailed phage (187) by studying their SDS polyacrylamide gel electrophoretic banding patterns (132) and to the location and nature of lipids in particular virions (20,26,106,214,237). Techniques of immuno-electron microscopy should also be more widely applied because they allow comparisons between capsids and not exclusively adsorption structures as studied by neutralization of infectivity (4).

Phage of identical morphology may be found in conjunction with different bacterial genera (3,151) and the B1 morphotype (Figure 2.1) has hosts in a spread of 51 of these (187). However, only phage that are morphologically identical and attack the same host may be serologically related (5). The direct correlation existing between characteristics such as capsid size, particle weight, and nucleic acid content and the relationship between particle buoyant density and its lipid content must be taken into account when considering independent characters suitable for an Adansonian classification

(5). A list of the various physical properties of virions that may be used for characterization has been given by Ackermann (4).

2.3.3. Nucleic Acid of Virions

In 1961 Cooper (67) first classified animal viruses by the nucleic acid in a virion, and Lwoff et al. (143,145) mentioned the "number of strands of nucleic acids" in their system of virus classification. The strandedness of the nucleic acid was also employed as a further distinguishing element by Gibbs et al. in 1966 (100), and Bradley, in 1967 (42), used these features to divide bacteriophages into three groups. Added to these are other fairly gross characteristics of the nucleic acid such as topology, length, and molecular weight of the molecules; whether the strands are composed of sets of permuted or unique base sequences; and the nature of the terminal sequences, whether cohesive or redundant (222). The molecular weight expressed as a percentage of that of the virion, presence of unusual bases, and the guanine–cytosine content of the nucleic acid are other distinguishing properties. In practice the latter characteristics are not as helpful as would appear at first sight: Tailed phage all have a DNA content of about 50%, unusual bases have been detected only in tailed phage (227), and phage and host DNA have similar base compositions (157).

Determination of molecular relatedness may now be made by examination of nucleotide sequences (180) or that of their encoded amino acids by using computers to detect homologies (98), and many phage have been compared by heteroduplex analysis (187). Restriction endonuclease patterns (217,229) serve as a useful complement to the latter, and nucleic acid hybridization affords a further means of comparing genomes of phage (122). Results of the latter test correlate well with morphological and serological properties (190) and hold great promise (107). Restriction endonuclease patterns of DNA and hybridization techniques are extremely sensitive and so can best be used to group closely related viruses (27,107,122). Because of their more general applicability some of the investigations listed above often replace those of *in vivo* cooperative interactions, produced as a result of mixed infection of a host, as indicators of degrees of relatedness among phage.

2.3.4. Strategies of Infection and Morphogenesis

Phage are not free-living organisms. They depend on bacterial hosts for multiplication and sometimes for prolonged accommodation during the lysogenic stage of some of them (113). The extracellular phase (the virion) is the dormant part of the cycle. Confining taxonomic efforts to only the latter structure may yield distorted results that could be compared to attempts at classifying multicellular organisms solely on the properties of their gametes (231). The variety of different strategies of infection (68,218)—the coordinated genetic and biochemical events that are elicited by the invading phage genome culminating in liberation of progeny phage—is an essential adjunct to the above studies.

After adsorption to the specific receptor(s) (Section 2.3.1) ejection and uptake (102) of the phage nucleic acid may occur (83). This mode of host cell infection differentiates phage (but see 108) from most animal viruses (218; Sections 2.1, 2.3.5, 2.4, 2.5). These processes may be very complex and are usually unique to a particular phage group (101). Ejection and uptake may be reasonably well distinguished in phage such as T4, P22, or λ—the irreversible complex varieties (53)—which pipe the ejected DNA and pilot protein (22,101,102) directly to the cytoplasmic membrane without disassembly of the "disposable" protein coat. With others such as the small isometric single-stranded (ss) DNA or RNA phage (101)—the reversible simple viruses (53)—ejection is more an uncoating or decapsidation. The genome comes out of the virion as a result of solubilization of some of the capsomeres at the cell membrane (63,84,172,179,199) with the pilot protein A of the capsid accompanying the nucleic acid into the cell (101,133). With the infecting genome now deposited in the periplasmic space [or potential space (30)] where it may be exposed to degradative enzymes, it must penetrate the cytoplasmic membrane through a pore. This is done differently by various phage with varying degrees of efficiency (93,101).

The filamentous phage are of interest because not only does the pilot protein, the product of gene 3 of the phage (105,180), accompany the DNA into the host cell (63) but also the major capsid protein of the infecting phage is incorporated in the bacterial cytoplasmic membrane (101).

The enveloped phage φ6 (226; Table 2.3) has a blunt taillike structure that possibly constitutes the host attachment apparatus of the phage (167; and see below). The phage attaches to host pili of uncertain nature (178) and also to somatic receptors. Attachment to the latter leads to fusion of the phage lipid-containing envelope with the bacterial outer membrane, resulting in the presence of phage nucleocapsids in the periplasmic space and even in the cytoplasm (21,26), where constituent proteins have also been detected (21,202). This penetration process may be analogous to that employed by some enveloped animal viruses (178) and has also been suggested for infection of *Acholeplasma laidlawii* by mycoplasma type 2 phage (150). Very little is known about the infection process of phage PM2 (178). The cellular receptor for attachment has not been identified but removal of the protein sp43 spikes from the capsid (see below) results in particles that cannot attach to host cells (46).

For the PRD1 group of phage (Table 2.3), which infect Gram-negative bacteria carrying certain plasmids (45), a two-step infection procedure has been proposed. First they attach to receptors on tips of pili coded for by the plasmids; these pili are then retracted and the phage reach secondary somatic receptors. It has not been ruled out, however, that attachment to pilus tips is merely coincidental and not required for infection (178). The lipid layer in these virions does not appear to have a role in the interaction with either of the above receptors but rather in DNA release: The tail appears to be an extension of the inner lipid-containing layer of the phage capsid (24) and emerges from within the virion only when the DNA is to be ejected (141). It may constitute a unique penetration device among phage (178).

Phage do not replicate their nucleic acid by repeatedly making copies identical to the original infecting genome. Replication invariably occurs by way of formation of intermediates whose structures differ from those of mature phage molecules. In the case of single-stranded genomes, the formation of templates with complementary base sequences is required for the synthesis of the next generation of phage chromosomes. For double-stranded (ds) DNA phage the replicating intermediates are often molecules of extended length. There is an overwhelming diversity in mechanisms of DNA replication (168) and each mechanism may have evolved for replication of a molecule with a particular and specific tertiary structure (77,134,174). It is plausible that other functions of phage DNA, such as recombination or the formation of prophage, have complicated the repliction process, and also that many of the control processes involved are exerted to maintain the DNA in a state suitable for replication or to counteract structural modifications imposed by the latter two processes. The RNA-containing phage illustrate a special aspect of control in which the genome and some of its products each assume widely differing tasks and where the distinction between replication and transcription or between structural protein and controlling protein becomes blurred (90).

The patterns of assembly of the complex dsDNA bacteriophages such as ϕ29, λ, T4, T7, and P22 are basically similar (63,86,118,126,172) and the morphogenesis of the small isomeric ssDNA phage ϕX-174 may follow the same general theme (94,170). This pattern of assembly comprising a bewilderment of detail and variety of protein–protein and protein–nucleic acid interactions (127) is worth attention in broad outline because individual idiosyncracies may be of use in future classification schemes (172).

The heads of these phage are all assembled by aggregation of the major head protein (plus a variable number of minor head proteins) with a core protein that often appears to fill the interior of the newly formed prohead. Apart from phage T4 (224) there is no evidence that the core can be preassembled to act as a true scaffold for subsequent assembly of the head around it (128). It seems that coaggregation of the proteins is the usual mode of formation of the prohead. The core may be highly structured, possibly lattice-like, and may well be head length-determining in prolate heads (85,224). For correct assembly of the proheads of nonisometric phage such as T4 or ϕ29 the presence of a protein located in the head–tail connecting region of the virion (named connector) is essential. It represents a symmetry transition between the fivefold vertex of phage head and the sixfold symmetry present in many phage tails. The assembly of proheads of some isometric phage such as P22, T7, and T3 is independent of this connector (61).

Bacterial host involvement in the morphogenesis of phage T4 is indicated by the unique assembly of the prohead and tail base plate on the bacterial membrane (200). Also mutations in the *Escherichia coli* locus *groE* block the development of a number of these phage by interfering with protein cleavage (96,97).

After assembly of the prohead the core protein(s) must be removed either before or during DNA packaging because DNA has to occupy the space inside the mature head. Core protein(s) may be removed by proteolysis, as happens in the case of T4, or they may be removed intact and then recycle to catalyze further assembly in the case of the scaffolding protein of phage P22 (219).

Suitable lengths of DNA concatemers are cleaved off as the DNA is packaged into the head (110). The cleavage can occur either at specific sites in the genome sequence as in bacteriophage λ, to produce the cohesive ends, or simply to package a headful of DNA: For a normal T4 particle this contains about 1.1 genome length of DNA with randomly permuted ends (172). During this process the prohead may undergo expansion (83). The newly packaged DNA is then stabilized by the incorporation of a series of proteins into the head (216). Any model for DNA packaging must be able to account for the subsequent ejection of the DNA from the head. There is great uncertainty about all these issues (83,102,191,197).

Only after completion and filling of the head are the various neck and tail appendages added (155). This addition may occur as a sequence of simple interactions, as happens in the unbranched pathway of assembly followed by phage φ29 (53). For more complex situations it may take place by addition of preformed assemblies in the branched pathway of phage λ or the multiply branched path of phage T4.

The assembly of the ssRNA-containing phage differs from that of the DNA-containing phage just discussed. A few molecules of the coat protein initially complex with the viral RNA strand to form a structure called complex I. Then additional molecules of coat protein are added to form the protein shell around the RNA (116,117) and the host cell membrane may also be involved (195).

The assembly process of filamentous phage is unique (53,77,189,230) and assumed to be similar for all members of the group (81). Briefly (53,169,230) the ssDNA molecule from the replicative form (162), attached to the pilot protein and coated with the gene 5 coded protein, the "intercellular coat protein," associates with the cell membrane. The outer surface of this membrane carries newly synthesized gene 8 coat protein as well as coat protein discarded by the original infecting virion (see above). As the DNA passes through the membrane to be extruded, the gene 5 protein is released inside the cell (where it can be reused) and the DNA (62) now becomes surrounded by the gene 8 coded coat protein present in the membrane (81,104). The virion ssDNA is thus always covered with protein and never appears to be naked even in the host cell (105).

   The assembly of these filamentous ssDNA phage thus differs markedly from that of the ssDNA icosahedral phage mentioned above. The mechanism of phage release, by not causing lysis of the cell membrane, also differs from that of other phage, apart from certain of the mycoplasma phage (151).

   The morphogenesis of the enveloped phage PM2 (Table 2.3) proceeds by the formation of a phospholipid bilayer membrane vesicle from the host bacterium cytoplasmic membrane as a result of a protein introducing a branch in the latter membrane (46). It is not an invagination (63) and the membrane formation does not require an outer scaffolding of coat protein or an inner core (47). Although proteolytic cleavages occur in the morphogenesis of many other phage (see above), proteolytic processing of PM2 proteins has not been detected (49). In this respect it resembles the assembly process of phage P22 (40; and see above). Phage proteins sp13 and sp 6.6 appear to have an as yet unspecified association with this membrane or possibly with the DNA of the phage (46,49,152). The phage DNA is then packaged along with encapsidation of the vesicle by the phage protein sp27 and addition of the spike protein sp43 (47). The phosphatidylglycerol concentration in phage membranes is more than double that of the host in which it matures (46). The specific enrichment of the lipid in the phage membrane is not yet explained but may require extraction of lipid from the host membrane by a protein with affinity for phosphatidylglycerol (48).

   As with phage PM2 the phospholipid membrane of the PRD1 group of phage (Table 2.3) is qualitatively similar to that of the bacterial host. Quantitatively, however, the phage membrane is enriched with respect to phosphatidylglycerol at the expense of phosphatidylethanolamine (46). The mechanism of this change is still unclear. It has been determined, however, that host mutants that are blocked in lipid synthesis still allow the assembly of phage PR4, a member of the PRD1 group. This suggests that the lipids of the virion are acquired from the host membrane and not directly from lipid biosynthesis (192). Also, Muller and Cronan (171) have modified the composition of phage lipids by changing the composition of the host membrane phospholipids. A pathway for assembly has been suggested (76,165) that involves the insertion of the virion membrane proteins into the host membrane and subsequent excision of this virus-specific patch from the membrane through mediation of the major capsid protein and some accessory proteins. The major protein tends to form a curved surface, and this leads to an invagination of the patch and ultimate budding of the area into the cytoplasm of the cell. These cytoplasmic particles now have all, apart from one, of the proteins present in the finished product. They are now filled with DNA (22,25) in the nuclear region of the cell and then move as intact virions to the cell periphery (141). The above picture best fits the present incomplete (24,165) experimental findings and differs from the assembly of large DNA phage mentioned above in at least two other aspects: There is no indication of a scaffolding protein and the host attachment (adsorption) apparatus is functional on particles before DNA is packaged.

The morphogenesis of phage $\phi6$ (Table 2.3) is very different from that of the above two lipid-containing phage. The nucleocapsid is assembled prior to envelopment (23). Assembly involves the formation of a dodecahedral (237) procapsid with four structural proteins, p1, p2, p4, and p7 (137), which then picks up the dsRNA to become a filled procapsid. Addition of a protein (p8) leads to formation of stable nucleocapsids. Phage protein p12 (166), which is not packaged, is required for envelopment of the nucleocapsid by phospholipids. This also involves addition of the cell wall lysin p5 and its precursor p11 and membrane proteins p9, p10, and p6. Addition of protein p3 renders the particle infectious. Protein p5 is not necessary for the membrane formation and resides between the capsid and the envelope (214). Proteins p3 and p6 are also not required for membrane formation. They constitute the host attachment apparatus of the phage (167,214), and it may be that p3 is anchored by p6. The exact function (or location) of membrane proteins p9 and p10 is not known (214), although both are required for nucleocapsid envelopment. The phospholipid composition of the membrane also shows quantitative differences from that of the host membrane (46), and the organization of synthesized phage membrane components as well as the mechanism of membrane formation around the nucleocapsid have not been resolved. What little is known about the maturation of the mycoplasma type 2 phage (Table 2.3) has been dealt with by Maniloff *et al.* (149).

Temporal patterns of phage-directed protein synthesis as determined by SDS polyacrylamide gel electrophoresis followed by autoradiography (111) is a useful method of distinguishing closely related phage, and has been employed to distinguish a number of morphologically T7-like phage (132). This latter procedure may be a method of delineating genera ("subgroups") from the families ("groups") of tailed phage (see Table 2.3; Section 2.5).

2.3.5. Phylogeny

Every organism has a history, and biological understanding should not ignore ancestry. But lack of a paleontological background has contributed to ignorance about the phylogeny of viruses of all affiliations and also bacteria: It has been the root cause of controversies regarding their arrangement (100,156), which (for phage) has been described as chaotic (8).

There is evidence that matters may be changing. Not only is the ontogeny of a number of phage known in some detail (37,46,83,85,86,126,128, 172,197,224) but techniques of present-day molecular biology are uncovering "memories" (187) of past relationships between phage embedded in phage (and bacterial) nucleic acids and in amino acid sequences of peptides they encode. Viruses possess mechanisms that enable them to adapt to changing environments. These include single nucleotide changes, frameshift mutations, gene duplication, reassortment of segmented genomes, and genetic exchanges with hosts.

As to the origins of phage, Bradley (42). Adams (12), and Campbell (58) have suggested that phage might have evolved from the bacterial chromosome as a means of genetic transfer. Zinder (238) has suggested that a primitive phage was simply a segment of bacterial chromosome producing a protein that could encapsidate DNA joined to the DNA encoding the origin of bacterial chromosomal replication. Some experimental support for this thesis exists: There is an amino acid sequence homology between domains of the *E. coli* polymerase I and that of phage T7 DNA polymerase (176); Fiddes et al. (89) have shown that the sequence corresponding to the origin of replication of the *E. coli* K12 chromosome and that of the tailed phage λ and the cubic ssDNA phage G4 are very similar. Also in this vein Bradley (42) proposed a hypothetical sequence of events relating the origin of filamentous phage of the fd type to Inc F plasmid-coded pili. The plasmid in question was assumed to have arisen from a portion of the bacterial chromosome that became detached (187). The pili coded for by this plasmid functioned as carriers of DNA in conjugation and later became autonomous phage.

There are strong indications, based on the nucleotide sequences and organization of the genomes (180), that the Inc N plasmid-specific filamentous phage IKe and Inc F filamentous phage such as M13, fd, or f1 evolved from a common ancestor. The sex plasmid F has maintenance genes that are organized in a similar manner to those of phage P1, and it is possible that this phage originated from plasmid genes (1). Bradley (43) also suggested that the RNA of phage such as MS-2 or Qβ is derived from transcripts of plasmid DNA coding for pili which render the host cell susceptible to infection by the phage. Reanney and Ackermann (187, and see 238) reviewed evidence that RNA-containing phage such as MS-2 and Qβ evolved from bacterial DNA through a recombination event linking genes for the capsid proteins to sequences encoding ribosomal RNA. Landy and Ross (136) have pointed out that sequences of the phage λ *att* site bear close resemblance to elements of the terminal repeat sections of IS1. They speculate that the λ *int* protein is related to the integrase protein(s) of IS1. Substance is lent to this belief by the fact that it is possible to integrate phage λ DNA into bacterial DNA by means of an insertion element (146).

In view of the above, it may well be that sometime in the future a major division will be made between bacteriophages and viruses infecting eukaryotic tissues (see Sections 2.1, 2.4, 2.5).

Reanney and Ackermann (187) considered the following: the survival value of extracellular genes, the relationship between phage and plasmids, and similarities between some bacterial and phage genes. They concluded that the genomes of temperate phage are part of the vertically inherited "selfish" bacterial DNA, that phage genes have probably been a factor in the evolution of host bacterial genomes, and that the evolution of phage (and bacteriocin), plasmid, and chromsomal DNAs should be regarded as a network of interacting processes. Some of the mechanisms involved could be transposition [possibly resulting in converting phage (28)] and fragmentation of DNA by restriction endonucleases aided by interbacterial transfer of nucleic acid affected by conjugation, transduction, or transformation.

When did this "selfish" bacterial nucleic acid (42,184,185), which ensured its future by assuming the extracellular encapsidated form of a phage, arise? Based on their present distribution and the coincidence of G–C values between the DNAs of tailed phage and their hosts, Reanney and Ackermann (187) deduce that the development of bacterial genes for these tailed particles antedated the divergence of many of the host taxa concerned. These genes must have arisen before division of the bacterial world into Grampositive and Gram-negative lines. In fact the genes must have been present in the common ancestor of the archaebacteria, eubacteria, and cyanobacteria at least 2×10^9 years ago. Using some of the arguments mentioned above and also morphology and nucleic acid data, the authors deduce that genes for cubic, filamentous, and pleiomorphic phage possible arose in bacteria at different but more recent times.

The number of different morphologies of tailed phage is best explained by rearrangements occurring in functionally and genetically interchangeable sets of ancestral genes called modules by Botstein and Herskowitz (39), and a scheme depicting interrelationships between tailed phage has been presented (187). The coding regions of two functionally compatible modules may be very different (115), but each is bounded by flanking regions that are homologous, thus ensuring proper placement, by recombination, in a phage genome (236). Viruses found in nature represent favorable combinations of such modules and evolution progresses through selection of modules that retain adequate flanking homology and functional compatibility with the maximum number of combinations of other modules (38,59).

The occurrence of "phage" modules not only in phage genomes but in bacterial host DNA is a distinct possibility (38,184). New conjunctions of phage modules may arise from time to time through genetic events mentioned above and may lead to the appearance of viable variants or various grades of defective bacteriocin-like particles (6,95,140,213). There is evidence that suggests that the R-type pyocins originated from a common temperate bacteriophage, which by mutation became defective in assembly of normal particles (131,193). This is supported by the finding that several *Pseudomonas aeruginosa* strains produce phage that display immunologic cross-reaction with R-type pyocins (124). Furthermore, the isolated tail of one of these phage has a pyocinlike bacteriocidal action (198).

The work on phylogeny discussed above is promising, but many areas are highly speculative. It certainly does not warrant the establishment of a "natural" (Section 2.2.1) system of arrangement for phage although the ICTV may soon have to modify its approach to evolution and phylogeny (Section 2.4).

2.3.6. Serology

Immune reactions of phage have been mentioned above in connection with electron microscopy, the fact that morphologically identical phage of the same host are serologically related, and the correlation that exists between nucleic acid hybridization and serological properties. It remains to be said that the tests employed (neutralization of infectivity, complement fixation, specific aggregation, or immune-electron microscopy) are still the main criteria for determining relatedness between strains and for making fine distinctions between similar phage. Both Adams (12) and Bradley (42) have raised the possibility of granting "low-level" or species ranking, in the same genus, to serological varieties of similar phage. This smacks of Kauffmanism, which was the cause of much absurdity in *Salmonella* taxonomy (73). The ICTV had provisionally approved of the division of adenoviruses into species on grounds of serological specificity (157), but this was disallowed at its Sendai meeting in 1984 (50,229).

2.3.7. Sensitivity to Physical and Chemical Agents

The various tests in use have been listed by Ackermann (4). The most useful are possibly sensitivity to enzymes, organic solvents, ultraviolet light, and temperature. Use of only the solvents chloroform and diethyl ether allows a nice distinction to be made between phage listed (Table 2.3) as *Leviviridae* and those named *Tectiviridae, Corticoviridae,* and *Cystoviridae* (65).

2.3.8. Host- and Environment-Dependent Properties

These are properties such as adsorption velocity, latent period, burst size, plaque morphology, and efficiency of plating of a phage system, which, owing to inherent variability, are not very useful taxonomic properties. Ackermann et al. (5) relegate them to low-level criteria.

2.4. DEVELOPMENT OF THE PRESENT BACTERIOPHAGE TAXONOMY

The development of phage taxonomy may best be seen as occurring in two phases: The first period, prior to 1966, was characterized by individuals promoting particular systems, followed by an era, partially overlapping, of international effort.

d'Hérelle in 1918 (78) and 1921 (79) named the phage he was working with *Bacteriophagum intestinale,* and in 1926 (80), *Protobios bacteriophagus.* This (or these) phage he thought demonstrated extreme powers of adaptive variation. He showed little insight into the implications of applying latinized binomials, and the names had no lasting significance. In the period from 1928 to 1955 various workers (12) used properties such as resistance to physical agents, host range, serology, plaque morphology, and eventually electron microscopy to classify limited groups of phage. These schemes were more like keys or catalogs (Section 2.2.1) but prompted Holmes in 1948 (119) to devise a system with an order *Phagineae,* a genus, and latinized binomial nomenclature to name 46 phage species. This system was thought to be presumptuous, "disastrous" (14), and ill-conceived, and was never applied (158).

Adams (11,12) was the first phage taxonomist of note. He considered all the available criteria for classification and concluded that a Linnaean system, although desirable, was premature. He also warned against grouping phage together with other viruses because doing so would imply a common origin (Sections 2.1, 2.3.5, 2.5).

In the meantime a Virus Subcommittee of the International Committee on Bacteriological Nomenclature (ICBN) had been established at the Fifth International Congress of Microbiology held in Rio de Janeiro in 1950 (158). It was dominated by animal and plant virologists, most of whom were opposed to "applying to viruses binomial names in the Linnaean sense" (14) because it was felt that "such names necessarily carried implications as to taxonomic relationships, and that sufficient evidence as to such relationships was not yet available" (15). The Virus Subcommittee also had a bacteriophage subsection under J. Craigie, who reported in 1953 that "Linnaean binomials could not at present be suitably applied in the phage field" (15). The deliberations of the Virus Subcommittee had very little effect on phage taxonomy of the time, and at a meeting of the subcommittee held in Montreal in 1962, phage workers were not even represented (158). Nevertheless this subcommittee set the pattern for future trends by recognizing the possible fundamental unity of all viruses and establishing the idea of specialist study groups of workers interested in particular sets of viruses who would report back to the subcommittee. Although this subcommittee actively opposed the application of binomials to viruses as a whole (14–16) some "non-Linnaean" binomial names were suggested during 1953 as a "compromise" (16). These names consisted of a "group" rather than genus name with the suffix "virus," followed by the second name, which would not be a species but rather the name of a "group member" (158). Some of the group names of animal viruses became widely used and the suffix "virus" is now in official use for viral generic names (157). The binomials were never popular and in 1962 the committee decided not to propose any further specific names for individual viruses.

In 1962 Lwoff et al. (143) proposed a taxonomy embracing all known viruses including three phage. The scheme was a hierarchical system based on arbitrarily chosen "essential integrants" involving the particular nucleic acid, capsid symmetry, presence of an envelope, and the diameter of helical capsids or the number of capsomeres of cubic capsids. In 1966 the system was expanded (145): Tailed phage were classified under the subphylum *De-*

oxyvira, class *Deoxybinala,* order *Urovirales,* and family *Phagoviridae.* Inexplicably, filamentous phage (family *Inophagoviridae*) were grouped under viruses with cubic capsid symmetry in the class *Deoxycubica* (Table 2.1).

Opposition to this classification as a whole came in the form of criticisms of the arbitrary decisions on the relative importance of different virus characters that were inherent in such a hierarchical classification. Also it was noted that the state of knowledge regarding viruses did not warrant such an all-embracing taxonomy. Gibbs et al. (100) and Gibbs and Harrison (99) were strongly opposed to Lwoff's classification. They proposed that virus taxonomy move toward an Adansonian system, that interim measures should not prejudice the attainment of this goal, and that virus names should consist of two parts: the vernacular name followed by a cryptogram consisting of a coded summary of eight characters.

In 1963 the Executive Committee of the International Association of Microbiological Societies agreed to a request by the Virus Subcommittee of the ICBN that the latter subcommittee be dissolved to make place for an International Committee on Nomenclature of Viruses (ICNV). A provisional committee (PCNV) functioned up to 1966, when the first series of meetings of the ICNV took place during the Ninth International Congress of Microbiology in Moscow (158). This may be regarded as the formal beginning of international viral taxonomy.

The ICNV rejected the bold proposal by the PCNV that an expanded Lwoff–Horne–Tournier taxonomic system (145; Table 2.1) be adopted in its entirety. It also did not accept the system of Gibbs et al. (100). Instead it set up four subcommittees to study groups of viruses. One of these was a bacteriophage subcommittee under the chairmanship of D. E. Bradley. However, interest in taxonomy of phage was limited and Bradley could appoint only three phage workers willing to serve on his subcommittee. Despite the opposing viewpoints of those who did (or did not) want a latinized binomial system, a hierarchical classification, or the use of "group" instead of genus, the ICNV had, by the end of the congress, adopted a code of nomenclature and a set of rules. Salient rules were (233):

3. Nomenclature shall be universally applied to all viruses.

4. An effort will be made towards a latinized binomial nomenclature.

5. Existing latinized names shall be retained whenever feasible.

7. New sigla shall not be introduced.

10. For pragmatic purposes the species is considered to be a collection of viruses with like characters.

11. A genus is a group of species showing certain common characters.

Overlapping with these developments were the 1965 and 1967 publications of Bradley (41, 42). Based on the use of six fundamental morphological

TABLE 2.1. The Lwoff and Tournier (1966) Taxonomic System for Phages (145)[a]

Common Name	Type Species	Genus	Subfamily	Family	Higher Taxa
RNA phage	bacterii	Androphagovirus	Androphagovirinae	Reoviridae	Subphylum: Ribovira Class: Ribocubica Order: Gymnovirales
Phage φX-174	monocatena	Microvirus	—	Microviridae	Subphylum: Deoxyvira Class: Deoxycubica Order: —
Phage fd	bacterii	Inophagovirus	—	Inophagoviridae	Subphylum: Deoxyvira Class: Deoxycubica
Phage T2	coli T secundus	Phagovirus	—	Phagoviridae	Subphylum: Deoxyvira Class: Deoxybinala Order: Urovirales[b]

[a] Relevant data extracted from tables representing the entire virus taxonomy.
[b] See Ref. 4.

phage types that he described (Figure 2.1) and properties of their nucleic acids, he could delineate six groups of phage. He did not go as far as naming specific phage but cast doubt on the validity of employing host range as a major criterion and mentioned the use of serology "at a very low level in the bacteriophage hierarchy." As mentioned (Section 2.3.2), this heavily weighted scheme still constitutes the basis of the present arrangement of phage. Tikhonenko in 1968 (223) proposed a similar scheme but it was less satisfactory because she grouped RNA- and DNA-containing cubic phage together. In 1971, Baltimore (19) divided viruses into classes I–VI on the basis of replication of genetic material and pathways followed for making messenger RNA. Double-stranded DNA-containing phage such as T4 fell in class I, ssDNA phage ϕX-174 belonged to class II, and the RNA-containing phage were placed in a class IV.

At the Tenth International Congress for Microbiology held in 1970 in Mexico City, Bradley's bacteriophage subcommittee submitted various proposals to the ICNV including a suggestion that phage families and genera should conclude with the suffix ". . . bactiviridae" and ". . . bactivirus," respectively. This was presumably thought necessary for the reason that no phage would be found to infect a eukaryote (88). The ICNV adopted a scheme of six genera based on morphology of the phage virions but did not approve of any of the names proposed by the phage subcommittee (Table 2.2; 232). In this scheme phage belonging to types B and C of Bradley were grouped together and a new morphological type (exemplified by the lipid-containing phage PM2) was introduced.

At this Mexico City meeting the ICNV also modified some of the rules made at its 1966 meeting in Moscow and laid down others:

13. The ending of the name of a virus genus is -virus.

14. To avoid changing accepted usage, numbers, letters or combinations may be accepted as names of species.

TABLE 2.2. Proposed Bacteriophage Taxonomic Groups: International Committee on Nomenclature of Viruses (ICNV)—1970[a]

Proposed Genus	Morphological Group of Bradley (25)	Vernacular Name	Type Species
Myovirus	A	T-even	Coliphage T4
Caudaevirus	B	λ phage	Coliphage λ
Lipovirus	—	Lipid phage PM2	Lipid phage PM2
Bullavirus	D	ϕX group	Coliphage ϕX-174
Masculovirus	E	Ribophage group	Coliphage f2
Inovirus	F	Filamentous phages	Coliphage fd

[a] See Ref. 8 and 232

This was official sanction for the use of cryptograms (100), which were subsequently employed (even for phage!) in the first (232) and second (88) reports of the above committee and its successor (see below).

15. These symbols may be preceded by an agreed abbreviation of the latinized name of a selected host genus or, if necessary, by the full name.

17. A family name will end in -idae.

18. A family is a group of genera with common characters.

The events bearing on phage taxonomy that took place between the 1970 Mexico City Congress and the Sixth International Virology Congress held in Sendai during 1984 will be treated together (158). In 1973 the ICNV changed its name to International Committee on Taxonomy of Viruses (ICTV) because the latter was more descriptive of its functions. Phage subcommittees were composed of more members than previously (see above). Their members represented groups of researchers active on particular phage–host systems. The chairmanship alternated between A. Eisenstark and H.-W. Ackermann for two periods each. Successive committees were faced with problems of arranging and naming phages with new morphologies that could not be accommodated in existing systems. Owing to the large numbers of different tailed phage described, the need to modify the Mexico City proposal of only two genera, namely, Myovirus for T-even types and Caudaevirus for phage λ types (Table 2.2; 232), was particularly acute. Bradley had predicted this (42) and Ackermann (2) also foresaw it. In 1974 Ackermann and Eisenstark (8), in a more utilitarian approach to classification, divided phage into 17 morphological types and recommended that tailed types A, B, and C (Figure 2.1; Table 2.2) be subdivided into three types each on the basis of head length. Their system also modified the D morphological group of Bradley: The original D became D1 and a larger type of DNA-containing phage with cubic symmetry was named D2; the latter subtype comprised only two phage (64,112); and the lipid-containing phage PM2 was designated D3. They also expanded Bradley's group E to include a phage with dsRNA having a lipid-containing envelope; this was named E2. Bradley's F group had also to accommodate a phage that appeared as short rods and was named F2. Finally, a phage found in *Acholeplasma* (103) that appeared as a round or oval, apparently enveloped particle was listed as a new morphology and named type G.

The second report of the ICTV embodying decisions made at the 1975 meeting in Madrid was issued in 1976 (88). It listed the phage subcommittee's proposals for Bradley's tailed phage A, B, C (Figure 2.1) as families *Myoviridae, Styloviridae,* or *Pedoviridae,* respectively. It also made recommendations for his type D by replacing *Bullavirus* with the genus *Morulavirus* in a family *Microviridae* and recommended that the genus *Lipovirus* be replaced by the family *Corticoviridae* to contain the type D3 of Ackermann and Eisenstark (8). The report also listed the E2 type of Ackermann and Eisenstark (8) as falling into a recommended monotypic family *Cystoviridae,* replaced the generic name *Masculovirus* (Bradley's type E) by a Ribophage group in a family named *Leviviridae,* and noted the proposal of the committee to establish the Ackermann and Eisenstark (8) type F2 as a second genus—*Plectrovirus* in the recommended family *Inoviridae.* No mention was made of types D2 or G mentioned above.

The third report of the ICTV in 1979 (156) listed approved family names for the PRD1 phage group and for phage belonging to the G morphological type as *Tectiviridae* and *Plasmaviridae,* respectively. Although the previously proposed family names *Corticoviridae, Microviridae, Inoviridae, Leviviridae,* and *Cystoviridae* were now approved, the family names for tailed phage mentioned above were still listed as "proposed." The only change of note in the fourth report of the ICTV (157) was that eight genus names for cubic, filamentous, and pleiomorphic phage mentioned above had been approved and that the family names for phage with contractile tails and those with short noncontractile tails, *Myoviridae* and *Podoviridae* (replacing *Pedoviridae*), respectively, had also been approved. The family name *Siphoviridae* (replacing the proposed *Styloviridae*) for phage with long noncontractile tails was approved by the ICTV at its meetings in Sendai in 1984 (50).

The same 10 phage families mentioned above were also accommodated in a new version, designed by Maurin et al. (160), of the 1962 and 1966 classifications of Lwoff et al. (143, 145; see above and Table 2.2).

The rules of the ICTV regarding viral taxonomy were under continuous scrutiny by that body in the period being surveyed (88,156–158) and some of the changes relevant to this chapter were as follows: Because there always was opposition to the use of a binomial system for names of virus species, the word "binomial" was deleted from rule 4 of the Moscow meeting (see above) at the 1975 meeting in Madrid (87). At the latter meeting rule 7 (see above) was also changed to allow the use of sigla under certain conditions. This was a concession especially to plant virologists, who frequently use this type of abbreviation in deriving names for viruses (157). The round of meetings held in Strasbourg during 1981 yielded a crop of new rules: New rule 11 defined a virus species as a "concept that will normally be represented by a cluster of strains from a variety of sources, or a population of strains from a particular source, which have in common a set or pattern of correlating stable properties that separate the cluster from other clusters of strains"; new rules 19 and 21 defined the genus and family concepts; new rules 13–16 provided guidance regarding the composition of binomial names (the word "binomial" had been deleted from rule 4 in 1975—see above), by stating that (in true Linnaean fashion!) genus should precede the species epithet and that the latter should preferably consist of a single word that could be followed by numbers or letters. The latter thus still sanctioned cryptograms although the ICTV had discontinued printing them in its reports because the cryptogram was thought to have outlived its usefulness (156; but see 54). Although newly designated numbers or letters alone were no longer allowed as species epithets, their long-standing equivalents already widely used were acceptable. This ensured a safe future for treasured names such as λ, T2, MS-2, and P22.

The contents of new rule 18 guarded against overhasty taxonomic decisions by allowing for a minimum period of 3 years to elapse between publication of provisionally approved species names and the definitive approval by ICTV. Finally, the ICTV felt it necessary to state that "virus taxonomy at its present stage has no evolutionary or phylogenetic implications." The present state of phage taxonomy is presented in condensed form in Table 2.3.

2.5. COMMENT

Defects in the arrangement depicted in Table 2.3 are immediately apparent. It represents an all-embracing defective-hierarchical system (see below)

TABLE 2.3. **The International Committee on Taxonomy of Viruses (ICTV) Taxonomy of Bacteriophages—1984**[a]

English Vernacular Family Name	Family	Genus	Type Species
Mycoplasma virus type 2 phage	*Plasmaviridae* (3)[b]	*Plasmavirus*	Phage MV-L2
PRD1 phage group	*Tectiviridae* (8)	*Tectivirus*	Phage PRD1
PM2 phage group	*Corticoviridae* (2)	*Corticovirus*	Phage PM2
Phage with contractile tails	*Myoviridae* (603)	T-even phage group[d]	Coliphage T2
Phage with long, noncontractile tails	*Siphoviridae*[c] (1050)	λphage group[d]	Coliphage λ
Phage with short tails	*Podoviridae* (372)	T7 phage group[d]	Coliphage T7
φX phage group	*Microviridae* (26)	*Microvirus*	Phage φX-174
Rod-shaped phage	*Inoviridae*		
Filamentous phage		*Inovirus* (20)	fd (proposed)
Mycoplasma virus type 1 phage		*Plectrovirus* (13)	MVL51 (proposed)
φ6 phage group	*Cystoviridae* (1)	*Cystovirus*	Phage φ6
ssRNA phage	*Leviviridae* (34)	*Levivirus*	Phage MS-2

[a] Extracted from relevant sections of ICTV reports (157,50).
[b] Figures in parentheses represent numbers of phages present in 1982 (157).
[c] See Ref. 160.
[d] English vernacular name. No approved genus name.

based on arbitrarily chosen weighted characters (almost exclusively virion morphology and nucleic acid type) while explicitly denying evolutionary or phylogenetic associations (Sections 2.3.5, 2.4). A similarly ambitious scheme proposed by Lwoff and Tournier (145) was rejected by the parent body of the ICTV a mere 20 years ago (Section 2.4). In the classical tradition of Aristotle and Linnaeus (55; Section 2.2.1) phage have been arranged into 10 families. The number of individual phage accommodated in families varies from an overwhelming more than 1000 (*Siphoviridae*) to the theoretically (55) embarrassing single entity for the *Cystoviridae*. Many of the families must still be *divided* into genera and all these must then be broken into specific modes (species) (55). Species names are not listed in the table—a "major gap" in the taxonomy (159). Type species are presented (with English vernacular names) but the type species concept may be criticized in that it is too static for viruses with very short generation times and fails to indicate the essential plasticity of the unit (Section 2.2.2). The type species should be replaced by several strains that represent the taxon thus giving some indication of its variability and spread (72).

In the three huge families of tailed phage, no provision is made for the morphological subtypes proposed by Ackermann and Eisenstark (8; Figure 2.1). These nine subtypes were derived from Bradley's A, B, and C morphological types on the basis of head length (Section 2.4). The resulting types are extremely heterogeneous (Section 2.3.2), and this mode of division does not appear to be satisfactory. Heterogeneity is acceptable at the family level but genera of a family should not group totally unrelated phage (Section 2.3.2). This is a major problem of phage arrangement at present, for more than 95% of described phage have tails (4). Here, therefore, are families without approved genera or species; the situation is reminiscent of that of monotypic genera (55).

The families of phage probably represent the most stable taxa established by the ICTV. Possibly because of the nature of the material being dealt with, or the approach to arrangement, it appears easier to categorize groups (families) than either subgroups or members (genera and species—see below). It may be reasonable to use stable properties such as virion morphology and nucleic acid structure (Sections 2.3.2, 2.3.3) to demarcate families or genera. However, attention may have to be given to aspects of phage existence such as the niceties of infection strategy (Section 2.3.4) or even host range and ultrafine structure (Sections 2.3.1, 2.3.2) as possible means of delineating "lower" taxa.

Mention is not made in the table of the D2 subtype of Ackermann and Eisenstark (8; Section 2.4). A possible reason is that, because no additions to this group of two phage have been made since 1974, the ICTV decided to delay consideration until independent verification of the morphology becomes available.

In the present taxonomy no provision is made for inclusion of one or more of the kinds of bacteriocins described (42,95,109,131,140,220; Section

2.3.5). A network of analogies has linked the study of bacteriocins with bacteriophages, and it could be argued that some of these should also be accommodated in a phage arrangement scheme (but see 42).

The present arrangement of phage resembles a diagnostic key in its exclusiveness and rigidity (Section 2.2.1) and the taxonomic structure depicted in Table 2.3 should not be allowed to impede the development of other more satisfactory systems of taxonomy: Taxonomies should be dynamic structures capable of adjustment as insight changes. A general classification would be preferable to a special one because of the weighting (114,173) caused by expressly using only a limited number of characters. Weighting is of course also a problem with general arrangements when decisions are made whether more importance should be attached to some characters of viruses than others (99,100). Such decisions have been made in the past (67,114,119,142–145) but I do not think there is a way of estimating the relative importance of different characters. I therefore agree with Gibbs et al. (100) and Gibbs and Harrison (99) that a bacteriophage classification incorporating the principles of Adanson (210) is the most desirable. Many plant viruses have been successfully classified by the use of an Adansonian method (92). Such an approach may not be possible for most phage at present owing to the large number of unit characters required (Sections 2.2.1, 2.3), but this is the goal to strive for. Even if results turn out to be very similar to the present classification the arrangement would have been achieved as objectively as possible.

The other objection is against the aura of phylogenetic classicism imparted to this taxonomy by the nomenclature used. This is denied by the ICTV (Sections 2.2.2, 2.4). For many—those "persons who are unduly influenced by a traditional definition of biological species which rests on barriers to sexual reproduction" (129)—the words "genus" and "species" are indelibly associated with interbreeding animals and plants in the hierarchical system of taxonomy devised by Linnaeus employing a latinized binomial nomenclature (92). These words also carry evolutionary overtones (164). The argument that Linnaeus applied the term "species" before Darwin's evolutionary theories were announced and could thus not have been infuenced by them (Section 2.2.2) is true: I maintain (in agreement with Milne, 164) that, owing to the breeding habits of the plants and animals he was working with, he *discovered* species. He did not create them. He did not even invent the name (56). This is no trick argument. There has been opposition to a latinized binomial system of nomenclature for viruses since the earliest days of international efforts to create a system (Section 2.4). It appears that the ICNV (later ICTV) has since its inception and, despite longstanding (15,16,87) and continued (164) opposition, favored a latinized binomial nomenclature. For reasons mentioned above (68; Sections 2.2.2, 2.4), this is tantamount to favoring a Linnaean classification. This attitude dates back to at least 1966 (16,233) *before* there was any formally approved virus classification (Section 2.4). It is a classical example of what nomenclature should not be. Nomenclature should not dictate arrangement. The latter should determine the former (208). Put in another way, "Nomenclature should be the handmaid of classification" (13). Cain (55) has also questioned the desirability of employing a binomial nomenclature in biological taxonomy.

It is accepted that there are different interpretations of the word "species" (70,159) and that the massive prokaryotic taxonomy is thoroughly enmeshed in a Linnaean system (33) although alternatives are available (207). It is argued that in these early days of phage taxonomy it is complicating matters unnecessarily to attempt the encasement of phage in a similar system and it is of little comfort that problems are also being encountered with the application of the above system to other organisms (159,164). Phage are subcellular particles of uncertain origin and phylogeny (Section 2.3.5) and are known to exchange genetic information with their hosts at times: They are not the candidates for a latinized binomial system of nomenclature, which implies a strict phylogenetic hierarchy (164). Plasmids resemble bacteriophages in many respects (1,52,75,121,175,187) but attempts to arrange plasmids into a system similar to that existing for phage (Table 2.3) would certainly not be taken seriously. Both are host-dependent for replication and are smaller and possibly less complex than their bacterial hosts. Why, then, employ a classical hierarchical taxonomy for phage and not for other non-free-living entities such as plasmids, transposons, or even insertion elements?

The categories of the "group" system discussed in Section 2.2.2—"virus group," "virus subgroup," and "member" (151)—are not seen as exact equivalents of families, genera, and species and must not be regarded as a vague nondescript dumping ground for entities that cannot be accommodated in the latter. The "group" system should be considered as an independent system of nomenclature that is more flexible and less committed and can better accommodate the nature of the variation encountered in clonally reproducing units. The groups may still be "loosely defined" (92) but this could be changed (if necessary) as implementation proceeds and discipline is applied. Little serious attention has yet been devoted to the "group" system of nomenclature, but see Table 1 in Milne (164) for an example. It must still be honed to suit requirements. Rules must be made. The beauty of the situation is that this is exactly what can be done because there are no existing procedures.

The "group" system of taxon nomenclature will not prejudice future development in phage taxonomy: When an Adansonian system of classification for phage eventually becomes feasible (Section 2.2.1) and the different levels of affinity are established, this numerical classification could simply be equated with the various ranks of the "group" system (Section 2.2.2). Alternatively, it could then be decided to abandon the "group" system of nomenclature altogether in favor of the various levels of phenons advocated by Sneath (207). Although it appears unlikely that new groups of phage will be detected in the well-studied eubacteria (but see 82), the archaebacteria could yield new types to be arranged and named. There is no reason why the "group" system could not handle such a development satisfactorily. Eventually when a phylogeny of phage may have to be taken into account (Section 2.3.5), the envisaged "group" nomenclature of these clonally reproducing entities would, in unchanged guise, still be the most correct system for accommodating a phylogeny.

The ICTV is committed by rule 3 (157; Section 2.4) to a universal nomenclature applied to all viruses, yet plant viruses constitute nearly half of the virus individual page entries in the above report, and most of these viruses are categorized in terms of a "group" nomenclature. Whereas in 1970 there were only 16 such groups, the number of approved groups had risen to 24 by 1982 (157). This may be taken as some measure of the support given to "group" nomenclature.

This appeal for a change in phage nomenclature is not heretical. Bacterial taxonomy has also undergone dramatic changes: As recently as the seventh edition of *Bergey's Manual* (31) the classification presented was claimed to be a natural (phylogenetic) scheme. For instance, the emphasis placed on the mode of flagellar insertion in the treatment of true bacteria reflected the evolutionary significance attached to this structural character by Orla-Jensen (177) and by Kluyver and van Niel (130). The eighth edition of the manual (32) abandoned all attempts at a phylogenetic approach and grouped organisms under vernacular headings for purposes of identification—a purely artificial arrangement. The present situation regarding bacterial taxonomy is mentioned in Section 2.2.1.

As mentioned above (also Section 2.2.2) the genus–species system is married to a latinized binomial nomenclature (55). The "group" system proposed will of course also require universal standard names for the phage. A change in the names of categories and perhaps those of scientific names of phage (208), incumbent on the adoption of the latter system of nomenclature, need not be a deterrent: The state of flux of the present system is indicated by the frequent substitution of "group" for the category "genus" in the present ICTV report (157); and proposed scientific names of phages and their categories have in the past been subject to drastic changes (Section 2.4; Tables 2.1–2.3). The novelty of the "group" system lies in the fact that no constraints need be placed on whether it is multi- or uninomial or even on the succession of words or symbols should a multinomial system be preferred. Linnaeus described a particular species by a *differentia* (55,221)—a brief descriptive phrase of no more than 12 words. He also produced a single catchword known as the *trivial epithet* for each species. With use, the latter replaced the former and became known as the *specific epithet* (56,70,73). There is thus ample precedence for naming phage with the use of more than two words! For plant viruses Milne (164) favors the English vernacular name followed by the group name as the possible basis of a binomial nomenclature.

The essentially arbitrary names of many phages are here to stay and must be accommodated in a proposed new system of nomenclature. Some reference to the propagating host is valuable for reasons mentioned (Section 2.3.1) but may be cumbersome, whereas a group name relates the phage to clusters of similar phage. Ackermann et al. (5) have suggested certain rules for designating an abbreviated genus–species name of the host bacterium followed by that of the phage, which usually consists of Latin capitals, Greek letters, or numerals (151). This is in agreement with new rule 15 of the ICNV (Section 2.4) and corresponds to the provisionally approved method of naming adenovirus species (157). Use of the host name may complicate matters because bacterial nomenclature is also in a state of flux (see above), but advantages nevertheless outweigh this objection. Perhaps it would be more suitable to be more elastic and not to specify the exact bacterial name component. The latter may even be omitted in the case of broad-host-range phage such as those that adsorb to bacterial appendages (Section 2.3.1). It should even be permissible to use mnemonics or sigla provided the final name of the phage is unique and informative. The names of many bacteriocins include the trivial designation of the producer strain (220).

The analogy with animal and plant viruses—the unity of virology—can be overplayed (see rule 3 of ICNV, 233; Sections 2.1, 2.3.5, 2.4) and it may be better to replace "virus" with "phage" in the "group" system advocated here (see above; also Section 2.2.2). The group (or subgroup, 164) name, followed by the vernacular phage name, could yield binomials such as Myocoli-, Myoeco-, or Myoentero-phage T4 (157), Podobruphage Bk (10), Siphopseudophage F116 (139), Myostrepphage RZh (7), Siphobacphage PBP1 (186), Podovibriophage 4996 (9), Plasmaacholephage L2 (but see 151), and Leviphage MS2 (157).

New names often create a psychological resistance to their use. The setting up of a phage nomenclature couched in terms of the "group" system, international in flavor, not unwieldy, and useful, will be a formidable task.

If the ICTV should ever support such a system of phage nomenclature the approach should then be cautious to retain consensus and avoid more mistakes.

REFERENCES

1. Abeles, A. L., K. M. Snyder, and D. K. Chattoraj. 1984. P1 plasmid replication:Replicon structure. J. Mol. Biol. *173*:307–324.

2. Ackermann, H.-W. 1969. Bactériophages propriétes et premières étapes d'une classification. Path.-Biol. *17*:1003–1024.

3. Ackermann, H.-W. 1975. La classification des bacteriophages des cocci gram-positifs: *Micrococcus, Staphylococcus* et *Streptococcus*. Path.-Biol. *23*:247–253.

4. Ackermann, H.-W. 1983. Current problems in bacterial virus taxonomy, pp. 105–121. *In* R. E. F. Matthews (ed.), A Critical Appraisal of Viral Taxonomy. CRC Press, Boca Raton, Florida.

5. Ackermann, H.-W., A. Audurier, L. Berthiaume, L. A. Jones, J. A. Mayo, and A. K. Vidaver. 1978. Guidelines for bacteriophage characterization. Adv. Virus Res. *23*:1–24.

6. Ackermann, H.-W., and G. Brochu. 1978. Particulate bacteriocins, pp. 691–737. *In* A. I. Laskin and H. I. Lechevalier (eds.), CRC Handbook of Microbiology, 2nd ed., Vol. 2. CRC Press, Cleveland, Ohio.

7. Ackermann, H.-W., E. D. Cantor, A. W. Jarvis, J. Lembke, and J. A. Mayo. 1984. New species definitions in phages of gram-positive cocci. Intervirology *22*:181–190.

8. Ackermann, H.-W., and A. Eisenstark. 1974. The present state of phage taxonomy. Intervirology *3*:201–219.

9. Ackermann, H.-W., S. S. Kasatiya, T. Kawata, T. Koga, J. V. Lee, A. Mbiguino, F. S. Newman, J.-F. Vieu, and A. Zachary. 1984. Classification of *Vibrio* bacteriophages. Intervirology *22*:61–71.

10. Ackermann, H.-W., F. Simon, and J.-M. Verger. 1981. A survey of *Brucella* phages and morphology of new isolates. Intervirology *16*:1–7.

11. Adams, M. H. 1953. Criteria for a biological classification of bacterial viruses. Ann. N. Y. Acad. Sci. *56*:442–459.

12. Adams, M. H. 1959. Bacteriophages. Interscience, New York.

13. Ainsworth, G. C. 1955. Nomenclature, the handmaid of classification. J. Gen. Microbiol. *12*:322–323.

14. Andrewes, C. H. 1953. The Rio Congress decisions with regard to study of selected groups of viruses. Ann. N.Y. Acad. Sci. *56*:428–432.

15. Andrewes, C. H. 1954. Nomenclature of viruses. Nature *173*:620–621.

16. Andrewes C. H., and P. H. A. Sneath. 1958. The species concept among viruses. Nature *182*:12–14.

17. Appelbaum, P. C., N. Hugo, and J. N. Coetzee. 1971. A flagellar phage for the Proteus-providence group. J. Gen. Virol. *13*:153–162.

18. Archibald, A. R. 1980. Phage receptors in Gram-positive bacteria, pp. 7–26. *In* L. L. Randall and L. Philipson (eds.), Virus receptors, Part I: Bacterial Viruses, Series B, Vol. 7. Chapman and Hall, London.

19. Baltimore, D. 1971. Expression of animal virus genomes. Bact. Rev. *35*:235–241.

20. Bamford, D. H. 1981. Lipid-containing bacterial viruses: Disruption studies on ϕ6, pp. 477–489. *In* L. DuBow (ed.), Bacteriophage Assembly. A. R. Liss, New York.

21. Bamford, D. H., and K. Lounatmaa. 1978. Freeze-fracturing of *Psuedomonas phaseolicola* infected by the lipid-containing bacteriophage ϕ6. J. Gen. Virol. *39*:161–170.

22. Bamford, D., T. McGraw, G. MacKenzie, and L. Mindich. 1983. Identification of a protein bound to the termini of bacteriophage PRD1 DNA. J. Virol. *47*:311–316.

23. Bamford, D. H., and L. Mindich. 1980. Electron microscopy of cells infected with nonsense mutants of bacteriophage ϕ6. Virology *107*:222–228.

24. Bamford, D. H., and L. Mindich. 1982. Structure of the lipid-containing bacteriophage PRD1: Disruption of wild-type and nonsense mutant phage particles with guanidine hydrochloride. J. Virol. *44*:1031–1038.

25. Bamford, D. H., and L. Mindich. 1984. Characterization of the DNA-protein complex at the termini of the bacteriophage PRD1 genome. J. Virol. *50*:309–315.

26. Bamford, D. H., E. T. Palva, and K. Lounatmaa. 1976. Ultrastructure and life cycle of the lipid-containing bacteriophage ϕ6. J. Gen. Virol. *32*:249–259.

27. Bamford, D. H., L. Rouhiainen, K. Takkinen, and H. Söderlund. 1981. Comparison of the lipid-containing bacteriophages PRD1, PR3, PR4, PR5 and L17. J. Gen. Virol. *57*:365–373.

28. Barksdale L., and S. B. Arden. 1974. Persisting bacteriophage infections, lysogeny and phage conversions. Ann. Rev. Microbiol. *28*:265–299.

29. Baumann, P., L. Baumann, M. J. Wooekalis, and S. S. Bang. 1983. Evolutionary relationships in *Vibrio* and *Photobacterium*: A basis for a natural classification. Ann. Rev. Microbiol. *37*:369–398.

30. Bayer, M. E. 1968. Adsorption of bacteriophage to adhesions between cell wall and membrane of *Escherichia coli*. J. Virol. *2*:346–356.

31. Bergey's Manual of Determinative Bacteriology, 7th ed. 1957. R. S. Breed, E. G. D. Murray, and N. R. Smith (eds.). Williams and Wilkins, Baltimore, Maryland.

32. Bergey's Manual of Determinative Bacteriology, 8th ed. 1974. R. E. Buchanan and N. E. Gibbons (eds.). Williams and Wilkins, Baltimore, Maryland.

33. Bergey's Manual of Systematic Bacteriology, Vol. 1, 9th ed. 1984. N. R. Krieg and J. G. Holt. (eds.). Williams and Wilkins, Baltimore, Maryland.

34. Bessey, C. E. 1908. The taxonomic aspect of the species question. Am. Nat. *42*:218–224.

35. Beumer, J., E. Hannecart-Pokorni, and C. Godard. 1984. Bacteriophage receptors. Bull. L'Inst. Pasteur *82*:173–253.

36. Bisset, K. A. 1962. The phylogenetic concept in bacterial taxonomy, pp. 361–373. *In* G. C. Ainsworth and P. H. A. Sneath (eds.). Microbial Classification, twelfth Symposium of the Society for General Microbiology. Cambridge University Press, Cambridge.

37. Black, L. W., and M. K. Showe. 1983. Morphogenesis of the T4 head, pp. 219–245. *In* C. K. Matthews, E. M. Kutter, G. Mosig, and P. G. Berget (eds.), Bacteriophage T4. American Society for Microbiology, Washington, DC.

38. Botstein, D. 1980. A theory of modular evolution for bacteriophages. Ann. N.Y. Acad. Sci. *354*:484–491.

39. Botstein, D., and I. Herskowitz. 1974. Properties of hybrids between *Salmonella* phage P22 and coliphage λ. Nature *251*:584–589.

40. Botstein, D., C. H. Waddell, and J. King. 1973. Mechanism of head assembly and DNA encapsulation in *Salmonella* phage P22. J. Mol. Biol. *80*:669–695.

41. Bradley, D. E. 1965. The morphology and physiology of bacteriophages as revealed by the electron microscope. J. R. Microsc. Soc. *84*:271–314.

42. Bradley, D. E. 1967. Ultrastructure of bacteriophages and bacteriocins. Bact. Rev. *31*:230–314.

43. Bradley, D. E. 1971. A comparative study of the structure and biological properties of bacteriophages, pp. 207–227. *In* K. Maramorosh and E. Kurstak (eds.), Comparative Virology. Academic, New York.

44. Bradley, D. E., J. N. Coetzee, T. Bothma, and R. W. Hedges. 1981. Phage X: A plasmid-dependent, broad host range, filamentous bacterial virus. J. Gen. Microbiol. *126*:389–396.

45. Bradley, D. E., and E. L. Rutherford. 1975. Basic characterization of a lipid-containing bacteriophage specific for plasmids of the P, N and W compatibility groups. Can. J. Microbiol. *21*:152–163.

46. Brewer, G. J. 1980. Control of membrane morphogenesis in bacteriophage. Int. Rev. Cytol. *68*:53–96.

47. Brewer, G. J. 1983. Mutant *ts1* of bacteriophage PM2 is defective in the major capsid protein and fails to package its DNA. J. Virol. *45*:226–232.

48. Brewer, G. J., and R. M. Goto. 1983. Accessibility of phosphatidylethanolamine in bacteriophage PM2 and in its Gram-negative host. J. Virol. *48*:774–778.

49. Brewer, G. J., and M. Singh. 1982. Kinetics and characterization of the proteins synthesized during infection by bacteriophage PM2. J. Gen. Virol. *60*:135–146.

50. Brown, F. 1986. The classification and nomenclature of viruses: Summary of results of meetings of the International Committee on Taxonomy of Viruses in Sendai, September 1984. Intervirology *25*:141–143.

51. Buck, K. W. 1983. Current problems in fungal virus taxonomy, pp. 139–176. *In* R. E. F. Matthews (ed.), A Critical Appraisal of Viral Taxonomy. CRC Press, Boca Raton, Florida.

52. Bukhari, A. I., J. A. Shapiro, and S. L. Adhya (eds.). 1977. DNA Insertion, Elements, Plasmids and Episomes. Cold Spring Harbor Laboratory, Cold Spring Harbor, New York.

53. Butler, P. J. G. 1979. Assembly of regular viruses, pp. 205–237. *In* R. E. Offord (ed.), International Review of Biochemistry; Chemistry of Macromolecules IIB, Vol. 25. University Park Press, Baltimore, Maryland.

54. Bystryi, N. F., and M. S. Drozhevkina. 1979. Basic criteria and classification scheme for bacteriophages acting against cholera virbrios of the classical and El Tor biotypes. J. Hyg. Epidem. Microbiol. Immunol. *23*:85–94.

55. Cain, A. J. 1958. Logic and memory in Linnaeus's system of taxonomy. Proc. Linn. Soc. Lond. *169*:144–163.

56. Cain, A. J. 1959. The post-Linnaean development of taxonomy. Proc. Linn. Soc. Lond. *170*:234–244.

57. Cain, A. J. 1962. The evolution of taxonomic principles, pp. 1–13. *In* G. C. Ainsworth and P. H. A. Sneath (eds.), Microbial Classification, Twelfth Symposium of the Society for General Microbiology. Cambridge University Press, Cambridge.

58. Campbell, A. 1961. Conditions for the existence of bacteriophage. Evolution *15*:153–165.

59. Campbell, A., and D. Botstein. 1983. Evolution of the Lambdoid phages, pp. 554–593. *In* R. W. Hendrix, J. W. Roberts, F. W. Stahl, and R. A. Weisberg (eds.), Lambda II. Cold Spring Harbor Laboratory, Cold Spring Harbor, New York.

60. Carr, N. G., and B. A. Whitton. 1982. The Biology of Cyanobacteria, Vol. 19. Blackwell, London.

61. Carrascosa, J. L., J. N. Carazo, C. Ibanez, and A. Santisteban. 1985. Structure of phage ϕ29 connector protein assembled *in vivo*. Virology *141*:190–200.

62. Casadevall, A., and L. A. Day. 1982. DNA packing in the filamentous viruses fd, Xf, Pf1 and Pf3. Nucleic Acids Res. *10*:2467–2481.

63. Casjens S., and J. King. 1975. Virus assembly. Ann. Rev. Biochem. *44*:555–611.

64. Clark-Walter, G. D., and S. B. Primrose. 1971. Isolation and characterization of a bacteriophage SiI for *Spirillum itersonii*. J. Gen. Virol. *11*:139–145.

65. Coetzee, J. N., D. E. Bradley, J. Fleming, L. du Toit, V. M. Hughes, and R. W. Hedges. 1985. Phage pilHα: A phage which adsorbs to IncHI and IncHII plasmid-coded pili. J. Gen. Microbiol. *131*:1115–1121.

66. Coetzee, J. N., D. E. Bradley, R. W. Hedges, J. Fleming, and G. Lecatsas. 1983. Bacteriophage M: an incompatibility group M plasmid specific phage. J. Gen. Microbiol. *129*:2271–2276.

67. Cooper, P. D. 1961. A chemical basis for the classification of animal viruses. Nature *190*:302–305.

68. Cooper, P. D. 1974. Towards a more profound basis for the classification of viruses. Intervirology *4*:317–319.

69. Cowan, S. T. 1955. The philosophy of classification. J. Gen. Microbiol. *12*:314–321.

70. Cowan, S. T. 1962. The Microbial species—a macromyth?, pp. 433–455. *In* G. C. Ainsworth and P. H. A. Sneath (Eds.), Microbial Classification, Twelfth Symposium of the Society for General Microbiology. Cambridge University Press, Cambridge.

71. Cowan, S. T. 1965. Principles and practice of bacterial taxonomy—a forward look. J. Gen. Microbiol. *39*:143–153.

72. Cowan, S. T. 1970. Heterical taxonomy for bacteriologists. J. Gen. Microbiol. *61*:145–154.

73. Cowan, S. T. 1978. L. R. Hill (ed.), A Dictionary of Microbial Taxonomy. Cambridge University Press, Cambridge.

74. Crowther, R. A. 1980. Structure of bacteriophage Pf1. Nature *286*:440–441.

75. Davey, R. B., and D. C. Reanney. 1980. Extrachromosomal genetic elements and the adaptive evolution of bacteria. Evol. Biol. *13*:113–147.

76. Davis, T. N., and J. E. Cronan. 1983. Nonsense mutants of the lipid-containing bacteriophage PR4. Virology *126*:600–613.

77. Denhardt, D. T. 1975. The single-stranded DNA phages. CRC Crit. Rev. Microbiol. *4*:161–223.

78. d'Herelle, F. 1918. Technique de la recherche du microbe filtrant bactériophage (*Bacteriophagum intestinale*). C.R. Soc. Biol. Paris *81*:1160–1167.

79. d'Herelle, F. 1921. Le bactériophage: son rôle dans l'immunité. Masson, Paris.

80. d'Herelle, F. 1926. The Bacteriophage and its Behavior. Williams & Wilkins, Baltimore, Maryland.

81. Dunker, A. K. 1980. Comments on filamentous phage assembly, pp. 383–388. *In* M. S. DuBow (ed.), Bacteriophage Assembly. Alan R. Liss, New York.

82. Dybvig, K., J. A. Nowak, T. L. Sladek, and J. Maniloff. 1985. Identification of an enveloped phage, mycoplasma virus L172, containing a 14 kilobase single-stranded DNA genome. J. Virol. *53*:384–390.

83. Earnshaw, W. C., and S. R. Casjens. 1980. DNA packaging by the double-stranded DNA bacteriophages. Cell *21*:319–331.

84. Edgell, M. H., and W. Ginoza. 1965. The fate during infection of the coat protein of the spherical bacteriophage R17. Virology *27*:23–27.

85. Eiserling, F. A., J. Corso, S. Feng, and R. H. Epstein. 1984. Intracellular morphogenesis of bacteriophage T4. II. Head morphogenesis. Virology *137*:95–101.

86. Feiss, M., and A. Becker. 1983. DNA packaging and cutting, pp. 449–484. *In* R. W. Hendrix, J. W. Roberts, F. W. Stahl, and R. A. Weisberg (ed.), Lambda II. Cold Spring Harbor Laboratory, Cold Spring Harbor, New York.

87. Fenner, F. 1976. The classification and nomenclature of viruses: Summary of results of meetings of the International Committee on Taxonomy of Viruses in Madrid, September 1975. Virology *71*:371–378.

88. Fenner, F. 1976. Classification and nomenclature of viruses: Second report of the International Committee on Taxonomy of Viruses. Intervirology *7*:1–115.

89. Fiddes, J. C., B. G. Barrell, and G. N. Godson. 1978. Nucleotide sequences of the separate origins of synthesis of bacteriophage G4 viral and complementary DNA strands. Proc. Natl. Acad. Sci. USA *75*:1081–1086.

90. Fiers, W. 1979. Structure and function of RNA bacteriophages, pp. 69–98. *In* H. Fraenkel-Conrat and B. Wagner (eds.), Comprehensive Virology, Vol. 13. Plenum, New York.

91. Fox, G. E., E. Stackebrandt, R. B. Hespell, J. Gibson, J. Maniloff, T. A. Dyer, R. S. Wolfe, W. E. Balch, R. S. Tanner, L. J. Magrum, L. B. Zablen, R. Blakemore, R. Gupta, L. Bonen, B. J. Lewis, D. A. Stahl, K. R. Luehrsen, K. N. Chen, and C. R. Woese. 1980. The phylogeny of prokaryotes. Science *209*:457–463.

92. Francki, R. I. B. 1983. Current problems in plant virus taxonomy, pp. 63–104. *In* R. E. F. Matthews (ed.), A Critical Appraisal of Viral Taxonomy. CRC Press, Boca Raton, Florida.

93. Fuchs, P., and A. Kohn. 1983. Changes induced in cell membranes adsorbing animal viruses, bacteriophages and colicins. Curr. Top. Microbiol. Immunol. *102*:57–99.

94. Fujisawa, H., and M. Hayashi. 1977. Assembly of bacteriophage ϕX174: Identification of

a virion capsid precursor and proposal of a model for the functions of bacteriophage gene products during morphogenesis. J. Virol. *24*:303–313.

95. Garro, A. J., and J. Marmur. 1970. Defective bacteriophages. J. Cell. Physiol. *76*:253–264.

96. Georgopoulos, C. P., R. W. Hendrix, S. R. Casjens, and A. D. Kaiser. 1973. Host participation in bacteriophage lambda head assembly. J. Mol. Biol. *76*:45–60.

97. Georgopoulos, C. P., R. W. Hendrix, A. D. Kaiser, and W. B. Wood, 1972. Role of the host cell in bacteriophage morphogenesis: Effect of a bacterial mutation on T4 head assembly. Nature New Biol. *239*:38–41.

98. Gibbs, A. J. 1980. How ancient are the Tobamoviruses? Intervirology *14*:101–105.

99. Gibbs, A. J., and B. D. Harrison. 1968. Realistic approach to virus classification and nomenclature. Nature *218*:927–929.

100. Gibbs, A. J., B. D. Harrison, D. H. Watson, and P. Wildy. 1966. What's in a virus name? Nature *209*:450–454.

101. Goldberg, E. 1980. Bacteriophage nucleic acid penetration, pp. 117–141. *In* L. L. Randall and L. Philipson (eds.), Virus Receptors, Part I: Bacterial Viruses, Series B, Vol. 7. Chapman and Hall, London.

102. Goldberg, E. 1983. Recognition, attachment and injection, pp. 32–39. *In* C. K. Mathews, E. M. Kutter, G. Mosig, and P. B. Berget (eds.), Bacteriophage T4. American Society for Microbiology, Washington, DC.

103. Gourlay, R. N. 1971. Mycoplasmatales virus-laidlawii 2, a new virus isolated from *Acholeplasma laidlawii*. J. Gen. Virol. *12*:65–67.

104. Grant, R. A., T.-C. Lin, R. E. Webster, and W. Koningsberg. 1980. Structure of filamentous bacteriophage: Isolation, characterization and localization of the minor coat proteins and orientation of the DNA, pp. 413–428. *In* M. S. DuBow (ed.), Bacteriophage Assembly. Alan R. Liss, New York.

105. Gray, C. W., G. G. Kneale, K. R. Leonard, H. Siegrist, and D. A. Marvin. 1982. A nucleoprotein complex in bacteria infected with Pf1 filamentous virus: Identification and electron microscopic analysis. Virology *116*:40–52.

106. Greenberg, N., and S. Rottem. 1979. Composition and molecular organization of lipids and proteins in the envelope of Mycoplasmavirus MVL2. J. Virol. *32*:717–726.

107. Grimont, F., and P. A. D. Grimont. 1981. DNA relatedness among bacteriophages of the morphological group C3. Curr. Microbiol. *6*:65–69.

108. Haberer, K., and J. Maniloff. 1982. Adsorption of the tailed mycoplasma virus L3 to cell membranes. J. Virol. *41*:501–507.

109. Hardy, K. G. 1975. Colicinogeny and related phenomena. Bact. Rev. *39*:464–515.

110. Harrison, S. C. 1983. Packaging of DNA into bacteriophage heads: A model. J. Mol. Biol. *171*:577–580.

111. Hausmann, R. 1976. Bacteriophage T7 genetics. Curr. Top. Microbiol. Immunol. *75*:77–110.

112. Hashimoto, T., D. L. Diedrich, and S. F. Conti. 1970. Isolation of a bacteriophage for *Bdellovibrio bacteriovorus*. J. Virol. *5*:97–98.

113. Hayes, W. 1968. Temperate bacteriophages and lysogeny, pp. 447–479. *In* The Genetics of Bacteria and their Viruses: Studies in Basic Genetics and Molecular Biology, 2nd ed. Blackwell, Oxford.

114. Heslop-Harrison, J. 1962. Purposes and procedures in the taxonomic treatment of higher organisms, pp. 14–36. *In* G. C. Ainsworth and P. H. A. Sneath (eds.), Microbial Classification, Twelfth Symposium of the Society for General Microbiology. Cambridge University Press, Cambridge.

115. Hilliker, S., and D. Botstein. 1976. Specificity of genetic elements controlling regulation of early functions in temperate bacteriophages. J. Mol. Biol. *106*:537–566.

116. Hohn, T. 1967. Self-assembly of defective particles of the bacteriophage fr. Eur. J. Biochem. *2*:152–158.

117. Hohn, T. 1976. Packaging of genomes in bacteriophages: A comparison of ss RNA bacteriophages and ds DNA bacteriophages. Phil. Trans. R. Soc. Lond. B *276*:143–150.

118. Hohn, T., and I. Katsura. 1977. Structure and assembly of bacteriophage Lambda. Curr. Top. Microbiol. Immunol. *78*:69–151.

119. Holmes, F. O. 1948. Order Virales. The filterable viruses, pp. 1125–1144. *In* R. S. Breed, E. G. D. Murray, and A. P. Hitchens (eds.), Bergey's Manual of Determinative Bacteriology, 6th ed. Williams and Wilkins, Baltimore, Maryland.

120. Iino, T., and M. Mitani. 1967. Infection of *Serratia marcescens* by bacteriophage ψ. J. Virol. *1*:445–447.

121. Ikeda, H. M. Inuzuka, and J.-I. Tomizawa. 1970. P1-like plasmid in *Escherichia coli* 15. J. Mol. Biol. *50*:457–470.

122. Jarvis, A. W. 1984. Differentiation of lactic streptococcal phages into phage species by DNA-DNA homology. Appl. Environ. Microbiol. *47*:343–349.

123. Jones, D., and P. H. A. Sneath. 1970. Genetic transfer and bacterial taxonomy. Bact. Rev. *34*:40–81.

124. Kageyama, M., T. Shinomiya, Y. Aihara, and M. Kobayaski. 1979. Characterization of a bacteriophage related to R-type piocins. J. Virol. *32*:951–957.

125. Kandler, O. 1981. Archaebakterien und Phylogenie der Organismen. Naturwissenschaften *68*:183–190.

126. Katsura, I. 1983. Tail assembly and injection, pp. 485–519. *In* R. W. Hendrix, J. W. Roberts, F. W. Stahl, and R. A. Weisberg (eds.), Lambda II. Cold Spring Harbor Laboratory, Cold Spring Harbor, New York.

127. Kellenberger, E. 1980. Control mechanisms in the morphogenesis of bacteriophage heads. Biosystems *12*:201–223.

128. King, J., C. Hall, and S. Casjens. 1978. Control of the synthesis of phage P22 scaffolding protein is coupled to capsid assembly. Cell *15*:551–560.

129. Kingsbury, D. W. 1984. Species classification problems in virus taxonomy, p. 388. *In* Abstracts of Sixth International Congress of Virology, Sendai.

130. Kluyver, A. J., and C. B. van Niel. 1936. Prospects for a natural system of classification of bacteria. Zbl. Bakt. II *94*:369–384.

131. Koninsky, J. 1982. Colicins and other bacteriocins with established modes of action. Ann. Rev. Microbiol. *36*:125–144.

132. Korsten, K. H., C. Tomkiewicz, and R. Hausmann. 1979. The strategy of infection as a criterion for phylogenetic relationships of non-coli phages morphologically similar to phage T7. J. Gen. Virol. *43*:57–73.

133. Krahn, P., R. O'Callaghan, and W. Paranchych. 1972. Stages in phage R17 infection. VI. Injection of a protein and RNA into the host cell. Virology *47*:628–633.

134. Krüger, D. H., and C. Schroeder. 1981. Bacteriophage T3 and bacteriophage T7 virus–host cell interactions. Microbiol. Rev. *45*:9–51.

135. Lamanna, C., and M. F. Mallette. 1965. The occurrence and taxonomy of bacteria, pp. 8–57. *In* Basic Bacteriology, 3rd ed. Williams and Wilkins, Baltimore, Maryland.

136. Landy, A., and W. Ross. 1977. Viral integration and excision: Structure of the Lambda *att* sites. Science *197*:1147–1149.

137. Lehman, J. F., and L. Mindich. 1979. The isolation of new mutants of bacteriophage φ6. Virology *97*:164–170.

138. Lindberg, A. A. 1973. Bacteriophage receptors. Ann. Rev. Microbiol. *27*:205–241.

139. Liss, A., H.-W. Ackermann, L. W. Mayer, and C. H. Zierdt. 1981. Tailed phages of *Pseudomonas* and related bacteria. Intervirology *15*:71–81.

140. Lotz, W. 1976. Defective bacteriophages: The phage tail-like particles. Prog. Mol. Subcell. Biol. *4*:53–102.

141. Lundström, K. H., D. H. Bamford, E. T. Palva, and K. Lounatmaa. 1979. Lipid-containing bacteriophage PR4: Structure and life cycle. J. Gen. Virol. *43*:583–592.

142. Lwoff, A. 1967. Principles of classification and nomenclature of viruses. Nature *215*:13–14.

143. Lwoff, A., R. Horne, and P. Tournier. 1962. A system of viruses. Cold Spring Harbor Symp. Quant. Biol. *27*:51–55.

144. Lwoff, A., R. W. Horne, and P. Tournier. 1962. Un système des virus. C. R. Acad. Sci. (Paris) Sèr. D *254*:4225–4227.

145. Lwoff, A., and P. Tournier. 1966. The classification of viruses. Ann. Rev. Microbiol. *20*:45–74.

146. MacHattie, L. A., and J. A. Shapiro. 1978. Chromosomal integration of phage Lambda by means of a DNA insertion element. Proc. Natl. Acad. Sci. USA *75*:1490–1497.

147. Mandel, M. 1969. New approaches to bacterial taxonomy: Perspective and prospects. Ann. Rev. Microbiol. *23*:239–274.

148. Maniloff, J. 1983. Evolution of wall-less prokaryotes. Ann. Rev. Microbiol. *37*:477–499.

149. Maniloff, J., S. P. Cudden, and R. M. Putzrath. 1980. Maturation of an enveloped budding phage: Mycoplasmavirus L2, pp. 503–513. *In* M. S. DuBow (ed.), Bacteriophage Assembly. Alan R. Liss, New York.

150. Maniloff, J., J. Das, and J. R. Christensen. 1977. Viruses of mycoplasmas and spiroplasmas. Adv. Virus Res. *21*:343–380.

151. Maniloff, J., K. Haberer, R. N. Gourlay, J. Das, and R. Cole. 1982. Mycoplasma viruses. Intervirology *18*:177–188.

152. Marcoli, R., V. Pirrotta, and R. M. Franklin. 1979. Interaction between bacteriophage PM2 protein IV and DNA. J. Mol. Biol. *131*:107–131.

153. Marvin, D. A., W. J. Pigram, R. L. Wiseman, E. J. Wachtel, and F. J. Marvin. 1974. Filamentous bacterial viruses: XII. Molecular architecture of the class I (fd, If1, IKe) virion. J. Mol. Biol. *88*:581–600.

154. Marvin, D. A., R. L. Wiseman, and E. J. Wachtel. 1974. Filamentous bacterial viruses: XI. Molecular architecture of the class II (Pf1, Xf) virion. J. Mol Biol. *82*:121–138.

155. Matsuo-Kato, H., H. Fujisawa, and T. Minagawa. 1981. Structure and assembly of bacteriophage T3 tails. Virology *109*:157–167.

156. Matthews, R. E. F. 1979. Third report of the International Committee on Taxonomy of Viruses: Classification and nomenclature of viruses. Intervirology *12*:132–296.

157. Matthews, R. E. F. 1982. Fourth report of the International Committee on Taxonomy of Viruses. *In* Classification and nomenclature of viruses. Karger, Basel.

158. Matthews, R. E. F. 1983. The history of viral taxonomy, pp. 1–35. *In* R. E. F. Matthews (ed.), A critical appraisal of viral taxonomy. CRC Press, Boca Raton, Florida.

159. Matthews, R. E. F. 1982. Future prospects for viral taxonomy, pp. 219–256. *In* R. E. F. Matthews (ed.), A critical appraisal of viral taxonomy. CRC Press, Boca Raton, Florida.

160. Maurin, J., H.-W. Ackermann, G. Lebeurier, and A. Lwoff. 1984. Un système des virus—1983. Ann. Virol. (Inst. Pasteur), *135E*:105–110.

161. Mayr, E. 1981. Biological Classification: Towards a Synthesis of Opposing Methodologies. Science *214*:510–516.

162. Meyer, T. F., and K. Geider. 1982. Enzymatic synthesis of bacteriophage fd viral DNA. Nature *296*:828–832.

163. Michener, C. G., and R. R. Sokal. 1957. A quantitative approach to a problem in classification. Evolution *11*:130–141.

164. Milne, R. G. 1984. The species problem in plant virology. Microbiol. Sci. *1*:113–117.

165. Mindich, L., D. Bamford, T. McGraw, and G. Mackenzie. 1982. Assembly of bacteriophage PRD1: Particle formation with wild-type and mutant viruses. J. Virol. *44*:1021–1030.

166. Mindich, L., and J. Lehman. 1983. Characterization of ϕ6 mutants that are temperature sensitive in the morphogenetic protein P12. Virology *127*:438–445.

167. Mindich, L., J. Lehman, and R. Huang. 1979. Temperature-dependent compositional changes in the envelope of ϕ6. Virology *97*:171–176.

168. Mitra, S. 1980. DNA replication in viruses. Ann. Rev. Genet. *14*:347–397.

169. Model, P., M. Russel, and J. D. Boeke. 1980. Filamentous phage assembly: Membrane insertion of the major coat protein, pp. 389–400. *In* M. S. DuBow (ed.), Bacteriophage Assembly. Alan R. Liss, New York.

170. Mukai, R., and M. Hayashi. 1977. Synthesis of infectious ϕX-174 bacteriophage *in vitro*. Nature *270*:364–366.

171. Muller, E. D., and J. E. Cronan. 1983. The lipid-containing bacteriophage PR4: Effects of altered lipid composition on the virion. J. Mol. Biol. *165*:109–124.

172. Murialdo, H., and A. Becker. 1978. Head morphogenesis of complex double-stranded deoxyribonucleic acid bacteriophages. Microbiol. Rev. *42*:529–576.

173. Murray, R. G. E. 1962. Fine structure and taxonomy of bacteria, pp. 119–144. In G. C. Ainsworth and P. H. A. Sneath (eds.), Microbial Classification, Twelfth Symposium of the Society for General Microbiology. Cambridge University Press, Cambridge.

174. Nossal, N. G. 1983. Prokaryotic DNA replication systems. Ann. Rev. Biochem. *53*:581–615.

175. Novick, R. P., R. C. Clowes, S. N. Cohen, R. Curtiss, N. Datta, and S. Falkow. 1976. Uniform nomenclature for bacterial plasmids: A proposal. Bacteriol. Rev. *40*:168–189.

176. Ollis, D. L., C. Kline, and T. A. Steitz. 1985. Domain of *E. coli* DNA polymerase I showing sequence homology to T7 DNA polymerase. Nature *313*:818–819.

177. Orla-Jensen, S. 1909. Die Hauptlinien des natürlichen Bacteriensystems. Zbl. Bakt. II *22*:305–311.

178. Palva, T., and D. Bamford. 1980. Attachment and penetration of lipid-containing bacteriophages, pp. 97–114. *In* L. L. Randall and L. Philipson (eds.), Virus receptors, Part I: Bacterial viruses, Series B, Vol. 7. Chapman and Hall, London.

179. Paranchych, W., P. M. Krahn, and R. D. Bradley. 1970. Stages in phage R17 infection. Virology *41*:465–473.

180. Peeters, B. P. H., R. M. Peters, J. G. G. Schoenmakers, and R. N. H. Konings. 1985. Nucleotide sequence and genetic organization of the genome of the N-specific filamentous bacteriophage IKe. J. Mol. Biol. *181*:27–39.

181. Pirie, N. W. 1955. Summing-up. J. Gen. Microbiol. *12*:382–386.

182. Pirie, N. W. 1962. Prerequisites for virus classification, pp. 374–393. *In* G. C. Ainsworth and P. H. A. Sneath (eds.), Microbial Classification, Twelfth Symposium of the Society for General Microbiology. Cambridge University Press, Cambridge.

183. Randall, L. L., and L. Philipson. 1980. Virus Receptors, Part I, Bacterial Viruses, Series B, Vol. 7. Chapman and Hall, London.

184. Reanney, D. C. 1976. Extrachromosomal elements as possible agents of adaption and development. Bacteriol. Rev. *40*:552–590.

185. Reanney, D. C. 1981. Evolutionary virology: A molecular overview, pp. 519–536. In A. J. Natromius, W. R. Dowdle, and R. Schinazi (eds.), The Human Herpesviruses. Elsevier, Amsterdam.

186. Reanney, D. C., and H.-W. Ackermann. 1981. An updated survey of *Bacillus* phages. Intervirology *15*:190–197.

187. Reanney, D. C., and H.-W. Ackermann. 1982. Comparative biology and evolution of bacteriophages. Adv. Virus Res. *27*:205–280.

188. Ross, R. 1962. Nomenclature and taxonomy, pp. 394–404. *In* G. C. Ainsworth and P. H. A. Sneath (eds.), Microbial Classification, Twelfth Symposium of the Society for General Microbiology. Cambridge University Press, Cambridge.

189. Russel, M., and P. Model. 1985. Thioredoxin is required for filamentous phage assembly. Proc. Natl. Acad. Sci. USA *82*:29–33.

190. Rutberg, L., R. W. Armentrout, and J. Jonasson. 1972. Unrelatedness of temperate *Bacillus subtilis* bacteriophages SPO2 and ϕ105. J. Virol. *9*:732–737.

191. Saiyo, K. 1975. Tail-DNA connection and chromosome structure in bacteriophage T5. Virology *68*:154–165.

192. Sands, J. A., and D. Auperin. 1977. Effects of temperature and host cell genetic characteristics on the replication of the lipid-containing bacteriophage PR4 in *Escherichia coli*. J. Virol. *22*:315–320.

193. Sano, Y., and M. Kageyama. 1984. Genetic determinant of pyocin AP41 as an insert in the *Pseudomonas aeruginosa* chromosome. J. Bact. *158*:562–570.

194. Santos, M. A., H. de Lencastre, and L. J. Archer. 1983. *Bacillus subtilis* mutation blocking irreversible binding of bacteriophage SPP1. J. Gen. Microbiol. *129*:3499–3504.

195. Schoulaker-Schwarz, R., and H. Engelberg-Kulka. 1981. *Escherichia coli traD* (Ts) mutant temperature sensitive for assembly of RNA bacteriophage MS2. J. Virol. *38*:833–871.

196. Schwartz, M. 1980. Interaction of phages with their receptor proteins, pp. 61–94. *In* L. L. Randall and L. Philipson (eds.), Virus Receptors, Part I: Bacterial Viruses, Series B, Vol. 7. Chapman and Hall, London.

197. Serwer, P., and R. Gope. 1984. Bacteriophage P22 capsids with a subgenome length of packaged DNA. J. Virol. *49*:293–296.

198. Shinomiya, T., and S. Shiga. 1979. Bactericidal activity of the tail of *Pseudomonas aeruginosa* bacteriophage PS17. J. Virol. *32*:958–967.

199. Silverman, P. M., and R. C. Valentine. 1969. The RNA injection step of bacteriophage f2 infection. J. Gen. Virol. *4*:111–124.

200. Simon, L. D. 1972. Infection of *Escherichia coli* by T2 and T4 bacteriophages as seen in the electron microscope: T4 head morphogenesis. Proc. Natl. Acad. Sci. USA *69*:907–912.

201. Simpson, G. G. 1961. Principles of Animal Taxonomy. Columbia University Press, New York.

202. Sinclair, J. F., and L. Mindich. 1976. RNA synthesis during infection with bacteriophage ϕ6. Virology *75*:209–217.

203. Skerman, V. B. D. 1959. A Guide to the Identification of the Genera of Bacteria. Williams and Wilkins, Baltimore, Maryland.

204. Sneath, P. H. A. 1957. Some thoughts on bacterial classification. J. Gen. Microbiol. *17*:184–200.

205. Sneath, P. H. A. 1962. The construction of taxonomic groups, pp. 289–332. *In* G. C. Ainsworth and P. H. A. Sneath (eds.), Microbial Classification, Twelfth Symposium of the Society for General Microbiology. Cambridge University Press, Cambridge.

206. Sneath, P. H. A. 1971. Theoretical aspects of microbiological taxonomy, pp. 581–593. *In* A. Pérez-Miravette and D. Peláez (eds.), Xth International Congress for Microbiology. Asociacion Mexicana de Microbiologia, Mexico, DF.

207. Sneath, P. H. A. 1978. Classification of microorganisms, pp. 9/1–9/31. *In* J. R. Norris and M. H. Richmond (eds.), Essays in Microbiology. Wiley, New York.

208. Sneath, P. H. A. 1984. Bacterial nomenclature, pp. 19–24. *In* N. R. Krieg and J. G. Holt (eds.), Bergey's Manual of Systematic Bacteriology, Vol. 1. Williams and Wilkins, Baltimore, Maryland.

209. Sokal, R. R. 1962. Typology and empiricism in taxonomy. J. Theoret. Biol. *3*:230–267.

210. Sokal, R. R., and P. H. A. Sneath. 1963. A critique of current taxonomy, pp. 5–59. *In* Principles of Numerical Taxonomy. W. H. Freeman, London.

211. Stackebrandt, E., and C. R. Woese. 1984. The phylogeny of prokaryotes. Microbiol. Sci. *1*:117–122.

212. Staley, J. T., and N. R. Krieg. 1984. Classification of procaryotic organisms: An overview, pp. 1–4. In N. R. Krieg and J. G. Holt (eds.), Bergey's Manual of Systematic Bacteriology, Vol. 1, 9th ed. Williams and Wilkins, Baltimore, Maryland.

213. Steensma, H. Y. 1981. Effect of defective phages on the cell membrane of *Bacillus subtilis* and partial characterization of a phage protein involved in killing. J. Gen. Virol. *56*:275–286.

214. Stitt, B. L., and L. Mindich. 1983. The structure of bacteriophage ϕ6: Protease digestion of ϕ6 virions. Virology *127*:459–462.

215. Stocker, B. A. D. 1955. Bacteriophage and bacterial classification. J. Gen. Microbiol. *12*:375–381.

216. Strauss, H., and J. King. 1984. Steps in the stabilization of newly packaged DNA during phage P22 morphogenesis. J. Mol. Biol. *172*:523–543.

217. Studier, F. W. 1979. Relationships among different strains of T7 and among T7-related bacteriophages. Virology *95*:70–76.

218. Subak-Sharpe, J. H. 1971. Chairman's conclusions, pp. 389–395. *In* G. E. W. Wolstenholme and M. O'Connor (eds.), Strategy of the Viral Genome. A Ciba Foundation Symposium, Churchill Livingstone, London.

219. Susskind, M. M., and D. Botstein. 1978. Molecular genetics of bacteriophage P22. Microbiol. Rev. *42*:385–413.

220. Tagg, J. R., A. S. Dajani, and L. W. Wannamaker. 1976. Bacteriocins of Gram-positive bacteria. Bact. Rev. *40*:722–756.

221. Thompson, W. R. 1952. The philosophical foundations of systematics. Can. Entomol. *84*:1–16.

222. Thomas, C. A., and L. A. MacHattie. 1967. The anatomy of viral DNA molecules. Ann. Rev. Biochem. *36*:485–518.

223. Tikhonenko, A. S. 1970. Ultrastructure of Bacterial Viruses. Plenum, New York.

224. Traub, F., and M. Maeder. 1984. Formation of the prohead core of bacteriophage T4 *in vivo*. J. Virol. *49*:892–901.

225. Turrill, W. B. 1952. Some taxonomic aims, methods and principles: Their possible applications to the algae. Nature *169*:388–393.

226. Vidaver, A. K., R. K. Koski, and J. L. van Etten. 1973. Bacteriophage ϕ6: A lipid-containing virus of *Pseudomonas phaseolicola*. J. Virol. *11*:799–805.

227. Warren, R. A. J. 1980. Modified bases in bacteriophage DNAs. Ann. Rev. Microbiol. *34*:137–158.

228. Watanabe, K., K. Ishibashi, Y. Nakashima, and T. Sakurai. 1984. A phage–resistant mutant of *Lactobacillus casei* which permits phage adsorption but not genome injection. J. Gen. Virol. *65*:981–986.

229. Whetstone, C. A. 1985. Should the criteria for species distinction in adenoviruses be reconsidered? Intervirology *23*:116–118.

230. Wickner, W. 1983. M13 coat protein as a model of membrane assembly. Trends Biochem. Sci. *8*:90–94.

231. Wildy, P. 1962. Classifying viruses at higher levels: Symmetry and structure of virus particles as criteria, pp. 145–163. *In* G. C. Ainsworth and P. H. A. Sneath (eds.), Microbial classification, Twelfth Symposium of the Society for General Microbiology. Cambridge University Press, Cambridge.

232. Wildy, P. 1971. Classification and nomenclature of viruses (first report of the ICNV), pp. 1–82. Monographs in Virology, Vol 5. Karger, Basel.

233. Wildy, P., H. S. Ginsberg, J. Brandes, and J. Maurin. 1967. Virus-classification, nomenclature and the international Committee on nomenclature of viruses. Prog. Med. Virol. *9*:476–482.

234. Wilson, J. J., and I. Takahashi. 1978. Adsorption of *Bacillus subtilis* bacteriophage PBS1. Can. J. Microbiol. *24*:1–8.

235. Wright, A., M. McConnell, and S. Kanegasaki. 1980. Lipopolysaccharide as a bacteriophage receptor, pp. 29–57. *In* L. L. Randall and L. Philipson (eds.), Virus receptors, Part I: Bacterial Viruses, Series B, Vol. 7. Chapman and Hall, London.

236. Yamamoto, N., P. Gemski, and L. S. Baron. 1983. Genetic studies of hybrids between coliphage ϕ80 and *Salmonella* phage P22. J. Gen. Virol. *64*:199–205.

237. Yang, Y., and D. Lang. 1984. Electron microscopy of bacteriophage ϕ6 nucleocapsid: Three-dimensional image analyses. J. Virol. *51*:484–488.

238. Zinder, N. D. 1980. Portraits of viruses: RNA phage. Intervirology *13*:257–270.

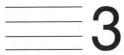

3

DISTRIBUTION OF COLIPHAGES IN THE ENVIRONMENT: GENERAL CONSIDERATIONS

K. FURUSE

Department of Microbiology, School of Medicine, Tokai University, Bohseidai, Isehara-shi, 259-11, Japan

3.1. INTRODUCTION

Starting with the discovery of the Micrococcus phage by F. W. Twort in 1915 (73) and the Shigella phage by d'Hérelle in 1917 (37), the bacteriophages were studied with great expectation as a possible mode of therapy in the treatment of bacterial diseases because of their host specificity, but to no avail. The reason underlying the ineffectiveness of the antibacterial action *in vivo* that is so prominent *in vitro* is as yet unknown. Possible reasons are the instability of the phage within the digestive system, inactivation by the host's immune system, and the development of phage-resistant bacteria. The 1940s brought about the practical use of antibiotics, and their effectiveness in the control of bacterial diseases brought with it a halt to all attempts at phage therapy.

Around that same time, however, the bacteriophage was adopted by Delbrück (10) and his associates as a model system for delving into genetic phenomena, and their studies became the foundation of present-day molecular biology. Utilizing certain coliphages (T-even group), they created a framework for the study of the replicatory mechanisms of the bacteriophage. From those initial experiments to this day, research has centered on the T-even coliphage and λ phage as models of the virulent and the temperate phage, respectively (49). As a result, great progress has been made in understanding the replication of viral nucleic acids, genetic recombination, mutation, and the fine structure of genes—the primary themes of molecular genetics. However, in the process of pursuing these studies on the phage, such aspects as their distribution in nature and their life cycle, for example, the biological and ecological aspects, were for the most part left unexplored. In fact, it would be more correct to say that such phenomenological studies were considered to be nothing more than mere enumerative biology and not a science, to be avoided at best. The RNA phage and filamentous phage with single-stranded DNA are exceptional cases, in that they have been classified and listed as a means of differentiating between the male and female strains of coliform bacteria. As a rule, the classification of phage has been limited to that arising from the practical consideration of classifying bacteria on the basis of their phage susceptibility, for example, phage typing. The advent of the 1970s marked the end of a phase in research involving the molecular biology of phage, and many phage researchers turned their attention to animal viruses. Studies on the phage after 1975 were narrowed to their significance as vectors [temperate phage λ, Mu (41), single-stranded DNA phage, M13 (53), etc.] in the realm of genetic engineering. As a result, we have almost no data that can answer questions on the ecological aspects of the phage with regard to the *E. coli*–coliphage system.

My colleagues and I have been interested in the RNA virus, not only in their ecology but also in their origin and diversification,* and hence I at-

---

* Diversification: The term "evolution" is generally used to denote the genetic relationships between strains of viruses (e.g., the rapid evolution of influenza virus), but the term "diversification" is employed here as a more suitable expression.

tempted a clarification of the phylogenetic relationships within a group of RNA viruses through comparison at the level of primary structure of their nucleotides. For this purpose, it was necessary to select an RNA virus that could be physicochemically studied with ease. We thus arrived at RNA phage and, as a primary step, set about isolating as many different RNA phage as possible. This selection was made because of the relative abundance of RNA phage satisfying this condition and the ease with which they could be isolated from domestic sewage. In planning this type of large-scale isolation of phage, it was necessary to conduct a systematic survey from an ecological standpoint. We limited the host bacteria for the isolation of RNA phage to male and female strains of *E. coli* K12 and the source of the phage to the two recognized as efficient sources for isolation, sewage and animal feces. Under these conditions, the distribution and influx of the phage over time was followed in the animal and sewage systems. We thus isolated several thousand strains of RNA phage and classified them into four groups according to their physicochemical and serological characteristics; in the process, several regularities in their distribution came to light. In this chapter, I discuss regularities in the geographical distribution of RNA phage and their host specificity, and I touch upon the phylogenetic relationships among the phage as related to their ecological distribution, as well as commenting on the origin and diversification of the RNA phage. Furthermore, I discuss briefly the ecology of DNA phage that arose as a spin-off from the RNA phage survey.

The bacteriophage can roughly be classified into three groups according to their mode of interaction with the surface structure of the host bacteria, that is, (*a*) those that infect by recognizing the appendages of the host bacteria such as pili and flagella (appendage phage), (*b*) those recognizing the outer layer such as the polysaccharide capsule (capsule phage), and (*c*) those recognizing the cell wall (somatic phage). From this standpoint, the RNA phage presented in this chapter can be classified as appendage phage, and the DNA phage as a type of somatic phage.

3.2. ECOLOGY OF RNA PHAGE

3.2.1. Historical Background

The RNA phage f2, which specifically infects the male strains of *E. coli*, was discovered by Loeb and Zinder in 1961 (48). Containing only three genes (2,32,39), f2 phage has been suggested as having the smallest genome in the virus kingdom. They can be prepared and handled easily in large-scale culture. Phage RNAs extracted from purified phage particles exhibit messenger activity in an *in vitro* protein-synthesizing system (59), as well as template activity in an *in vitro* RNA-synthesizing system by way of the RNA replicase prepared from RNA phage-infected cells (34,70). Thus RNA phage have been used as highly efficient and ideal material for the analyses of protein synthesis and RNA replication at the molecular level.

All RNA phage isolated in America in the early 1960s [e.g., f2 (48), MS-2 (9), and R17 (64)] belong to group I (serological group), whereas those isolated in Japan, for example, Qβ (74), GA (74), and SP (58), belong to groups III, II, and IV, respectively, suggesting the inclusion of fairly large numbers of serologically different phage in this group.

Fiers et al. (17) determined the complete nucleotide sequence of MS-2 RNA in 1976. This was the first instance of a complete analysis of the entire genome of the virus. By the methods developed by Maxam and Gilbert (1977) (50) and by Sanger et al. (1977) (65) for the determination of the nucleotide sequence with relative ease, it became possible to compare data on phage grouping obtained by the serological and physicochemical methods with those obtained from primary structure analysis.

Bearing in mind the fundamental problems in virology—origin and diversification of RNA viruses—we planned to complete the classification of this group of phage together with an ecological study which, in itself, is a new area in virology.

3.2.2. Isolation and Grouping of RNA Phage

3.2.2.1. Isolation of RNA Phage.   Approximately 1 g of sewage sample or feces was taken from domestic drainage, or animals and man, and suspended in 5 mL of peptone–glucose (PG) medium. Samples of raw sewage (about 2.5 mL) from sewage treatment plants were also suspended in the same volume of PG medium. The samples were treated with 0.5 mL of chloroform to kill bacteria and centrifuged to remove bacterial debris and certain other precipitates. The supernatant fractions were stored in a cold room (4°C) as original phage samples. The sample numbers, site, source, environmental conditions, dates of collection, pH, and temperature of the sewage were recorded in each survey.

RNA phage specifically infect only the male strains of *E. coli* (F$^+$, F′, or Hfr) but not female strains. They also cannot make plaques on male strains in the presence of RNAase (100 μg per plate). Because of these characteristics, phage that lysed male strains but did not lyse female strains or male strains with RNAase were picked up as RNA phage (23, 27, 54).

3.2.2.2. Serological Classification.   Newly isolated RNA phage were subjected to spot tests on male hosts with (K = 1 per plate) or without antiphage serum of four standard phage [MS-2 (group I), GA (II), Qβ (III), and SP (IV)]. After this preliminary classification by the spot test method, each RNA phage was subjected to further analysis for subgrouping by the plating method on male host strains with (K = 1 per plate) or without antiserum. The degree of neutralization [plaque-forming units (PFU) on serum plate per PFU on serumfree plate] was measured and the values for the newly isolated phage were compared with those of standard ones (Table 3.1). By this method, several thousand strains of RNA phage isolated from

TABLE 3.1. Serological Grouping of RNA phage

Group	Phage	MS-2 (I)	GA (II)	$Q\beta$ (III)	SP (IV)
		\multicolumn{4}{c}{Degree of inactivation by an antiphage serum[a]}			
I	MS-2	1×10^{-4}	0.7	1.0	1.3
II	GA	0.6	1×10^{-4}	1.3	1.3
III	$Q\beta$	1.1	1.0	1×10^{-5}	0.9
IV	SP	1.0	1.2	1.2	8×10^{-4}
Serological	JP34	2×10^{-3}	0.1	1.0	1.0
intermediates[b]	MX1	1.0	1.0	0.1	1×10^{-3}
	ID2	0.4	0.5	0.7	0.6

Source: Reproduced with permission from Kodansha Scientific (24).

[a] Calculated as PFU on serum plate (K = 1/plate)/PFU on serumfree plate. A standard phage of each group gave values of about 1×10^{-4} with the homologous antiserum and of about 1 with the heterologous serum. A phage giving a value of >0.5 was judged to belong to another group.
[b] JP34 (serological intermediate between groups I and II) (23), MX1 (between groups III and IV) (22), and ID2 (among the four groups) were classified into groups II, III, and IV, respectively, according to several other physicochemical criteria.

various sources were classified into one of the known four groups of RNA phage. Phage of each group were further subdivided into three to seven subgroups according to the specific inactivation pattern resulting from given antisera within a group (23,27,28,54,55).

In the course of the serological classification, we found a few unique phage that cross-react with two or more antisera of different groups. As shown in Table 3.1, JP34 was neutralized by the antiserum of MS-2 (I), but also inactivated significantly by that of GA (II) (23). Likewise, MX1 was neutralized by SP (IV) antiserum, but also inactivated significantly by that of $Q\beta$ (III) (22). ID2 was neutralized by the antisera of group IV phages, but also inactivated significantly by those of phage groups I, II, and III, although to a lesser extent. Thus the existence of these intermediate phage suggested a close relationship between groups I and II, and between groups III and IV. ID2 was seen to share characteristics common to groups I, II, III, and IV, and was thought to be significant from the phylogenetic point of view (27).

3.2.2.3. *Physicochemical and Biological Properties of RNA Phage.*
As shown in Table 3.2, one critical difference among the four RNA phage groups is the absence of read-through (RT) protein in groups I and II (see Figure 3.1). We can easily distinguish these RNA phage groups by the density profiles from cesium chloride density gradient centrifugation (25). JP34 gave essentially the same density reading as that of group II, and was classified under group II rather than group I by this criterion.

The lowest optimum temperature for phage growth (30°C) was exhibited by the group II phages in *in vitro* propagation experiments (18). At 37°C,

TABLE 3.2. Physicochemical and Biological Properties of Various RNA Phage

Major group	Group	Phage	Buoyant density in CsCl (g/cm³)[a]	Number of virion proteins[b]	Optimum temperature for phage growth[c]		
					Temperature (°C)	Latent period (min)	Burst size
A	I	MS-2, B01, JP501	1.457	2 (M, C)	37	28	4500
	II	GA, BZ13, KU1, JP34, JP500	1.442	2 (M, C)	30	25	2000
B	III	Qβ, VK, TW18, MX1	1.469	3 (M, C, RT)	37	33	2000
	IV	SP, TW19, TW28	1.469	3 (M, C, RT)	37	39	2000

[a] For method, see Ref. 2. Two or three phage of different groups were centrifuged in the same tube. Identification of each phage was made by adding appropriate antisera (K = 1/plate for each antiserum) to the plate or by using appropriate hosts according to the phage combination employed (25).
[b] From the electrophoretogram of the proteins from purified phage particles (25). M, maturation protein; C, coat protein, RT, read-through protein.
[c] One-step growth experiments were performed at 20, 25, 30, 37, 40, and 42°C (18).

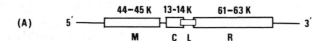

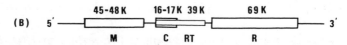

Figure 3.1. Structure of phage genomes of major groups A and B. Letters under phage genome show each cistron: M, maturation protein gene; C, coat protein gene; L, lysis protein gene; R, replicase (II or β subunit) gene; RT, read-through protein gene. RT protein (40) was detected only in the phage of major group B (25). L protein was detected first in group I phage MS-2 (4); phage of the other three groups have not yet been examined. Recently Inokuchi et al. visualized this in the GA genome by complete nucleotide sequence analysis (44). Group II phage have the smallest molecular weights both in RNA and protein, and group IV phage have the largest. Figures over phage genome show molecular weights of each protein. Data from Federoff and Zinder (16), Kamen (46), Furuse et al. (25), and Yonesaki and Haruna (77).

which is the optimum temperature for phage of the other three groups, the burst size of GA decreased appreciably (one-fourth of the optimal condition). Phages belonging to other subgroups of group II showed higher temperature sensitivities than GA. Thus group II phage are thought to be the representatives adapted to lower temperature conditions in the environment.

3.2.2.4. *Molecular Weights of Phage RNAs.*

According to the sucrose density gradient centrifugation method, the relative S values for the phage RNAs were determined as shown in Table 3.3 (group I/II/III/IV:24S/23S/ 27S/28S), in which ^{32}P-labeled *E. coli* RNA (4S tRNA, and 16 and 23S rRNAs) was used as internal marker in one series of experiments and ^{3}H-Qβ RNA (27S) in the others. The phage RNAs of JP34 and MX1 had S values of 23 and 27, respectively, and were classified under groups II and III, respectively, by this criterion (25). By comparing the electrophoretic mobilities of the 20 phage RNAs in agarose–acrylamide composite gel with those of the standard ones (TMV RNA, 16 and 23S *E. coli* rRNAs), the average molecular weights of the RNAs of phage groups I, II, III, and IV were calculated as 1.21×10^6, 1.20×10^6, 1.39×10^6, and 1.42×10^6, respectively. Slight differences in molecular weights of the RNAs between groups I and II, and between groups III and IV were also observed in these experiments as in the case of sedimentation analysis. The phage RNAs of JP34 and MX1 had molecular weights of 1.19×10^6 and 1.43×10^6, and were classified into groups II and IV, respectively (25).

The molecular size of the RNAs of group III and IV phages was about 20% greater than that of groups I and II, and the difference between them (about 600 nucleotides) corresponds well with the presence of read-through protein in the former and its absence in the latter. Based on these results, we believe that groups I and II can be incorporated into one major group A and groups III and IV into another major group B (25). Direct evidence for this was given by the complete nucleotide sequence analysis of the RNAs of MS-2 (I) (17), GA (II) (44), Qβ (III) (52), and SP (IV) (Inokuchi et al., personal communication), as shown in Table 3.3.

3.2.2.5. *Nucleotide Sequences of the Phage RNAs.*

In order to clarify the phylogenetic relationships among the four RNA phage groups, it would be best to compare the complete nucleotide sequences of the phage RNAs. Because the only data available for this were those on MS-2, Inokuchi et al. (43) analyzed the 200–250 nucleotide sequence located at the 3′-terminal region of 15 RNA phage. Although the precise nature of the 3′-terminal region that constitutes an untranslated region in the phage genome is not fully understood at present, it is considered that it may be playing important roles in the propagation of RNA phage as a recognition site for the RNA replicase and/or a binding site for the protein related to phage assembly.

The data collected revealed first that the nucleotide sequences were

TABLE 3.3. Molecular Weights of Various Phage RNAs

Major group	Group	Phage RNA	S value measured by sucrose density gradient centrifugation[a]	Mol. Wt. measured by polyacrylamide gel electrophoresis	Number of nucleotide determined by sequence analysis[b]
A	I	MS-2, f2, R17, JP501	24S	1.21×10^6	3569 (MS2)
A	II	GA, SD, TH1, BZ13, KU1, JP34	23S	1.20×10^6	3466 (GA)
B	III	Qβ, VK, ST, TW18	27S	1.39×10^6	4218 (Qβ)
B	IV	SP, FI, TW19, TW28, MX1, ID2	28S	1.42×10^6	4279 (SP)

Source: Reproduced with permission from Kodansha Scientific (24).

[a] Data from Furuse et al. (25).

[b] Data for MS-2, GA, Qβ, and SP from Fiers et al. (17), Inokuchi et al. (44), Mekler (52), and Inokuchi et al. (personal communication), respectively.

highly conserved (more than 90% homology) among subgroups within a group (except for group IV). According to this criterion, JP34 was shown to be closer to GA than to MS-2, whereas MX1, which belongs primarily to group III (66% homology), was placed in the most remote position within the group. If the data on group IV are compared with those from the other three groups, lower sequence homology (70%) in this region was observed in group IV phage, coinciding with the serologically polymorphic nature of this group (27). Secondly, when the sequences were compared among the four groups, they were well conserved between groups I and II (about 50% homology) and between groups III and IV (about 40% homology) (Figure 3.2) (42).

The 34-nucleotide sequence at the 3′-terminal region is well conserved among the four RNA phage groups, as shown by the frames in Figure 3.2, supporting the idea that all RNA phage share a common ancestor.

A fairly long nucleotide sequence (35 nucleotides) existing at the position between the 100th and 140th nucleotide in the MS-2 RNA could not be found at the corresponding region of the GA genome, but a similar nucleotide sequence reappeared after the 140th nucleotide in both RNAs (Figure 3.2). The simplest explanation for this is to assume a deletion (from MS-2) or insertion (into GA) of the relevant sequence in the genome. Comparing the complete nucleotide sequence of GA with that of MS-2, Inokuchi et al. (44) found 60% homology between the two genomes. He also pointed out a deletion of 45 bases in the replicase cistron that resides at the 5′-terminal region in the GA genome.

These results correspond well to those obtained from the other aforementioned criteria and support our previous proposal that groups I and II can be united into one major group A and groups III and IV into another major group B (2,25,27,43,56,67).

3.2.3. Distribution of RNA Phage in the Environment

In the course of our systematic surveys on RNA phage, the sources for the isolation of RNA phage checked initially were the following: (*a*) sewage from domestic drainage, (*b*) raw sewage from treatment plants, (*c*) animal feces including those of man, cows, pigs, and several animals in zoological gardens, (*d*) river water, (*e*) pond or lake water, (*f*) irrigation water, and (*g*) seawater. The results indicated that the first three (*a*–*c*) represent the most suitable sources for the isolation of RNA phage among the materials tested (23,27,54,55). Dhillon et al., (11,13) have also demonstrated the widespread distribution of RNA phage in Hong Kong sewage. We thus focused on these three materials, made extensive and systematic efforts to isolate as many different kinds of RNA phage as possible, and found two prominent features in the distribution of RNA phage in the environment, one in geographical distribution and the other in preferential distribution in animals.

Figure 3.2. Nucleotide sequences at the 3'-terminal region of the RNAs from four RNA phage groups. The sequences marked with the dotted and solid lines in the MS-2 sequence correspond with those in the GA sequence. Thirty-five bases are observed between two sequences (dotted and solid lines) in the MS-2 RNA. The common sequences at the extreme 3'-terminal region in the four RNA phage groups are enclosed within boxes. Higher homologies between groups I and II and between groups III and IV are also observed in this region. Data from Inokuchi (42) and Inokuchi et al. (43). Reproduced with permission from Kodansha Scientific (24).

96

3.2.3.1. *Distribution in Domestic Drainage.*

We first analyzed the sewage samples collected from domestic drainage in Japan (23) and other Asian countries (27,54,62). A sample that contained one or more coliphages (RNA phage) per 0.1 mL of original phage sample was judged to be a coliphage positive (RNA phage positive) sample. As shown in Table 3.4, coliphages were detected in almost all sewage samples collected from the domestic drainage examined. The amounts of total coliphages in this material were fairly high, ranging from 10 to 10^7 PFU/mL (many between 10^2 and 10^5 PFU/mL). The relative amounts of RNA phages in the sewage samples were high, occupying 10–90% of the total coliphages isolated. Thus RNA phage were isolated at high frequency from samples from Taiwan (28%) (54), Japan (39–51%) (21,23,31), the Philippines (48%), Singapore (35%), Indonesia (21%) (27), and Korea (56%) (62). The sewage samples from India and Thailand, however, had few RNA phage (3.4 and 5.3%, respectively) (27), a result consistent with data for Brazil (5%) (55) and Mexico (2%) (22). No significant differences in sewage temperature (20–30°C), volume of sewage water, or apparent environmental conditions were detected among the domestic drainage in these Asian countries.

When two or more RNA phage isolated from the same original sample exhibited similar inactivation patterns, they were considered to be the same strain (the majority of samples conformed to this case). The results of serological grouping are summarized in Table 3.4. All the RNA phage tested were inactivated markedly by one of the four standard antisera except for JP 34 and ID2. These two phage were shown to be the serological intermediates but were classified finally into groups II (JP 34) and IV (ID2), respectively, by several physicochemical properties. We thus classified the isolated 1020 strains into one of the four known RNA phage groups (I/II/III/IV: 18/495/504/3) (Table 3.4). As shown in the table, group II phage predominated in mainland Japan and in the northern and southern islands in the seas adjacent to Japan (II/III: 3/1), whereas the most prevalent RNA phage in the southwest islands of Japan (Amamiohshima, mainland Okinawa, Ishigakijima, and Iriomotejima), Taiwan, the Philippines, Singapore, and Indonesia were those of group III. The distribution pattern of RNA phage in Korea (II/III: 1/1) was an intermediate type between those of mainland Japan and Southeast Asia. Phage belonging to groups I and IV were noticeably few in domestic drainage in these Asian countries. It is worth noting the existence of a gradual increase in group III over group II phage according to separation from Japan proper toward the south as follows: Rebun and Rishiri (mostly group II) → Japan proper (north of Kyushu) (II/III: 3/1) → Iki and Tsushima (II/III: 3/2) → Korea (II/III: 1/1) → Southeast Asia (south of Amamiohshima) (mostly group III). We therefore proposed a borderline between Kyushu and Amamiohshima Island for the geographical distribution of RNA phage in domestic drainage from South and East Asia (Figure 3.3) (27).

Although it is not yet possible to elucidate the underlying causes of this gradual change in distribution of RNA phage over the global Asian area, our

TABLE 3.4. Distribution Pattern of RNA Phage in Sewage from Domestic Drainage in Asian Countries

Country (area)	Survey year	No. of samples collected	Coliphage positive[a]		RNA phage positive			No. in RNA phage group[b]				Refs.
			No. of samples	%	No. of samples	%	No. of strains[c]	I	II	III	IV	
Taiwan	1970	121	107	88.5	34	28.1	38	6	8	22	2	54
Japan (proper)	1972–1976	736	505	68.6	373	50.7	433	2	331	100	0	19,23
Japan (Island)												19,22
North	1975	13	6	46.2	5	38.5	5	0	5	0	0	
South	1973–1975	157	118	75.2	76	48.4	80	0	34	46	0	
West	1973–1977	150	117	78.0	63	42.0	75	1	44	30	0	
Southwest	1973–1975	413	366	88.6	216	52.3	231	4	22	205	0	
Philippines	1976	61	61	100	29	47.5	30	1	1	28	0	27
Singapore	1976	29	20	69.0	10	34.5	11	0	1	10	0	27
Indonesia	1976	34	30	88.2	7	20.6	7	0	1	6	0	27
India	1976	59	58	98.3	2	3.4	2	0	0	1	1	27
Thailand	1976	38	34	89.5	2	5.3	2	0	1	1	0	27
Korea	1979	132	125	94.7	74	56.1	106	4	47	55	0	62
Grand total		1943	1547	79.6	891	45.9	1020	18	495	504	3	

Source: Reproduced with permission from the American Society for Microbiology (27,62).

[a] Bacterial strains used are the derivatives of *E. coli* K12 strains, A/λ(F⁺), Q13 (RNase I⁻, Hfr), and W3110 (F⁻).

[b] Antiphage sera of groups I (MS-2, BO1, JP501), II (GA, BZ13, TH1, KU1, JP34), III (Qβ, VK, ST, TW18, MX1), and IV (SP, FI, TW19, TW28) were used for the serological grouping or subgrouping of newly isolated RNA phage.

[c] Number of RNA phage exhibiting different serological properties in the same original phage sample. Usually we picked up 3–10 plaques/sample and analyzed them serologically.

98

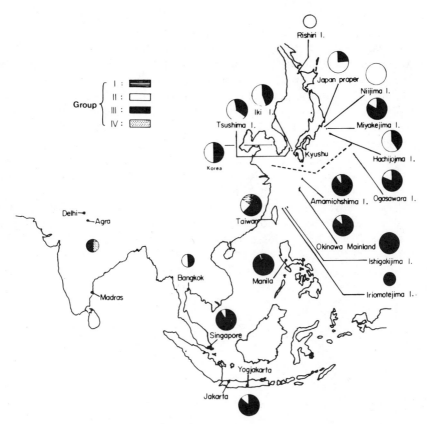

Figure 3.3. Distribution of RNA phage in South and East Asia. The data refer to Table 3.4. Smaller circles indicate a fewer number of RNA phage strains isolated. The dotted line shows a borderline between Kyushu and Amamiohshima Islands. Reproduced with permission from the American Society for Microbiology (27) and Kodansha Scientific (24).

data appear to reflect more strongly the influence of general climate (i.e., integrating or maximum–minimum temperatures, or both) surrounding the domestic drainage, because (*a*) sewage samples were obtained exclusively from domestic drainage whose physical conditions appeared to be directly controlled by the local climate; (*b*) domestic drainage seems to suffer little influence from contamination by animal sources such as human or domestic animal feces, insofar as we have been able to observe, and seems to construct in itself one stable and favorable natural habitat for this phage (see below); and (*c*) group II phage, which are the most prevalent RNA phage in mainland Japan, display the lowest optimal temperature (30°C) for growth among the four known groups of RNA phages in *in vitro* experiments (Table 3.2). They can propagate fairly well even at 20°C, which represents a limiting temperature for group I, III, and IV phage. Furthermore, phage of groups I,

III, and IV can propagate almost normally at 40°C, whereas group II phage can produce few progenies at this temperature (18).

Little is known about the stability and continuity of RNA phage in domestic drainage over moderately long periods of time. In order to investigate this problem, systematic studies were made on sewage samples collected from fixed domestic drainage in three cities of mainland Japan—Choshi, Niigata, and Toyama—at frequencies of two to four times per year between 1973 and 1977. The average isolation frequencies of total coliphages (% of coliphage-positive samples in collected materials) and RNA phage (% of RNA phage-positive samples in collected materials) in sewage samples obtained from Choshi, Niigata, and Toyama for the 5 years were 88 and 77%, 65 and 44%, and 54 and 30%, respectively. The amounts of total coliphages and RNA phage in the sewage samples were fairly high and constant (many showing 10^2–10^5 PFU/mL) (19). As shown in Figure 3.4, the average ratios of group II to III phage over the 5-year period in the cities of Choshi, Niigata, and Toyama were 2:1, 3.6:1, and 8:1, respectively, and the values for each domestic drainage were fairly stable, although some seasonal fluctuations

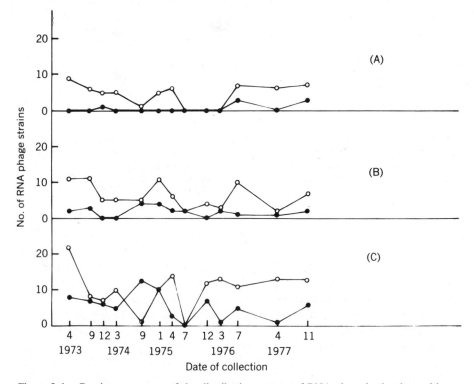

Figure 3.4. Continuous survey of the distribution pattern of RNA phage in the three cities Toyama (A), Niigata (B), and Choshi (C) of Japan proper (from 1973 to 1977). Open circles, group II; solid circles, group III. Reproduced with permission from Gakkaishi Kanko Centre (19) and Kodansha Scientific (24).

were observed over relatively short periods of time. A survey in Choshi city also revealed a similar result: II/III: 6/1 in 1981 compared with II/III: 4/1 in 1977. Insofar as domestic drainage is concerned, it can thus be said that RNA phage are highly stable in terms of their quantity and group types over relatively long periods of time (at least for 5 years in Niigata and Toyama, and for 10 years in Choshi). From this, it appears that the previous results of systematic surveys in various countries can be deemed as fairly reliable comparable data, although they were obtained only from single surveys in each country.

3.2.3.2. *Distribution in Animal and Human Sources.*

In an attempt to clarify the distribution pattern of RNA phage in animals and in man, and to determine the extent of the contribution of RNA phage from animal sources to the propagation and transmission cycles of RNA phage in their natural habitats, we investigated the numbers and types of RNA phage in animal sources such as (*a*) fecal samples from domestic animals (cows, pigs, horses, and fowl), some other animals in zoological gardens, and humans, (*b*) the gastrointestinal contents of cows and pigs, and (*c*) sewage samples from treatment plants in slaughterhouses. The concentration of RNA phage from the first and second sources was fairly low ($10–10^3$ PFU/mL of original phage sample), whereas that from the third source was fairly high ($10^3–10^5$ PFU/g). Concerning the group types of RNA phage from the first and second sources, human feces were found to contain RNA phage of groups II and III in almost equal proportions, the gastrointestinal contents of pigs included those of groups I and II equally, and the feces and gastrointestinal contents of mammals other than man and pigs had those of group I exclusively. From the third source, we found mostly group I phage, with a minor fraction of group II phage (Table 3.5) (63). Thus the prominent feature of the distribution pattern of RNA phage in animals is the existence of preferential relationships between RNA phage groups and their host animals (more precisely, the microbial flora in the animals). Group III phage were isolated exclusively from human feces, and group I phage were isolated from the feces and gastrointestinal contents of mammals other than man. Group II phage are thought to belong primarily to humans, although they were also found in the gastrointestinal contents of pigs and in raw sewage collected from treatment plants of slaughterhouses. We were unable to determine whether the group II phage observed in pigs were intrinsic to them or had been introduced from human sources by chance. In any case, pigs appear to be able to support the propagation of RNA phage of group II as well as group I under the usual breeding conditions, suggesting some similarities of gastrointestinal conditions between humans and pigs. Group IV phage could be isolated from the feces of animals and humans, sewage from domestic drainage, and raw sewage from treatment plants, showing the broadest habitat among the four RNA phage groups, although the isolation frequencies of these phage were very low. Group I phage could be isolated only from

TABLE 3.5. Distribution Pattern of RNA Phage in Animal and Human Sources

Source	Material	No. of samples	Coliphage positive		RNA phage positive			RNA phage group			
			No. of samples	%	No. of samples	%	No. of strains[a]	I	II	III	IV
Birds (zoo)[b]	Feces	25	23	92.0	0	0	0	0	0	0	0
Fowls	Feces	30	9	30.0	0	0	0	0	0	0	0
Mammals (zoo)[c]	Feces	97	72	74.2	5	5.2	5	3[d]	0	0	2[e]
Horses	Feces	30	3	10.0	1	3.3	1	1	0	0	0
Cows	Gastrointestinal contents	20	13	65.0	4	20.0	4	4	0	0	0
Cows	Feces	20	6	30.0	0	0	0	0	0	0	0
Pigs	Gastrointestinal contents (pooled)	3	3	100	1	33.3	3	1	1	0	1
Pigs	Large-intestinal contents	30	30	100	6	20.0	8	3	5	0	0
Pigs	Feces	11	10	90.9	0	0	0	0	0	0	0
Humans	Feces	597	140	23.5	14	2.3	18	0	9	7	2
Total		863	309	35.8	31	3.6	39	12	15	7	5

Source: Reproduced with permission from the American Society for Microbiology (63).

[a] Number of phage exhibiting different serological properties in the same original phage sample. Usually we picked up 5–20 plaques/sample in the case of fecal and gastrointestinal materials and 20–60 plaques/sample for sewage from treatment plants and analyzed them serologically.

[b] Aves (number of samples in parentheses): Struthiniformes—ostrich (2); Cinoniiformes—black-headed ibis (3), scarlet ibis (3), greater flamingo (3), Chilean flamingo (3); Gilliformes—Japanese pheasant (3), copper pheasant (2); Gruiformes—Manchurian crane (3); Strigiformes—feather-toed scops owl (3).

[c] Mammalia (number of samples); Marsupialia—white-throated wallaby (3); Primates—common tree shrew (2), show loris (3), thick-tailed bush baby (3), common squirrel monkey (3), douroucouli (3), Japanese macaque (5), siamang (5), chimpanzee (5), lowland gorilla (2); Lagomorpha—Etigo hare (2); Rodentia—white-cheeked flying squirrel (3); Carnivora—Japanese red fox (3), raccoon-like dog (3), Himalayan black bear (3), Japanese black bear (1), Yezo brown bear (3), tiger (3), lion (5); African cheetah (3); Proboscidae—African elephant (3); Perissodactyla—domestic horse (1), Chapman's zebra (2), Malayan tapir (4), Indian rhinoceros (3); Artiodactyla—Japanese wild boar (3), Bacterian camel (4), Yaku Island sika deer (3), reticulated giraffe (3), domestic yak (2), scimitar-horned oryx (3), Japanese serow (2), mouflon (3).

[d] These three strains were isolated from a fox, an elephant, and a horse, respectively.

[e] These two strains were isolated from two tigers.

animals such as foxes, elephants, horses, cows, and pigs, and from sewage from slaughterhouses, but not from humans or sewage from domestic drainage. Group I phage are therefore thought to belong primarily to animals. This is supported by the fact that although all the animals tested were bred in zoological gardens or breeding farms, where they are thought to have ample chances of becoming contaminated by human phages through their food, they revealed no group III phage, which are believed to be specific to man.

Dhillon et al. (14) reported an example of a similar "habitat preference" in the virulent phages ϕX-174 and S13, which contain single-stranded DNA and are closely related. They found that S13 type phage were isolated only from the feces of pigs, whereas ϕX-174 type phage were isolated only from cow dung.

The fact that the feces of some domestic animals (pigs and cows) contained fairly large amounts of RNA phage suggests that the RNA phage are capable of propagating in the intestines of these animals. Experiments were designed to clarify this point using germfree mice as the test animals (3). After establishment of gnotobiotic mice infected by an appropriate strain of *E. coli,* the group I RNA phage MS-2 was orally inoculated into the gnotobiotic mice, and the amounts of bacteria and RNA phage in the fresh feces of the mice were monitored continuously over a period of 1–3 months.

Failures of RNA phage propagation in germfree and F^--established gnotobiotic mice are shown in Figure 3.5a and b. Mice were inoculated with the group I RNA phage MS-2 (10^9 PFU/mouse). Phage MS-2 excreted in the feces of mice reached a maximum level of 10^4–10^6 PFU/g of feces within 24 hr after phage inoculation. It then decreased gradually for 3–5 days and finally became undetectable. The bacterial level was kept constant in the gnotobiotic mice at around 10^9 colony-forming units (CFU) per gram of feces throughout the experiment. It can be concluded, therefore, that RNA phage cannot propagate or colonize in the intestines of germfree or F^--established gnotobiotic mice.

Propagation of RNA phage in F^+-established gnotobiotic mice is shown in Figure 3.5c. The gnotobiotic mice were inoculated with phage MS-2 (10^9 PFU/mouse). At 6–24 hr after phage inoculation, the MS-2 excreted in the feces of the mice reached a maximum level of 10^{10}–10^{11} PFU/g of feces. At 2–4 days after phage inoculation, the titer of MS-2 decreased appreciably, but the PFU levels were found to stabilize at between 10^5 and 10^9 PFU/g of feces thereafter. The bacterial level was also kept constant at around 10^9 CFU/g of feces throughout the experiment.

Analysis of the propagation of RNA phage in a double infection system between two RNA phage groups revealed that the best colonization response was exhibited by the group I phage MS-2 (3). This is consistent with the aforementioned data that the feces and gastrointestinal contents of mammals other than man contained group I phages exclusively (63).

The effect of antibiotics on the propagation of RNA phages in the mouse intestine is shown in Figure 3.5 d. At 75 days after inoculation of MS-2, F^+-

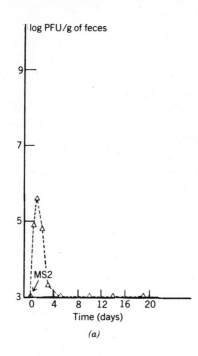

(a)

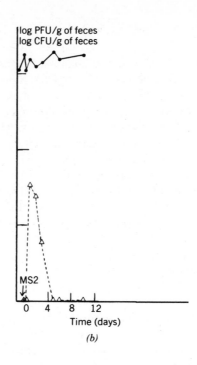

(b)

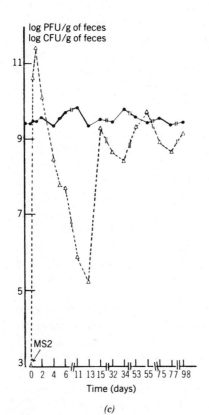

(c)

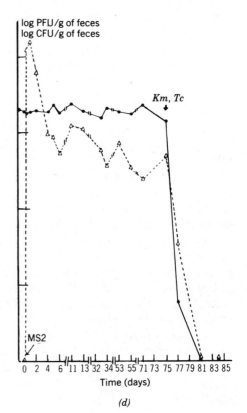

(d)

established (excreting 10^9 CFU/g of feces) and MS-2-established (excreting 10^8 PFU/g of feces) mice were administered tetracycline and kanamycin, as described in the legend to Figure 3.5. The bacteria decreased rapidly upon administration of the antibiotics and disappeared from the feces of the mice within 6 days. Phage MS-2 excreted in the feces of mice decreased concomitantly with the decrement in number of bacteria and finally disappeared from the feces of the mice within 6 days after administration of the antibiotics. These results suggest that bacterial (male strain) growth or colonization in the intestines of germfree mice is necessary for the propagation of RNA phage in the mice.

In the course of this experiment, it was observed that many of the bacteria isolated from the feces of the mice had undergone changes in their susceptibility to RNA phage with respect to lytic response or phage-producing capacity irrespective of phage infection. Thus the observed changes in phage sensitivity were thought to be independent of the challenge with RNA phage. One group was still sensitive to MS-2 in lytic response and was characterized as F^+ cells. The second was not lysed by MS-2 but retained the ability to produce progeny phage (though in smaller burst size) and was characterized as modified F^+ cells. The third group was neither sensitive to MS-2 nor able to produce progeny phages, was able to accept F factor from other male strains, and was characterized as F^- cells. It should be noted that RNA phage were excreted continually in the feces of F^+-established mice in spite of such drastic changes in bacterial character.

Changes in bacteriophage sensitivity of Streptococcus in sheep rumen have been observed by Iverson and Millis (45) in the absence of detectable bacteriophages. Effects of bacteriophages on rumen organisms have also been reported by Orpin and Munn (61), although the role of bacteriophages in relation to the rumen organisms is not thoroughly understood.

At present, it is not possible to conclude whether the majority of the bacterial population having smaller burst sizes support the propagation of RNA phage or whether the smaller fractions of bacteria that constitute only minor fractions in the population but have normal burst sizes do so. It is of interest to determine the exact nature of these various bacteria because we have little knowledge of the true hosts for RNA phage and of the mode of existence of plasmids such as the F and R factors in their natural habitats (11).

Recently, Havelaar et al. (35) developed a highly selective method to enumerate F-specific phage by constructing a *Salmonella typhimurium* har-

Figure 3.5. Propagation of RNA phage in germfree and gnotobiotic mice. Five-week-old germfree mice (ICR male strains purchased from CLEA Japan, Tokyo) were used. Propagation of MS-2 in germfree mice (*a*), in F^--established gnotobiotic mice (*b*), in F^+-established gnotobiotic mice (*c*), and the effect of antibiotics on the propagation of MS-2 (*d*). Arrows indicate the time of phage inoculation. The time of administration of kanamycin (Km) and tetracycline (Tc) is indicated by a bold arrow. Symbols: ($\triangle$) log PFU of MS-2/g of feces; ($\bullet$) log CFU of *E. coli*/g of feces. Reproduced with permission from the American Society for Microbiology (3).

boring F plasmid. Because of the lower concentration of *Salmonella* phage in sewage samples, plaques produced on these host strains are derived exclusively from F-specific phage. Using this system, Havelaar et al. isolated many RNA phage from the fecal samples of various domestic animals in The Netherlands. Serological analysis demonstrated that group I phage were isolated exclusively from animal sources, and group II and III phage from human sources. Thus our previous results obtained in Japan were reconfirmed in The Netherlands. Another and most prominent feature of their study was the abundance of group IV phage isolated from animal feces (36), and it is of interest to pursue these data in terms of overall distribution.

3.2.3.3. *Distribution in Raw Sewage from Treatment Plants.*

To determine the transmission cycle of RNA phages in their natural habitats, we investigated the distribution patterns of RNA phage in raw sewage (which is thought to manifest the overall distribution pattern) collected from treatment plants in various localities in Japan and several other countries (20). As shown in Table 3.6, the densities of total coliphages were variable, ranging from 10 to 10^8 PFU/mL (with many showing 10^2–10^4 PFU/mL). There appears to be a tendency toward somewhat higher phage densities in summer (10^4–10^6 PFU/mL) than in winter (10–10^3/mL) from the continuous surveys (two to four times per year) over 3–5 year periods on raw sewage from several treatment plants in Japan. Almost all of the samples contained RNA phage at relatively high densities, that is, 5–90% of the total coliphages present (with many at 30–60%). Thus RNA phage constituted the predominant phage species in raw sewage from treatment plants, as in the case of sewage from domestic drainage (23,27,54). By serological analysis, 1832 RNA phage strains isolated from Japan were classified into one of the four known groups (I/II/III/IV: 221/487/1122/2). Most of the samples contained only group II and III phage. Samples from the treatment plants in Sapporo, Tokyo, and Toyama contained appreciable amounts of group I phage. As a whole, raw sewage from treatment plants in Japan contained RNA phage belonging to the three groups in a ratio of I/II/III: 1/2/5. Thus two prominent features gathered from this material were the predominance of group III over group II phage and the existence of group I phage in certain treatment plants (the treatment plant in Tokyo received sewage from slaughterhouses together with that from human sources). The continual occurrence of RNA phage in raw sewage from treatment plants was also confirmed from continuous surveys over 3–5 year periods, although the density of RNA phage fluctuated more widely than that in domestic drainage (19).

Most of the sewage samples contained RNA phage of group II and/or III at fairly high densities comparable with those from the treatment plants in Japan. However, sewage samples from Recife (Brazil) and Würzburg (West Germany) contained group I phage only, although we were unable to ascertain whether or not the samples included sewage from slaughterhouses. Based on the distribution patterns of RNA phage in sewage from domestic

TABLE 3.6. Distribution Pattern of RNA Phage in Raw Sewage Collected from Treatment Plants in Various Countries

Country	City	Density of total coliphages[a]	No. of phage tested	No. of RNA phage strains[b]	RNA phage group (no.)				Date of collection	Refs.
					I	II	III	IV		
Mexico	Campeche	5×10^2	8	0	0	0	0	0	3 April 1973	22
Saudi Arabia	Kahfuji	1×10^3	43	0	0	0	0	0	March 1973	22
Australia	Sydney	2×10^4	70	10	0	10	0	0	19 Aug. 1973	22
U.S.	Oakland (CA)	10^2–10^8	198	154	0	52	102	0	29 Aug. 1973	22
	Oakland (CA)	3×10^2	50	28	1	27	0	0	7 Dec. 1973	22
	Richmond (CA)	5×10^4	52	51	0	0	51	0	7 Dec. 1973	22
	New York (NY)	2×10^2	48	45	0	45	0	0	Jan. 1976	20
	San Elijo (CA)	5×10^3	28	7	0	1	6	0	10 Aug. 1977	20
Brazil	Recife	3×10^2	22	4	4	0	0	0	27 Nov. 1975	20
Philippines	Manila	1×10^3	33	3	0	0	3	0	3 Nov. 1976	27
FRG	Würzburg	5×10^3	58	17	17	0	0	0	June 1978	20
	Würzburg (University Hospital)	5×10^4	51	6	0	6	0	0	June 1978	20
Japan	Sapporo, Tokyo, Fukuoka, etc.	10–10^6	3243	1832	221	487	1122	2	1973–1977	20
	Naha (Okinawa)	3×10^3	70	56	0	0	56	0	13 Sept. 1974	20

Source: Reproduced with permission from the American Society for Microbiology (20).

[a] PFU/mL of original phage sample.

[b] Number of phage exhibiting different serological properties in the same original phage sample. Usually we picked up 30–70 plaques per sample and analyzed them serologically.

drainage in Japan proper (II/III: 3/1), in animal feces and sewage from slaughterhouses (mostly group I), and in human feces (III/II: 1/1), it can be reasonably assumed that group I phage observed in the raw sewage from treatment plants are most likely introduced from animal sources, and group II and group III phage from human sources.

An RNA phage amplified selectively in the suitable alimentary tract of mammals (and also presumably in sewage systems) reinfects and propagates again in the alimentary tract of mammals through food and water contaminated with the RNA phage excreted from animals and man in the external environment. In this manner, the transmission cycle of RNA phage would be completed.

Concerning the circulation of RNA phage in sewage systems, the following observations are worth noting: (*a*) From the 5-year period of continuous survey of the distribution pattern of RNA phage in fixed domestic drainages in Japan, the qualitative and quantitative stabilities of these phage in sewage were demonstrated (19). The same was true in raw sewage from treatment plants in Japan over a 5-year survey period (20). (*b*) RNA phage could be isolated at high frequency from sewage of domestic drainage even in winter (most sewage temperatures were below 10°C) as well as in the three other seasons (31). (*c*) Prevalence of group II phage in the temperate zone and that of group III in the tropical and subtropical zones was observed in domestic drainage throughout the global Asian district (23, 27). These observations suggest strongly that sewage of domestic drainage constitutes one of the suitable natural habitats for RNA phage, allowing or supporting their proliferation through unknown mechanisms that can only be explored through the accumulation of experimental data on the propagation of bacteriophages under limited nutritional and temperature conditions.

3.2.4. Phylogenetic Relationships among the Four RNA Phage Groups

According to the serological characteristics, which are thought to reflect mainly the structure of the coat proteins together with several physicochemical characteristics of the phage RNAs, RNA phage can be classified into four groups (I–IV), in which groups I and II can be further assembled into major group A and groups III and IV into major group B.

In order to clarify the phylogenetic relationships among the four RNA phage groups, it would be necessary to compare not only the structural aspects of the phage particles but also the functional ones such as degree of genetic complementation, formation of hybrid particles, and template activity of the RNA replicase.

3.2.4.1. Genetic Complementation Tests.   Genetic recombination and/ or complementation are good indications for judging the distance of relationships between two groups. No genetic recombination has been observed in RNA phage.*

In order to check whether intergroup complementation does occur between mutants of any two RNA phage groups, we isolated a set of amber mutants from the representative phage of each group [MS-2 (I), GA (II), Qβ (III), SP (IV) and ID2 (IV)] by treating phage particles with hydroxylamine or nitrous acid. Mutants thus obtained were classified into three cistrons (genes 1, 2, and 3) by intragroup complementation tests. The function of genes 1, 2, and 3 were identified by physiological study as maturation protein (M) gene, coat protein (C) gene, and RNA replicase (R) gene, respectively (2,18). Several amber mutants of MS-2 and Qβ used in this experiment were provided by Drs. Horiuchi and Zinder of the Rockefeller University. We observed significant complementations between groups III (M mutant) and IV (C mutant), and between groups I (M mutant) and IV (M mutant), although the phage yields recovered from the intergroup complementation tests were about one-tenth of those from the intragroup complementation tests (Table 3.7). When the replicase mutant was used as one parent in the cross, no complementation was observed in any cross combination in the intergroup complementation tests. This would be due to the strict template specificity manifested by the replicase in *in vivo* RNA replication. Hence the production of infective phage particles in the intergroup complementation test suggests strongly the close relationships between the two relevant groups.

In contrast to the apparent complementation between groups III and IV, which constitute one major group B, no complementation was observed in any cross combination between groups I and II, which constitute the other major group A. Progeny phage recovered unexpectedly from the crosses between maturation protein mutants of groups I and IV showed the nature of SP in both the coat protein and RNA, indicating asymmetric complementation. In any case, this unusual complementation would suggest some close relationship between the maturation proteins of groups I and IV. ID2, which cross-reacts with antisera of the other three groups, was seen to effect good complementation with the mutants of group IV phage SP, but not with mutants of any of the other three groups.

3.2.4.2. Formation of Hybrid Particles. Miyake and Shiba (57) reported that progeny phage recovered from the cells infected with two serologically

---

* Recombination in RNA viruses: Evidence of genetic recombination in RNA viruses is limited to segmented viruses such as the influenza virus or reovirus. In these cases, however, recombination occurs through reassortment of the segments but not through interstrand genetic recombination (66). No recombination has been observed in RNA phage containing single-stranded positive strand RNA (38). Earlier data on the frequency of recombination of nonsegmented RNA viruses such as polio virus or foot-and-mouth disease virus are very scarce, variable, and not so convincing (7). Recently, several reports on the recombination of picorna viruses have been published (1,47,51,72). However, it seems necessary to confirm this with further study.

TABLE 3.7. Intergroup Complementation[a] between Several Amber Mutants of Groups I, III, and IV

		MS-2 (I)[b]			Qβ (III)					
		M	C	R	M		C		R	
SP(IV)		sus 1	am 9	sus 10	sus 5	am 205	am 9	am 12	sus 51	am 1
M	sus 246	11.0	0.5	1.1	0.4	0.8	0.2	0.2	0.2	1.0
	sus 261	19.7	0.3	0.4	1.4	0.1	0.1	0.1	0.5	0.1
	sus 262	35.1	0.2	2.0	1.8	0.6	1.2	0.4	2.7	0.6
C	sus 6	4.6	0.6	0.3	81.0	45.0	0.8	2.0	2.4	1.8
	sus 228	1.8	0.2	0.1	25.2	41.0	0.1	0.5	0.1	0.1
R	sus 259	1.3	0.1	0.2	1.0	0.1	0.1	0.1	0.1	0.1
	sus 264	3.0	0.2	0.3	1.0	1.5	0.3	0.3	1.7	1.5

Source: Reproduced with permission from The Keio Medical Society (18).

[a] Quasi-single burst technique (32) was used for the complementation test. The values indicate the numbers of viable phage (PFU/0.1 mL) averaged for 10 tubes. These values were about 1/10 those obtained from intragroup complementation. Positive results that gave significantly larger values than those of single infection are underlined.

[b] MS-2 *sus 1* has a mutation in maturation protein (M) gene, *am 9* in coat protein (C) gene, and *sus 10* in replicase (R) gene, respectively.

different phage of group IV (SP and FI) contained appreciable amounts of hybrid particles which were composed of the SP RNA and FI coat protein and vice versa, together with the normal parental SP and FI phage. Furthermore, they also found mosaic particles that contain the coat protein subunit of both phages at various ratios. Formation of such hybrid particles was observed between group III and IV phage but not between groups I or II and IV.

Hybrid particles can be produced in an *in vitro* reconstitution system in which phage RNA, coat protein, and maturation protein isolated from the purified phage particles are mixed under appropriate conditions. Using this *in vitro* system, Shiba (67) reported the efficient formation of infective phage particles between MS-2 and GA, and between Qβ and SP, supporting the close relationships between groups I and II and between groups III and IV.

3.2.4.3. *Template Specificity of the RNA Replicase.* Close relationships among the four RNA phage groups can also be deduced from template specificities of the RNA replicase in an *in vitro* RNA-synthesizing system.

The RNAs from group I and II phage showed no template activity with Qβ and SP replicase, although the RNAs from group IV phage showed reduced but significant template activity (one-fifth of the extent of Qβ RNA) with Qβ replicase, and those from group III phage did the same (one-fifth of the value of SP RNA) with SP replicase (Figure 3.6) (56). At present, how-

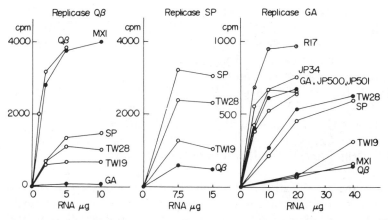

Figure 3.6. Template activities of various phage RNAs with Qβ, SP, and GA replicase. RNA synthesis by GA replicase was carried out in a reaction mixture (total 0.25 mL) containing 20 μmol Tris–HCl (pH 7.4), 2 μmol MgCl$_2$, 0.2 μmol each of ATP, CTP, GTP, and [^{3}H]-UTP (specific activity 5 Ci/mol), various concentrations of template RNAs, and 75 μg enzyme. After incubation at 30°C for 20 min, the reaction mixture was transferred to an ice bath, precipitated with trichloroacetic acid, collected on membrane filters, washed, dried, and counted for radioactivity. RNA synthesis by Qβ or SP replicases was carried out under essentially the same condition used for GA replicase except for the incubation temperature (37°C). Reproduced with permission from Academic Press (76).

ever, the RNA replicases of group I phage are unstable and cannot be used in this type of experiment (16). Recently, Yonesaki and Haruna (77) isolated and purified an RNA replicase of group II phage GA. Certain host factor(s) for GA RNA replication (GA HF) derived from uninfected *E. coli* stimulate RNA synthesis by GA replicase (75). In the presence of GA HF, GA replicase can synthesize both the viral (positive) strand and its complementary (negative) strand. In this GA HF-dependent RNA-synthesizing system, group I and II phage RNAs can serve almost equally as template, but group III and IV phage RNAs cannot. Consistent with previous data, we can thus distinguish groups I and II from groups III and IV by this GA RNA replication system. On the other hand, the purified GA replicase alone can synthesize only limited amounts of the complementary RNA strand (77). Using this GA HF-independent complementary RNA-synthesizing system, we checked the template activities of the various phage RNAs with GA replicase (76). As shown in Figure 3.6, the highest template activities were obtained with the RNAs from group I (R17 and JP501) and II (GA, JP34, and JP500) phage, and next with those from group IV (SP, TW19, and TW28) phage. No template activity was exhibited by the RNA from Group III phage Qβ. Thus group-specific RNA synthesis in which group IV phage RNAs showed significant template activities, though lower than those of group I and II phage RNAs, was observed. Though phage MX1 was first classified into group IV by the serological method (22) and on the basis of several physicochemical criteria, it was also shown to have an S value characteristic of group III by sucrose density gradient centrifugation (25). In the present study, MX1 RNA revealed almost the same template activity as Qβ RNA with Qβ replicase and no template activity as seen in Qβ RNA with GA replicase. A low template activity of MX1 RNA with SP replicase was also observed as in the case of Qβ RNA. Accordingly, phage MX1 was classified into group III under these criteria.

As was the case with phage MX1, group IV phage tend to share several characteristics with phage of the other three groups and are thought to be an intermediate between the two major groups A and B.

3.2.5. Origin and Diversification of RNA Phage

Based on the classification of RNA phage by several biological and physicochemical criteria and also consideration of the known natures of virus RNA, I present a "deletion hypothesis" on the origin and diversification of RNA phage.

The RNA phage that have been classified into the four serological groups have revealed high homology in their 3'-terminal nucleotide sequences (43). This supports the idea that the present-day RNA phage originated from a common ancestor. Because the RNAs of major group B containing a read-through protein cistron additionally are about 20% larger than those of major group A, some deletion or insertion of the fairly long stretch of nucleotide sequences, together with the substitution of the nucleotides, is assumed to have played an important role in the course of diversification of the RNA phage. When the nucleotide sequence (35 bases) found at the 100th–140th position at the 3'-terminal of the MS-2 RNA was removed from that position, the sequence homology with GA RNA increased markedly. This suggests a diversification through deletion or insertion between phage of major group A.

In considering the direction of diversification of the RNA virus, it becomes of essential importance to decide which group, the smaller or the larger, is closer to the prototype.

All cellular organisms include DNA as genetic material. Evolution or diversification by increasing genetic information through gene duplication is the widely accepted theory in this system. However, is it possible to apply such a duplication theory to the virus system, especially to the RNA virus, which lacks recombination and repair replication systems (29,30,38)?

On the basis of the above-mentioned classification, group IV phage have been found to exhibit more diversity in several respects compared to the phage of the other three groups, namely: (a) this group includes seven serological subgroups, the largest number of subgroups among the four groups (28); (b) the serological affinities among these seven subgroups tend to be relatively weak compared to those in the other three groups; (c) they can be isolated from various sources such as animal feces, human feces, raw sewage from treatment plants, and sewage from domestic drainage; and (d) the molecular weights of the virion proteins (25), electrophoretic mobilities of the phage particles (33), and the primary structure of the 3'-terminal region of the phage RNAs (43) tend to be relatively variable within this group. Furthermore, some phage of group IV exhibit intermediate characteristics in several respects involving the four RNA phage groups. First, one of the group IV phage (ID2) shows some serological cross-reactions with phage of the other three groups (27). Second, some amber mutants of the group IV phage SP produce viable progenies in intergroup genetic complementation tests with MS-2 (group I) and with Qβ (group III) (2,18). Third, on the basis of template activity of various phage RNAs with GA replicase in an *in vitro* complementary RNA-synthesizing system, an affinity between group II and IV phage is observed (76). This biological and physicochemical diversity and the intermediate character exhibited by the group IV phage may have some significance in relation to the phylogenetic relationships among the four groups of RNA phage. These observations led me to consider that group IV phage may have a closer relationship to the possible progenitor of RNA phage than do phage of any of the other three groups. According to this hypothesis, RNA phage might have diverged from group IV (which contains the largest RNA) to group III through base changes and/or minor deletions in one direction, and from group IV to groups I and II through fairly large deletion(s) in the other direction. In the course of the process of diversification, the function to which host they colonize would also be acquired.

RNA phage have three unique points different from DNA phage as follows:

1. The RNA replicase necessary for RNA replication is composed of four subunits, though the phage genome codes only one cistron for subunit II or β of the replicase and the remaining three are derived from the host proteins (S1 protein from ribosomes, elongation factors EF-Ts and EF-Tu) (46).

2. RNA phage do not have any error-correcting mechanisms such as recombination, host cell reactivation, ultraviolet light reactivation, and multiplicity reactivation, which are thought to accompany breakage and reunion of the genome (29,38).

3. The mutation frequency of RNA phage is about 10^3 times higher than that of DNA phage (5,38).

Because of these characteristics of the RNA phage, it should be desirable for the RNA phage to have smaller genomes containing only the minimum of information indispensable for phage reproduction. Thus it is not unreasonable to assume that a prototype of group IV phage, which contain the largest genome, might have diverged into groups I and II by deleting unessential regions from the genome to minimize the replication error caused by lower replication fidelity and higher mutation frequency (15) of the RNA phage.

In addition, from the fact that the molecular weight of any virus RNA is between 1×10^6 and 7×10^6 per segment, showing remarkable contrast with the broader variations seen in the DNA virus (10^6–10^8), the probability of diversification through insertion or duplication in the RNA genome would seem to be the lesser possibility. Besides this, deletion from the RNA genome has been widely observed in the influenza virus, which contains segmented negative strand RNA (8), and in the polio virus, which has nonsegmented positive strand RNA (60). From these viewpoints, the largest segment deleted from the genome of group IV phage would be judged to be the read-through protein cistron. Group I and II phage thus established have acquired new functions enabling them to complete the propagation cycles in the absence of read-through protein.

3.3. ECOLOGY OF DNA PHAGE

It has been found that the intestinal contents or feces of certain domestic animals (pigs and cows) and humans contain fairly large amounts of coliphages (phage) (14,63), and that RNA phage that specifically infect the male strains of *E. coli* could continue to propagate in the intestines of gnotobiotic mice established with male strains of *E. coli* but not those established with female strains of *E. coli*. Furthermore, administration of antibiotics gradually led to the disappearance of bacteria and, concomitantly, of the RNA phage (3). These facts support the idea that the intestines of mammals may constitute one of the natural habitats of bacteriophages given the existence of suitable microbial hosts. In the course of systematic surveys on RNA phage in fecal samples from man, we also found some regularities in the distribution pattern of DNA phage depending on the sources from which they were derived.

E. coli represents only a minor constituent of the normal bacterial flora in the human alimentary tract. However, we employed coliphages with their host *E. coli* as a representative microbial system of the human intestine, because the phage–host relationships of this system have been well characterized compared to those of other systems, at least in *in vitro* experiments. We can isolate various DNA phage from sewage samples from domestic drainage or raw sewage from treatment plants, but do not know anything about the quantity or quality of coliphages in such materials. In contrast to this, human feces contain fairly simple phage populations (many fecal samples contain phage of one or two serological groups) compared to those found in sewage systems.

In order to elucidate the ecological role of phage in the human intestine, especially to determine the effects of phage on the succession of bacterial flora, we analyzed, as a first step, the amounts of total phage and numbers of phage strains (serotypes) present in fecal samples collected from both healthy individuals and patients with diarrhea (26). The data obtained indicate qualitative and quantitative differences in the distribution of phage existing between healthy subjects and diarrheal patients, and the existence of an intimate correlation between bacterial and phage densities.

We initially used 10 laboratory strains of *E. coli* K12 [W3110 (F$^-$) and (F$^+$), A(F$^+$), BE110 (Hfr, λ^- derivative of S26), Q13 (Hfr), and 594 (F$^-$)], B [B(B)(F$^-$) and B(H)(F$^-$)], and C [C(F$^-$) and C (*Syn$^-$*) (F$^-$)] as host bacteria. We came to realize, however, that strain C was the most suitable for the isolation of DNA phage (especially for temperate phage, and this may depend on the absence of any known restriction systems in this strain); subsequently we adopted this strain as the main host and one strain from each of K12 and B in addition. Usually, these laboratory strains give much higher titers than the wild-type strains (many as much as 10–10^4 times higher). However, all these laboratory strains are rough types, none of them having capsules (K antigen) and some of them with pili. So the phage recovered from these host strains are thought to be composed mostly of somatic phage and some appendage phage. Collection of fecal samples, preparation of original phage samples, isolation, and grouping of DNA phage by the serological method were undertaken in essentially the same manner as in the RNA phage study (27) with the exception of the selection of the strain of *E. coli* to serve as the host strain.

3.3.1. Phage Distribution in Healthy Subjects and Diarrheal Patients

We collected 607 samples from eight populations of healthy subjects and 140 samples from three populations of diarrheal patients (all were cases of trav-

TABLE 3.8. Quantitative and Qualitative Analysis of Fecal Samples Collected from Both Healthy Subjects and Diarrheal Patients

Source	No. of samples			No. of total coliphages per g of feces		Phage types by Serological grouping[a]	
	Collected	Phage present (%)[b]	Classified	<10^5 (%)	≥10^5 (%)	Temperate (%)	Virulent (%)
Healthy subjects	607	209 (34.4)	175	600 (98.8)	7 (1.2)	156 (89.1)	19 (10.9)
Diarrheal patients	140	98 (70.0)	100	115 (82.1)	25 (17.9)	51 (51.0)	48 (48.0)

[a] Serological grouping was accomplished as described previously (27). See also Table 3.1, footnote *a*.
[b] Number of samples containing ≥10 coliphages/g of feces.

eler's diarrhea). From among the 607 samples from healthy subjects, only 209 samples (34%) were phage positive (samples containing more than 10 PFU/g of feces) and the phage concentrations were relatively low, with very few samples (1%) containing phage at higher concentrations (more than 10^5 PFU/g of feces). Most of the phage were classified as temperate phage. In contrast to this, 98 (70%) out of 140 samples from diarrheal patients were phage positive, and the phage concentrations were much higher. A fairly large number of samples (18%) were found to contain higher phage concentrations (more than 10^5 PFU/g of feces), with many of them being virulent phage (Table 3.8). Such were the quantitative and qualitative differences existing between phage isolated from fecal samples from healthy subjects and diarrheal patients. Most of the phage isolated from healthy subjects could be classified into temperate phages related to $\phi 80$ (36%), λ (27%), and 28 (17%). Contrasted to this, about half the isolates from the diarrheal patients were composed of virulent phage related to T4, T5, and TU23, with many of them at higher titers. Temperate phage at lower titers were also observed in this case. Appearance of virulent phage with higher titers in diarrheal patients seems to reflect some disturbance of their intestinal bacterial flora.

3.3.2. Continuous Survey in Healthy Subjects

We collected about 9 serial fecal samples each from 19 healthy subjects at 2-week intervals over approximately 4 months, and analyzed the isolated phage serologically. Samples collected from the same individual at different times contained almost the same numbers of phage, belonging to the same serological types. Many of them were phage related to $\phi 80$ and λ (both temperate phage). It can thus be said that temperate phage in the feces of healthy subjects tend to be fairly stable with regard to their numbers and serological types, at least over periods of about 4 months. Similar results were obtained in analyses of other populations of healthy adults and newborns. In these cases, we selected *E. coli* strains from the fecal samples before chloroform treatment and analyzed their phage-producing ability upon induction by ultraviolet irradiation. Phage isolated from *E. coli* after induction by ultraviolet irradiation showed fundamentally the same serological properties as those isolated directly from the original phage sample. Thus fecal temperate phage appear to be released through spontaneous induction of corresponding lysogenic bacteria (Furuse et al., unpublished data). Stability of *E. coli* in the human intestine was also observed in a continuous survey, whereas abrupt changes of the bacterial flora including *E. coli* were observed upon administration of antibiotics (71). Efficient isolation of temperate phage from human feces in Hong Kong was reported by Dhillon et al. (14), who demonstrated that many of them (80%) were temperate phage related serologically to HK022. Recently, HK022 was confirmed to cross-react with anti-$\phi 80$ serum and was thought to belong to the $\phi 80$ type. Thus the predominating phage species in healthy subjects in Hong Kong were temperate phage of the $\phi80$ type as well as in Japan.

In contrast to sewage samples from domestic drainage or raw sewage from treatment plants (20,23), human fecal samples contained only limited numbers of RNA phage (12 RNA phage strains among 747 fecal samples). Other authors have reported similar results (14,63). We followed samples collected from individuals whose feces had been shown to contain RNA phage at the time of the first survey, but were unable to detect any RNA phage in the samples obtained from the same subjects on the second and third occasions (Furuse et al., unpublished data).

3.3.3. Continuous Survey in Leukemic Patients

We attempted to determine the distribution patterns of coliphages and their host *E. coli* in fecal samples collected from patients with leukemia nursed in a protective environmental ward and undergoing antileukemic and antibacterial chemotherapy. Samples with phage densities of less than 10 PFU/g of feces comprised about 85% (60/71) of the total samples, as was expected from the successive chemotherapy applied. Most of the samples obtained from patients under chemotherapy had few *E. coli,* in parallel with the absence of coliphages. In contrast to this, three (60%) out of five samples collected from patients before they entered a protective environmental unit contained phage, and all five samples (100%) collected from patients who had left the unit and terminated antibacterial chemotherapy contained phage of appreciable amounts, indicating the rapid recovery of phage flora in the intestines once free from antibacterial chemotherapy (Furuse et al., unpublished data).

One prominent feature observed in patient KT in this continuous survey was the simultaneous change in the distribution pattern of coliphages and the clinical symptoms. As shown in Figure 3.7, after several weeks of the initial febrile period, the condition of the patient stabilized. In week 17, W31-type coliphages were isolated suddenly at a titer of 10^2 PFU/g of feces. After 1 week, the phage titer reached a maximum level of 10^9 PFU and remained constant at that high level until week 21, the time of our last sampling. The clinical symptoms of this patient began to take a serious turn in the 16th week, shortly before the outbreak of the phage. He died of septicemia by *Serratia* in week 26. All the phage isolated from the fecal samples collected at different times were W31-related phage (designated as KT1 through KT4 serially). In contrast to the appearance of high titers of W31-related phage, we could not detect *E. coli* in the same fecal samples. The contradiction between the appearance of coliphages and the absence of *E. coli* was resolved by using *Serratia marcescens* isolated from the patient as host bacteria.

One step-growth experiment of KT3 with *E. coli* C gave a latent period of 14 min and a burst size of 500 under ordinary experimental conditions. In

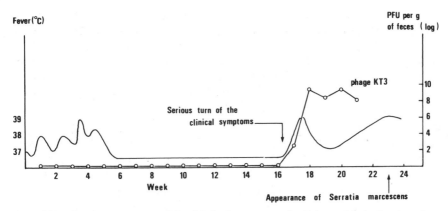

Figure 3.7. Continuous survey of the distribution pattern of coliphages in a leukemic patient.

contrast to this, *S. marcescens* infected with KT3 could divide by itself and produce progeny phage in liquid culture, as in the case of male-specific filamentous phage. That is, they constitute a persistent infection system. About 10% of the KT3-infected cells could produce colonies on agar plates, whereas *E. coli* C could not produce any colonies under the same condition. Thus virulence of KT3 is strong against *E. coli* C but moderate to *Serratia*. From these results, it can be reasonably understood that the unusually high titer of KT3 phage observed in patient KT could be supported by *Serratia* through persistent infection of this phage. Bertani et al. (6) has reported a similar phenomenon observed between coliphage P2 and *Serratia*.

These two examples (one diarrhea and the other sepsis) caused by bacterial infection suggest the applicability of phage examinations in the diagnosis of bacterial diseases, although the data presented here are too limited for generalization. We must accumulate more data of this kind to augment our limited knowledge of coliphages in their natural habitats (12,27).

In order to bring phage therapy, which constitutes one aim of phageology, closer to realization, we would need to (*a*) complete the classification of coliphages into the aforementioned three categories (appendage, capsule, and somatic phage); (*b*) prepare complete phage sets responsible for bacterial populations which would contain bacteria resistant to a given phage; and (*c*) develop a method for effectively introducing such phage sets into the intestines of animals, although two successful reports on phage therapy in experimental infections have been reported by Smith and Huggins (68,69).

3.4. CONCLUSIONS

We isolated several thousand strains of RNA phage from their natural habitats, for example, sewage and animal feces, and classified them into four serological groups. Regularity in the geographical distribution of RNA phage, for example, prevalence of group II phage in the temperate zone and group III phage in the tropical and subtropical zones, was observed in domestic drainage throughout the global Asian district. Preferential relationships between RNA phage groups and their host animals, for example, exclusive isolation of group II and III phage from human feces and group I phage from the feces of mammals other than humans, were observed. However, group IV phage were isolated from both humans and animals at relatively low frequencies.

Based on the classification of RNA phage by several biological and physicochemical criteria, analysis of phylogenetic relationships among four RNA phage groups, and consideration of the known nature of virus RNA, a "deletion hypothesis" on the origin and diversification of RNA phage was presented.

Analysis of total coliphages in fecal samples collected from healthy subjects, diarrheal patients, and leukemic patients under chemotherapy revealed qualitative and quantitative differences in the distribution of temperate and virulent phage in these materials, suggesting an intimate correlation between bacterial and phage densities.

REFERENCES

1. Agol, V. I., S. G. Drozdov, V. P. Grachev, M. S. Kolesnikova, V. G. Kozlov, N. M. Ralph, L. I. Romanova, E. A. Tolskaya, A. V. Tyufanov, and E. G. Viktorova. 1985. Recombinants between attenuated and virulent strains of poliovirus Type 1: Derivation and characterization of recombinants with centrally located crossover points. Virology *143*:467–477.

2. Ando, A., K. Furuse, T. Miyake, T. Shiba, and I. Watanabe. 1976. Three complementation subgroups in group IV RNA phage SP. Virology *74*: 64–72.

3. Ando, A., K. Furuse, and I. Watanabe. 1979. Propagation of ribonucleic acid coliphages in gnotobiotic mice. Appl. Environ. Microbiol. *37*:1157–1165.

4. Atkins, J. F., J. A. Steitz, C. W. Anderson, and P. Model. 1979. Binding of mammalian ribosomes to MS2 phage RNA reveals an overlapping gene encoding a lysis function. Cell *18*:247–256.

5. Batschelet, E., E. Domingo, and C. Weissmann. 1976. The proportion of revertant and mutant phage in a growing population as a function of mutation and growth rate. Gene *1*:27–32.

6. Bertani, G., B. Torheim, and T. Laurent. 1967. Multiplication in *Serratia* of a bacteriophage originating from *Escherichia coli*: lysogenization and host-controlled variation. Virology *32*: 619–632.

7. Cooper, P. D. 1977. Genetics of picornaviruses, pp. 133–207. *In* H. Fraenkel-Conrat and R. Wagner (eds.), Comprehensive Virology, Vol. 9. Plenum, New York.

8. Davis, A., A. L. Hiti, and D. P. Nayak. 1980. Influenza defective interfering viral RNA is formed by internal deletion of genomic RNA. Proc. Natl. Acad. Sci. USA *77*:215–219.

9. Davis, J. E., J. H. Strauss, and R. L. Sinsheimer. 1961. Bacteriophage MS2. Another RNA phage. Science *134*:1427.

10. Delbrück, M. 1940. The growth of bacteriophage and lysis of the host. J. Gen. Physiol. *23*:643–660.

11. Dhillon, E. K. S., and T. S. Dhillon. 1974. Synthesis of indicator strains and density of ribonucleic acid-containing coliphages in sewage. Appl. Microbiol. *27*:640–647.

12. Dhillon, E. K. S., T. S. Dhillon, Y. Y. Lam, and A. H. C. Tsang. 1980. Temperate coliphages: Classification and correlation with habitats. Appl. Environ. Microbiol. *39*:1046–1053.

13. Dhillon, T. S., Y. S. Chan, S. M. Sun, and W. S. Chan. 1970. Distribution of coliphages in Hong Kong sewage. Appl. Microbiol. *20*:187–191.

14. Dhillon, T. S., E. K. S. Dhillon, H. C. Chan, W. K. Li, and A. H. C. Tsang. 1976. Studies on bacteriophage distribution: virulent and temperate bacteriophage content of mammalian feces. Appl. Environ. Microbiol. *32*:68–74.

15. Domingo, E., D. Sabo, T. Taniguchi, and C. Weissmann. 1978. Nucleotide sequence heterogeneity of an RNA phage population. Cell *13*:735–744.

16. Fedoroff, N. V., and N. D. Zinder. 1971. Structure of the poly(G) polymerase component of the bacteriophage f2 replicase. Proc. Natl. Acad. Sci. USA *68*:1838–1843.

17. Fiers, W., R. Contreras, F. Duerinck, G. Haegeman, D. Iserentant, J. Merregaert, W. Min Jou, F. Molemans, A. Raeymaekers, A. Vandenberghe, G. Volekaert, and M. Ysebaert. 1976. Complete nucleotide sequence of bacteriophage MS2 RNA: Primary and secondary structure of the replicase gene. Nature *260*:500–507.

18. Furuse, K. 1982. Phylogenetic studies on RNA coliphages. J. Keio Med. Soc. *59*:265–274.

19. Furuse, K., A. Ando, S. Osawa, and I. Watanabe. 1979. Continuous survey of the distribution of RNA coliphages in Japan. Microbiol. Immunol. *23*:867–875.

20. Furuse, K., A. Ando, S. Osawa, and I. Watanabe. 1981. Distribution of ribonucleic acid coliphages in raw sewage from treatment plants in Japan. Appl. Environ. Microbiol. *41*:1139–1143.

21. Furuse, K., A. Ando, and I. Watanabe. 1975a. Isolation and grouping of RNA phages. V. A survey in the islands in the adjacent seas of Japan. J. Keio Med. Soc. *52*:259–263.

22. Furuse, K., A. Ando, and I. Watanabe. 1975b. Isolation and grouping of RNA phages. VII. A survey in Peru, Bolivia, Mexico, Kuwait, France, Australia, and the United States of America. J. Keio Med. Soc. *52*:355–361.

23. Furuse, K., T. Aoi, T. Shiba, T. Sakurai, T. Miyake, and I. Watanabe. 1973. Isolation and grouping of RNA phages. IV. A survey in Japan. J. Keio Med. Soc. *50*:363–376.

24. Furuse, K., and A. Hirashima. 1982. Origin and diversification of RNA phages, pp. 1–25. *In* Y. Okada and A. Ishihama (eds.), RNA as genetic material. Kodansha Scientific, Tokyo.

25. Furuse, K., A. Hirashima, H. Harigai, A. Ando, K. Watanabe, K. Kurosawa, Y. Inokuchi, and I. Watanabe. 1979. Grouping of RNA coliphages based on analysis of the sizes of their RNAs and proteins. Virology *97*:328–341.

26. Furuse, K., S. Osawa, J. Kawashiro, R. Tanaka, A. Ozawa, S. Sawamura, Y. Yanagawa, T. Nagao, and I. Watanabe. 1983. Bacteriophage distribution in human faeces: Continuous survey of healthy subjects and patients with internal and leukaemic diseases. J. gen. Virol. *64*:2039–2043.

27. Furuse, K., T. Sakurai, A. Hirashima, M. Katsuki, A. Ando, and I. Watanabe. 1978. Distribution of ribonucleic acid coliphages in South and East Asia. Appl. Environ. Microbiol. *35*:995–1002.

28. Furuse, K., T. Sakurai, Y. Inokuchi, H. Inoko, A. Ando, and I. Watanabe. 1983. Distribution of RNA coliphages in Senegal, Ghana, and Madagascar. Microbiol. Immunol. *27*:347–358.

29. Furuse, K., T. Sakurai, and I. Watanabe. 1967. The effects of ultraviolet irradiation on various RNA phages. Virus *17*:159–164.

30. Furuse, K., and I. Watanabe. 1971. Effects of ultraviolet light (UV) irradiation on RNA phage in H$_2$0 and in D$_2$0. Virology *46*:171–172.

31. Furuse, K., and I. Watanabe. 1973. Alteration in the distribution pattern of RNA phages. I. Patterns in summer and in winter. J. Keio Med. Soc. *50*:437–443.

32. Gussin, G. N. 1966. Three complementation groups in bacteriophage R17. J. Mol. Biol.*21*:435–453.

33. Harigai, H., A. Hirashima, K. Furuse, and I. Watanabe. 1981. Electrophoretic properties of RNA coliphages. Microbiol. Immunol. *25*:965–968.

34. Haruna, I., K. Nozu, Y. Ohtaka, and S. Spiegelman. 1963. An RNA "replicase" induced by and selective for a viral RNA: Isolation and properties. Proc. Natl. Acad. Sci. USA *50*:905–911.

35. Havelaar, A. H., W. M. Hogeboom, and R. Pot. 1984. F-specific RNA bacteriophages in sewage: methodology and occurrence. Wat. Sci. Tech. *17*:645–655.

36. Havelaar, A. H., K. Furuse, and W. M. Hogeboom. 1986. Bacteriophages and indicator bacteria in human and animal faeces. J. Appl. Bacteriol. *60*:255–262.

37. d'Hérelle, F. 1917. Sur un microbe invisible antagoniste des bacilles dysentériques. C.R. Acad. Sci. Paris. *165*:373–375.

38. Horiuchi, K. 1975. Genetic studies of RNA phages, pp. 29–50. In N. D. Zinder (ed.), RNA phages. Cold Spring Harbor Laboratory, Cold Spring Harbor, N.Y.

39. Horiuchi, K., and S. Matsuhashi. 1970. Three cistrons in bacteriophage Qβ. Virology *42*:49–60.

40. Horiuchi, K., R. Webster, and S. Matsuhashi. 1971. Gene products of bacteriophage Qβ. Virology *45*:429–439.

41. Howe, M. M., and E. G. Bade. 1975. Molecular biology of bacteriophage Mu. Science *190*:624–632.

42. Inokuchi, Y. 1981. Analysis of the primary structure of the gene of RNA coliphage. II. Homology of RNA sequences in the 3'-terminal region of RNAs from RNA coliphages. J. Keio Med. Soc. *58*:165–177.

43. Inokuchi, Y., A. Hirashima, and I. Watanabe. 1982. Comparison of the nucleotide sequences at the 3'-terminal region of RNAs from RNA coliphages. J. Mol. Biol. *158*:711–730.

44. Inokuchi, Y., R. Takahashi, T. Hirose, S. Inayama, A. Jacobson, and A. Hirashima. 1986. The complete nucleotide sequence of the group II RNA coliphage GA. J. Biochem. *99*:1169–1180.

45. Iverson, W. G., and N. G. Millis. 1977. Succession of *Streptococcus bovis* strains with differing bacteriophage sensitivities in the rumens of two fistulated sheeps. Appl. Environ. Microbiol. *33*:810–813.

46. Kamen, R. 1970. Characterization of the subunits of Qβ replicase. Nature *228*:527–533.

47. King, A. M. Q., D. McCahon, W. R. Slade, and J. W. I. Newman. 1982. Recombination in RNA. Cell *29*:921–928.

48. Loeb, T., and N. D. Zinder. 1961. A bacteriophage containing RNA. Proc. Natl. Acad. Sci. USA *47*:282–289.

49. Lwoff, A. 1953. Lysogeny. Bacteriol. Rev. *17*:269–337.

50. Maxam, A. M., and W. Gilbert. 1977. A new method for sequencing DNA. Proc. Natl. Acad. Sci. USA *74*:560–564.

51. McCahon, D., and W. R. Slade. 1981. A sensitive method for the detection and isolation of recombinants of foot-and-mouth disease virus. J. Gen. Virol. *53*:333-342.

52. Mekler, P. 1981. Ph.D. thesis. University of Zürich.

53. Messing, J., B. Gronenborn, B. Muller-Hill, and P. H. Hofschneider. 1977. Filamentous

coliphage M13 as a cloning vehicle: insertion of a *Hin*d II fragment of the *lac* regulatory region in M13 replicative form *in vitro*. Proc. Natl. Acad. Sci. USA *74*:3642–3646.

54. Miyake, T., K. Furuse, T. Shiba, T. Aoi, T. Sakurai, and I. Watanabe. 1971. Isolation and grouping of RNA phages in Taiwan. J. Keio Med. Soc. *48*:25–34.

55. Miyake, T., K. Furuse, T. Shiba, T. Aoi, T. Sakurai, and I. Watanabe. 1973. Isolation and grouping of RNA phages. II. A survey in Brazil. J. Keio Med. Soc. *50*:353–362.

56. Miyake, T., I. Haruna, T. Shiba, Y. H. Itoh, K. Yamane, and I. Watanabe. 1971. Grouping of RNA phages based on the template specificity of their RNA replicases. Proc. Natl. Acad. Sci. USA. *68*:2022–2024.

57. Miyake, T., and T. Shiba. 1970. Formation of hybrid particles in RNA phages. I. Hybrid particles between RNA phages SP and FI. Virology *43*:675–684.

58. Miyake, T., T. Shiba, T. Sakurai, and I. Watanabe. 1969. Isolation and properties of two new RNA phages SP and FI. Japan. J. Microbiol.*13*:375–382.

59. Nathans, D., G. Notani, J. H. Schwartz, and N. D. Zinder. 1962. Biosynthesis of the coat protein of coliphage f2 by *E. coli* extracts. Proc. Natl. Acad. Sci. USA *48*:1424–1431.

60. Nomoto, A., A. Jacobson, Y. F. Lee, J. J. Dunn, and E. Wimmer. 1979. Defective interfering particles of poliovirus: Mapping of the deletion and evidence that the deletions in the genomes of DI(1), (2) and (3) are located in the same region. J. Mol. Biol. *128*:179–196.

61. Orpin, C. G., and E. A. Munn. 1974. The occurrence of bacteriophages in the rumen and their influence on rumen bacterial population. Experientia *30*:1018–1020.

62. Osawa, S., K. Furuse, M. S. Choi, A. Ando, T. Sakurai, and I. Watanabe. 1981. Distribution of ribonucleic acid coliphages in Korea. Appl. Environ. Microbiol. *41*:909–911.

63. Osawa, S., K. Furuse, and I. Watanabe. 1981. Distribution of ribonucleic acid coliphages in animals. Appl. Environ. Microbiol. *41*:164–168.

64. Paranchych, W., and A. F. Graham. 1962. Isolation and properties of an RNA containing bacteriophage. J. Cell. Comp. Physiol. *60*:199–208.

65. Sanger, F., S. Nicklen, and A. R. Coulson. 1977. DNA sequencing with chain-terminating inhibitors. Proc. Natl. Acad. Sci. USA *74*:5463–5467.

66. Scholtissek, C. 1979. Influenza virus genetics. Adv. Genetics *20*:1–36.

67. Shiba, T., 1974. *In vitro* assembly of virus particles. II. Formation of hybrid particles between various RNA phage *in vitro*. J. Keio Med. Soc.*51*:305–314.

68. Smith, H. W., and M. B. Huggins. 1982. Successful treatment of experimental *Escherichia coli* infections in mice using phage: its general superiority over antibiotics. J. Gen. Microbiol. *128*:307–318.

69. Smith, H. W., and M. B. Huggins. 1983. Effectiveness of phages in treating experimental *Escherichia coli* diarrhoea in calves, piglets and lambs. J. Gen. Microbiol. *129*:2659–2675.

70. Spiegelman, S., I. Haruna, I. B. Holland, G. Beaudreau, and D. Mills. 1965. The synthesis of a self-propagating and infectious nucleic acid with a purified enzyme. Proc. Natl. Acad. Sci. USA *54*:919–927.

71. Tabaqchali, S., A. Howard, C. H. Teoh-Chan, K. A. Bettelheim, and S. L. Gorbach. 1977. *Escherichia coli* serotypes throughout the gastrointestinal tract of patients with intestinal disorders. Gut *18*:351–355.

72. Tolskaya, E. A., L. I. Romanova, M. S. Kolensnikova, and V. I. Agol. 1983. Intertypic recombination in poliovirus: Genetic and biochemical studies. Virology *124*:121–132.

73. Twort, F. W. 1915. An investigation on the nature of the ultramicroscopic viruses. Lancet *189*(2):1241–1243.

74. Watanabe, I., T. Miyake, T. Sakurai, T. Shiba, and T. Ohno. 1967. Isolation and grouping of RNA phages. Proc. Jap. Acad. *43*:204–209.

75. Yonesaki, T., and A. Aoyama. 1981. *In vitro* replication of bacteriophage GA RNA. Involvement of host factor(s) in GA RNA replication. J. Biochem. *89*:751–757.

76. Yonesaki, T., K. Furuse, I. Haruna, and I. Watanabe. 1982. Relationships among four groups of RNA coliphages based on the template specificity of GA replicase. Virology *116*:379–381.

77. Yonesaki, T., and I. Haruna. 1981. *In vitro* replication of bacteriophage GA RNA: Subunit structure and catalytic properties of GA replicase. J. Biochem. *89*:741–750.

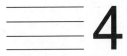

4

ECOLOGY OF PHAGE IN FRESHWATER ENVIRONMENTS

SAMUEL R. FARRAH

Department of Microbiology and Cell Science, University of Florida, Gainesville, Florida 32611

4.1. INTRODUCTION

Freshwater environments contain a variety of macroorganisms and microorganisms. They may interact in parasitic, symbiotic, or commensal relationships. These environments may also contain subcellular entities, the viruses. Viruses for plants, insects, cyanobacteria, and bacteria have been found and may influence these organisms in freshwater environments.

This chapter considers the ecology of viruses that are specific for bacteria (bacteriophages, or phage). Bacteriophages differ from cellular organisms in two important ways. First, they are smaller (20–100 nm diameter) and contain few types of macromolecules. They generally contain only an outer protein coat surrounding a core of either DNA or RNA. Some phage also contain lipid. Second, they are obligate intracellular parasites. As such, they require the presence of viable host bacteria for their replication. In many cases, replication of virulent phage leads to lysis and death of the host bacterium. In other cases, lysogenic phage may form stable associations with the host bacteria and may be responsible for introduction of new and beneficial genes to the host bacterium (21). Either type of interaction is likely to affect the host bacterium's ability to survive and therefore the total numbers of the host bacterial species in a particular environment.

Because phage have the potential to influence the numbers of bacteria in a given freshwater environment, their interactions with their host bacteria are of interest to microbial ecologists.

4.2. METHODS FOR DETECTING THE PRESENCE OF BACTERIOPHAGES IN WATER

Much of the work on the ecology of bacteriophages has been directed toward their study during wastewater treatment and following their release to receiving waters (14,22,28,29,36). Studies on the total number of bacteriophages in sewage and wastewater have been conducted using electron microscopy (1). Also, studies have been done to determine the number of specific bacteriophages such as those of indicator or pathogenic bacteria that are present in raw and treated wastewater (12,16,17,19,37,39).

Therefore, many procedures for detecting bacteriophages were developed for use with wastewater. Moreover, much work has been done on the development of procedures for detecting human enteroviruses in natural environments (4,8,10). Many of these procedures cannot be applied to the study of bacteriophages in freshwater. Some of the previously described procedures that can and cannot be used to detect phage in freshwater are discussed below.

Because the nutrient level and concentration of bacteria in wastewater are generally higher than those found in freshwater, the levels of bacteriophages are also generally higher. Therefore, direct analysis of freshwater using electron microscopy (1) as has been applied to the study of bacteriophages in sewage is not practical for most freshwater environments. Several of the procedures developed for recovering human enteroviruses from water and other environmental samples are not suitable for phage recovery. The extremes of pH encountered in these procedures tend to inactivate bacteriophages and greatly reduce the efficiency of virus recovery (20,25,27,32,34). Adsorption–elution procedures using positively charged filters have been shown to be suitable for recovering a variety of laboratory and indigenous phage from fresh waters. These filters are capable of adsorbing viruses at the ambient pH of natural waters or after slight acidification of the water. The adsorbed viruses can be eluted from the filters with solutions of beef extract or solutions containing salts and detergents (11,20,36). In many cases, a one-step concentration procedure is sufficient to detect phage. In other cases, the number of phage in the eluting solution is below the level of detectability. Further concentration of phage can be accomplished using a second adsorption–elution procedure (36). A recently described modification of the organic flocculation procedure suitable for concentrating both enteric viruses and bacteriophages in beef extract has been described (35). In this procedure, ammonium sulfate at pH 7–9 is used in place of low pH to flocculate the beef extract. Viruses adsorb to the flocs and are recovered by centrifugation. The viruses are then recovered by dissolving the floc in a small amount of deionized water. A procedure that is capable of concentrating phage from turbid as well as clear samples uses a combination of flocculated casein and magnetite (5). In this magnetite organic flocculation procedure (MOF), casein and magnetite are added to samples and the pH is lowered to 4.5. The resulting flocs can be recovered using a magnet. This procedure does not require special equipment or centrifugation and can therefore be used under field conditions. These one- and two-step procedures have been used to detect small numbers of bacteriophages present in several liters of water. It is possible that some of the concentrated samples contain enough phage for direct analysis using electron microscopy. These concentrated samples should be subjected to direct examination because electron microscopy is capable of detecting phage in the absence of a suitable host bacterium.

Most procedures for detecting phage use a double-agar overlay, which permits detection of viable phage by their ability to form plaques on a susceptible host (2). Havelaar and Hogeboom (14) found that the concentration of calcium and magnesium ions in the media as well as the strain of *E. coli* used as the host influenced the sensitivity of the plaque assay. These workers also developed bacterial strains that were highly specific for detecting male-specific bacteriophages (15). Primrose et al. (24,25) described several procedures for making phage assays specific for particular phage types. By using mixed indicators, equilibrium buoyant-density centrifugation, and specific hybridization probes and enzymes, these workers were able to enumerate bacteriophages in a eutrophic lake specifically. Isometric male-specific phages, filamentous phages, different classes of P-group plasmid-specific phages, and phages resembling ϕX-174 and λ were separated and identified using these procedures. Further application of these and other procedures could be used to provide a more detailed description of phage in freshwater.

Indigenous bacteria in concentrated samples from freshwater may overgrow host bacteria and make assays for phage difficult. Chloroform can be used to kill bacteria without affecting non-lipid-containing phages. However, Kennedy et al. (18) found that plating of samples on selective media such as EC reduced contamination problems when *E. coli* was used as the host.

In summary, procedures have been modified or developed that permit low levels of viruses in large volumes of water to be detected. These procedures have been used to study phage in a variety of freshwater environments and may be modified to be selective for certain types of phage.

4.3. DISTRIBUTION OF PHAGE IN FRESHWATER ENVIRONMENTS

In most of the studies on the distribution of phage in freshwater environments strains of *E. coli* have been used as host bacteria. This likely reflects the interest expressed in the possibility of using *E. coli* phage (coliphages) as indicators of enteric viruses or bacteria. As shown in Table 4.1, coliphages have been isolated from lakes, rivers, and other types of freshwater environments in several parts of the world. Numbers up to 10^5/L have been found.

Enteric bacteria other than *E. coli* and indigenous coliform bacteria may also be present in freshwater environments. Therefore, it would be expected that phage active against these bacteria would also be recovered from these waters. Although relatively few workers have studied these phage, phage active against *Klebsiella pneumoniae* and *Salmonella* spp. have been recovered from freshwater. They have been found in numbers comparable to those of *E. coli* (Table 4.2).

Understanding the role of bacteriophages in the ecology of freshwater environments will require, at a minimum, knowledge of the numbers of individual bacteria and their phage. Observations on changes in the numbers of different bacteria and their respective phage might then be used to determine the relation between numbers of bacteriophages and their host bacte-

TABLE 4.1. Levels of *Escherichia coli* Phage in Freshwater Environments

Source of water	Host bacteria	Phage/liter[a]	Ref.
Polishing pond, Florida	*E. coli* B	1.9×10^4	5
Lake Alice, Florida	*E. coli* B	1.1×10^3	5
Lake Alice, Florida	*E. coli* C	1.5×10^2	18
Lake Alice, Florida	*E. coli* B	1.5×10^1	18
River Avon, England	*E. coli* W3110	6.5×10^5	20
River water, England	*E. coli* HfrH	1.8×10^4	23
River water, England	*E. coli* W3110	3.5×10^3	31
Lake Alice, Florida	*E. coli* C	1.3×10^4	35
Apies river, South Africa	*E. coli* K12Hfr	1.7×10^5	12
Vaal River, South Africa	*E. coli* K12Hfr	4.0×10^3	12
Hudson River, New York	*E. coli*	5.0×10^1	16
Potomac River, Washington, DC	*E. coli* C	1.0×10^4	17
Danube River, Czechoslovakia	*E. coli* B-39	2.0×10^4	37
Cypress strand, Florida	*E. coli* C-3000	2.4×10^5	30

[a] Values are either representative data or mean values calculated from data presented in the indicated reference.

TABLE 4.2. Phage of Enteric Bacteria Other Than *Escherichia coli* in Freshwater Environments

Source of water	Host bacteria	Phage/liter[a]	Ref.
River Avon, England	*Klebsiella pneumoniae* 889	1.0×10^6	20
River Avon, England	*Salmonella typhimurium* his G46	8.5×10^3	20
River water, England	*S. typhimurium* G46	8.2×10^2	23
River water, England	*K. pneumoniae*	1.2×10^3	23
River water, England	*K. pneumoniae* 889	1.1×10^3	31

[a] Values are either representative data or mean values calculated from data presented in the indicated reference.

ria. Unfortunately, few data on the levels of phage specific for indigenous bacteria in freshwater are available. Logan et al. (20), Primrose and Day (23), and Seeley et al. (31) recovered phage active against *Aeromonas hydrophila* in several samples of river water (Table 4.3). The phage were found at approximately 10^3 PFU/L. The concentrations of the host bacterium, *A. hydrophila,* were not determined. *A. hydrophila* is indigenous to many freshwater environments and may be an indicator of the trophic level of these environments (26). Therefore, it should be possible to study changes in the number of phage active against *A. hydrophila* that occur during variations in the numbers of *A. hydrophila* following seasonal or long-term changes in the trophic level of freshwater environments. Primrose and Day (23) also recovered phage active against *Pseudomonas aeruginosa*. This organism is capable of metabolizing a wide variety of organic compounds and is widely distributed in natural environments. The variation in numbers of *P. aeruginosa* and its phage during changes in the nutrient levels of freshwaters caused by human pollution or natural changes in nutrient levels may also be a useful model for studying the phage–host relationship under natural conditions.

TABLE 4.3. Phage of Indigenous Bacteria in Freshwater

Source of water	Host bacteria	Phage/liter[a]	Ref.
River Avon, England	*Aeromonas hydrophila*	1.4×10^3	20
River Avon, England	*A. salmonicida*	1.6×10^3	20
River Avon, England	*Hyphomicrobium*	6.4×10^3	20
River Avon, England	*Rhodopseudomonas blastica*	3.8×10^2	20
River water, England	*Pseudomonas aeruginosa*	3.5×10^3	23
River water, England	*A. hydrophila*	2.3×10^3	23
River water, England	*A. hydrophila*	2.3×10^3	31

[a] Values are either representative data or mean values calculated from data presented in the indicated reference.

TABLE 4.4.   Survival of Bacteriophages in Freshwater

Source of water	Bacteriophage	Decay rate $(k)^a$ hr^{-1}	Ref.
River water, Finland	T7	0.0061	22
Ground water, Florida	f2	0.059	6
Cypress strand water, Florida	MS-2	0.019	30
Ground water, Wisconsin	MS-2	0.005	42

^{a} Typical decay rates for samples at approximately 20°C; data from the indicated references have been used to calculate k values according to the equation: $k = [\ln (C_o/C_t]/t$, where C_o = the concentration of phage initially present, t = time, in hours, and C_t = the concentration of phage at t hr.

4.4.   FACTORS INFLUENCING THE SURVIVAL AND MULTIPLICATION OF PHAGE IN FRESHWATER

Bacteriophage survival in freshwater has been studied using source water seeded with bacteriophages and incubated under laboratory conditions and *in situ* using survival chambers (6,22,30,42). As shown in Table 4.4, similar decay rates were obtained for different viruses in several types of freshwater. Niemi (22) studied survival of phage T7 in different types of water, including river and lakewater and groundwater, and concluded that the temperature of incubation and the type of water influenced phage survival. Yates et al. (42) found that viral survival in groundwater was not correlated with the pH, nitrate concentration, turbidity, or hardness of groundwater samples obtained from several sources in the United States. These authors found that the temperature of incubation was the only variable that correlated with the survival of phage MS-2. Scheuerman et al. (30) reported that the inactivation rate of phage MS-2 was similar to that of poliovirus and was lower than the rates for indicator bacteria in survival chambers placed in a cypress strand under natural conditions.

Based on these and other studies, it is likely that several factors have the potential to influence the numbers and activities of bacteriophages in freshwater. These include the densities of both host bacteria and their phage; the association of phage and bacteria with solids; the presence of organic matter, especially organic matter that influences the metabolic activity of the host bacteria; ultraviolet and visible light; temperature; pH; the concentration and type of ions, and the metabolic activities of microorganisms other than the host bacteria. The effects of several of these variables on phage survival have been studied under laboratory conditions. Fewer studies on survival under natural conditions have been reported.

Little is known about the effect of light on phage in freshwater. It has been shown that visible light along with certain dyes or natural colored compounds can cause photooxidation of viruses (40). Along with the lethal effects of ultraviolet (UV) light, near ultraviolet (NUV) light has been shown to affect the attachment of bacteriophage T7 to its host (13). Although visible, UV, and NUV light all have the capacity to inactivate or alter phage attachment, the role that they play in freshwater has not been determined.

Enteroviruses have been found in higher numbers in both marine and freshwater sediments than in the corresponding waters (4). It has been suggested that adsorption to sediment aids the survival of viruses (4,9). The limited data available do not show that increased numbers of bacteriophages are present in freshwater sediments. Scheuerman et al. (30) found higher levels of coliphages in waters of a cypress strand than in the underlying sediment. In contrast, these workers found similar levels of human enteroviruses in the sediment and in the water following passage of wastewater through experimental corridors in the cypress strand.

Humic and fulvic acids are naturally present in many water environments. Humic acids have been shown to influence enterovirus attachment to microporous filters (38). Fulvic acid was found to complex with viruses and reduce their ability to adsorb to soils (7). Because some freshwater environments contain highly colored water with relatively high concentrations of humic and fulvic acids, it might be expected that these compounds would influence the attachment of bacteriophages with sediments and other solids and possibly with their host bacteria. The presence of organic matter may also influence phage numbers by influencing the metabolic rate of the host bacterium. Anderson (3) found that the titer of phage produced by *Salmonella typhi* growing in sewage was similar to titers produced by its growth in laboratory media, 10^9–10^{10} particles/mL. The presence of certain metallic ions, particularly calcium and magnesium ions, has been found to influence phage production under laboratory conditions (14). The concentration and nature of ions has also been shown to influence the adsorption of bacteriophages to microporous filters (4,8). Based on the results of laboratory studies, it would be expected that ions would influence the adsorption of phages to solids and influence their replication. Data on the influence of ions on bacteriophages in freshwater environments are not available.

In addition to organic matter and inorganic ions, the pH also influences bacteriophage adsorption (4,8). It might be expected that the pH of natural waters would influence the distribution of bacteriophages between sediments, other solids, and water. Studies on the distribution of phage and its relationship to the pH of the water have not been reported.

The concentrations of bacteriophage and its host bacterium are likely to be important factors in the ecology of phage in freshwater environments. At low levels of host bacteria, the probability of a phage encountering a susceptible host are reduced and productive infection may not occur. Anderson (3) concluded that phage replication did not occur until the concentration of host bacteria (*Salmonella typhi*) was 3×10^4 mL. At lower concentrations of bacteria, phage multiplication could not be detected. A similar threshold value of approximately 10^4/mL *Staphylococcus aureus, Bacillus subtilis,* or *E. coli* was found to be required for replication of their respective phage by Wiggins and Alexander (41). Anderson (3) also investigated the level of bacteriophage that would be required to eliminate completely its host bacteria. Multiple tubes were inoculated with *Salmonella typhi* at a concentration of one or two cells/tube. Increasing concentrations of bacteriophages were added to the tubes. The tubes were incubated under conditions that support phage multiplication and were examined for bacterial growth. Anderson (3) found that relatively high concentrations of phage (10^6) were required to influence the number of bacteria. Even with an initial population of one or two bacteria, an initial concentration of 10^8 phages/mL was required to completely eliminate the host bacteria. Based on these results, Anderson concluded that complete elimination of a host bacteria by phage in natural environments was not likely. The effect of bacteriophages on mineralization of glucose by *E. coli* was studied by Wiggins and Alexander (41). Mineralization of glucose by $1.1. \times 10^2$ CFU/mL of *E. coli* was not affected by a concentration of 50 PFU/mL of T4. Concentrations of $1.1. \times 10^3$ and 1.0×10^4 did reduce the rate of mineralization of glucose in samples inoculated with 1.3×10^2 CFU/mL of *E. coli.* Although both these studies suggest that a concentration of approximately 10^4 bacteria/mL is required for phage multiplication, it should be noted that both studies were conducted under laboratory conditions. Bacteria in freshwater environments may be under more nutrient-limiting conditions and may be concentrated at the surface of solids rather than in the overlaying water. Studies on the minimal number of bacteria required to support phage replication under natural conditions would be a welcome addition to the laboratory studies.

Studies on the effect of temperature on bacteriophage adsorption to host bacteria and on bacteriophage replication have yielded interesting and potentially important results. Anderson (3) found that high titers of phage against *Salmonella typhi* were produced when the phage and bacteria were incubated at 38.5°C. No phage were produced when the cultures were incubated at room temperature, 18–20°C.

Seeley and Primrose (33) reported that bacteriophages could be classified by the effect of temperature on their efficiency of plating. These authors found three types of coliphages. Low-temperature (LT) phage had optimum plating temperatures at or below 30°, mid-temperature (MT) phage in the range of 15–42°, and high-temperature (HT) phage in the range of 25–42° or 30–45°C. As shown in Table 4.5, human and cow feces contained HT and MT phage but no LT phage. Raw sewage contained mostly MT and LT types. In contrast, water from the Swift river in England contained few HT type phage and a large percentage of LT phage. These authors found that the proportion of LT phage was greatest near the source of the river and decreased as the river received pollution from sewage treatment plants. With self-purification, the percentage of LT phage again increased. When *Aeromonas hydrophila* was used as the host bacterium, only LT phage were

TABLE 4.5. Temperature Profiles of Bacteriophages from Different Habitats[a]

Source of phage	Number of samples tested	% of phage giving lysis only in the range:			
		15–30°C LT type	25–42°C HT type	30–45°C HT type	15–45°C MT type
Human feces	50	0	54	0	46
Cow feces	7	0	0	72	28
Raw sewage	49	7	0	9	84
River Swift	19	53	0	5	42

[a] Modified from Seeley and Primrose (33).

isolated. These results suggest that indigenous bacteriophages in freshwater environments are capable of replicating at temperatures near those of the water. In contrast, phage introduced with fecal pollution are not capable of replicating at the low temperatures found in some environments.

Anderson (3) found that the titer of phage added to sewage declined, but no decline was observed in the titer of phage in nutrient broth incubated under similar conditions. Wiggins and Alexander (41) found little or no inactivation of phage added to autoclaved lake water or sewage over 30 days. In contrast, phage titers in unsterilized samples of lake water or sewage declined to less than 0.1% of the initial values over the same time period. These results suggest that the activity of indigenous microorganisms, not the temperature of incubation, was responsible for the observed inactivation in the unsterilized samples.

4.5. DISCUSSION

Procedures have been developed for recovering low numbers of phage in relatively large volumes of water, in some cases from several hundred liters. Methods for characterizing the recovered phage or for isolating specific phage are also available. Although many studies have investigated phage active against *E. coli* and other enteric bacteria, phage active against bacteria indigenous to freshwater environments have also been detected. Because it is likely that these phage are inactivated with time in freshwater environments, it is reasonable to assume that the phage replicate under natural conditions. The conditions required for and the consequences of this replication have not been established. It is likely that growth of indigenous bacteria above a threshold value required for phage multiplication would lead to an increase in the number of phage. Laboratory studies suggest that at some concentration of phage, the numbers would be sufficient to decrease the number of bacteria or to reduce their metabolic activity. These interactions would be complicated by mutations that lead to the emergence of phage-resistant bacteria and then mutations of phage to produce varieties that are capable of attacking the new strains of bacteria.

In summary, the interactions of phage and their host bacteria, with each other and with their environment are likely to be complex and influenced by several variables. Future studies on the relationship between indigenous bacteria and their phage under different environmental conditions should provide a better understanding of the role of these interactions in freshwater environments.

REFERENCES

1. Ackermann, Hans-W., and T. Nguyen. 1983. Sewage coliphages studied by electron microscopy. Appl. Environ. Microbiol. *45*:1049–1059.

2. Adams, M. H. 1959. Bacteriophages. Interscience, New York.

3. Anderson, E. S. 1957. The relations of bacteriophages to bacterial ecology. Seventh Symp. of the Soc. for Gen. Microbiol, Royal Institution, London.

4. Bitton, G. 1980. Adsorption of viruses to surfaces: Technological and ecological implications, pp. 331–374. *In* G. Bitton and K. C. Marshall (eds.), Adsorption of Microorganisms to Surfaces. Wiley, New York.

5. Bitton, G., L. T. Chang, S. R. Farrah, and K. Clifford. 1981. Recovery of coliphages from wastewater effluents and polluted lake water by the magnetite–organic flocculation method. Appl. Environ. Microbiol.*41*:93–96.

6. Bitton, G., S. R. Farrah, R. H. Ruskin, J. Butner, and Y. J. Chou. 1983. Survival of pathogenic and indicator organisms in ground water. *21*:405–410.

7. Bixby, R. L., and D. J. O'Brien. 1979. Influence of fulvic acid on bacteriophage adsorption and complexation in soil. Appl. Environ. Microbiol. *38*:840–845.

8. Gerba, C. P. 1984. Applied and theoretical aspects of virus adsorption to surfaces. Adv. Appl. Microbiol.*30*:133–168.

9. Gerba, C. P., and G. E. Schaiberger. 1975. Effect of particulates on virus survival in seawater. J. Water Pollut. Control Fed. *47*:93–103.

10. Goyal, S. M., and C. P. Gerba. 1982. Concentration of viruses from water by membrane filters. pp. 59–116. In C. P. Gerba and S. M. Goyal (eds.), Methods in Environmental Virology. Dekker, New York.

11. Goyal, S. M., K. S. Zerda, and C. P. Gerba. 1980. Concentration of coliphages from large volumes of water and wastewater. Appl. Environ. Microbiol. *39*:85–91.

12. Grabow, W. O. K., P. Coubough, E. M. Nupen, and B. W. Bateman. 1984. Evaluation of coliphages as indicators of the virological quality of sewage-polluted water. Water SA *10*:7–14.

13. Hartman, P. S., and A. Eisenstark. 1981. Alteration of bacteriophage attachment capacity by near-UV irradiation. J. Virol. *43*:529–532.

14. Havelaar, A. H., and W. M. Hogeboom. 1983. Factors affecting the enumeration of coliphages in sewage and sewage-polluted waters. Antonie van Leeuwenhoek *49*:387–397.

15. Havelaar, A. H., and W. M. Hogeboom. 1984. A method for the enumeration of male-specific bacteriophages in sewage. J. Appl. Bacteriol. *56*:439–447.

16. Hilton, M. C., and G. Stotzky. 1973. Use of coliphages as indicators of water pollution. Can. J. Microbiol.*19*:747–751.

17. Kenard, R. P., and R. S. Valentine. 1974. Rapid determination of the presence of enteric bacteria in water. Appl. Microbiol. *27*:484–487.

18. Kennedy, Jr., J. J., G. Bitton, and J. L. Oblinger. 1985. Comparison of selective media for assay of coliphages in sewage effluent and lake water. Appl. Environ. Microbiol. *49*:33–36.

19. Kott, Y., N. Roze, S. Sperber, and N. Betzer. 1974. Bacteriophages as viral pollution indicators. Water Res. *8*:165–171.

20. Logan, K. B., G. E. Rees, N. D. Seeley, and S. B. Primrose. 1980. Rapid concentration of bacteriophages from large volumes of freshwater: evaluation of positively charged, microporus filters. J. Virol. Methods *1*:87–97.

21. Luria, S. E., J. E. Darnell, Jr., D. Baltimore, and A. Campbell. 1978. General Virology. Wiley, New York.

22. Niemi, M. 1976. Survival of *Escherichia coli* phage T7 in different water types. Water Res. *10*:751–755.

23. Primrose, S. B., and M. Day. 1977. Rapid concentration of bacteriophages from aquatic habitats. J. Appl. Bacteriol. *42*:417–421.

24. Primrose, S. B., N. D. Seeley, and K. B. Logan. 1981. Recovery of viruses: methods and applications, pp. 211–234. *In* M. Butler and M. Goddard (eds.), Viruses in Water and Wastewater. Pergamon, Oxford, England.

25. Primrose, S. B., N. D. Seeley, K. B. Logan, and J. W. Nicolson. 1982. Methods for studying aquatic bacteriophage ecology. Appl. Environ. Microbiol. *43*:694–701.

26. Rippey, S. R., and V. J. Cabelli. 1980. Occurrence of *Aeromonas hydrophila* in limnetic environments: relationship of the organism to trophic state. Microbiol. Ecol. *6*:45–54.

27. Sabitino, C. M., and S. Maier. 1980. Differential inactivation of three bacteriophages by acid and alkaline pH used in the membrane adsorption-elution method of virus recovery. Can. J. Microbiol. *26*:1403–1407.

28. Safferman, R. S., and M. E. Morris. 1976. Assessment of virus removal by a multistage activated sludge process. Water Res.*10*:413–420.

29. Saleh, F. 1977. Recovery and susceptibility patterns of faecal streptococci bacteriophages. Water Res. *11*:403–409.

30. Scheuerman, P. R., S. R. Farrah, and G. Bitton. 1987. Reduction of microbiological indicators and viruses in a cypress strand. Water Sci. Technol. *19*:539–546.

31. Seeley, N. D., G. Hallard, and S. B. Primrose. 1979. A portable device for concentrating bacteriophages from large volumes of freshwater. J. Appl. Bacteriol. *47*:145–152.

32. Seeley, N. D., and S. B. Primrose. 1979. Concentration of bacteriophages from natural waters. J. Appl. Bacteriol. *46*:103–116.

33. Seeley, N. D., and S. B. Primrose. 1980. The effect of temperature on the ecology of aquatic bacteriophages. J. Gen. Virol. *46*:87–95.

34. Seeley, N. D., and S. B. Primrose. 1982. The isolation of bacteriophages from the environment. J. Appl. Bacteriol. *53*:1–17.

35. Shields, P. A., and S. R. Farrah. 1986. Concentration of viruses in beef extract by flocculation with ammonium sulfate. Appl. Environ. Microbiol. *51*:211–213.

36. Shields, P. A., T. F. Ling, V. Tjatha, D. O. Shah, and S. R. Farrah. 1986. Comparison of positively charged membrane filters and their use in concentrating bacteriophages in water. Water Res. *20*:145–151.

37. Simkova, A., and J. Cervenka. 1981. Coliphages as ecological indicators of enteroviruses in various water systems. Bull. World Health Org. *59*:611–618.

38. Sobsey, M. D., and A. R. Hickey. 1985. Effects of humic and fulvic acids on poliovirus concentration from water by microporous filtration. Appl. Environ. Microbiol. *49*:259–264.

39. Stetler, R. E. 1984. Coliphages as indicators of enteroviruses. Appl. Environ. Microbiol. *48*:668–670.

40. Wallis, C., and J. L. Melnick. 1965. Photodynamic inactivation of animal viruses: A review. Photochem. Photobiol. *4*:159–170.

41. Wiggins, B. A., and M. Alexander. 1985. Minimum bacterial density for bacteriophage replication:implications for significance of bacteriophages in natural ecosystems. Appl. Environ. Microbiol. *49*:19–23.

42. Yates, M. V., C. P. Gerba, and L. M. Kelley. 1985. Virus persistence in groundwater. Appl. Environ. Microbiol. *49*:778–781.

5

ECOLOGY OF MARINE BACTERIOPHAGES

KARLHEINZ MOEBUS

Biologische Anstalt Helgoland, Postfach 850, D-2192 Helgoland, Federal Republic of Germany

5.1. HISTORICAL NOTES

Soon after their detection in 1916 by d'Herelle (17), bacteriophages were isolated from inshore waters by numerous authors, using enteric bacteria as hosts (43,57). Using these bacteria, no phage could be detected in offshore

137

samples, leading those authors to conclude that the phage they obtained from inshore waters were not indigenous to the marine environment but were transported to the sea by rivers or sewage.

Without specifying the bacteria used, ZoBell (57) in 1946 wrote "that bacteriophage . . . occurs in sea water along the coast [of California], but it is rather difficult to demonstrate its presence. The infrequency with which positive results are obtained indicates that bacteriophage is not present in high concentrations or that it occurs only sporadically." He also noted that phage never were found in water collected beyond the littoral zone. From his findings ZoBell concluded that "since bacteriophage is generally found associated with large numbers of rapidly multiplying bacteria, it is very doubtful if the sparse bacterial population characteristic of the open ocean is conducive to the development or activity of bacteriophage."

Kriss and Rukina (34) in 1947 reported the formation of plaques on bacteria isolated from water and sediment of the Black Sea. Their findings, however, were not accepted as a demonstration of marine bacteriophages because no attempt was made to isolate the lytic principles observed (48,52). It remains questionable if bacteriophages were involved at all, because plaque formation was also observed with several controls (bacteria spot-tested with sterile broth instead of sample). This was ascribed to lysogeny by Kriss (33), contradicting the concept of this bacterial trait. Furthermore, the host bacteria used were not typical marine species.

In 1954 a bacteriophage–host system was described by Smith and Krueger (46), with the host depending on about 3% NaCl for optimal growth. This sytem was isolated from mud collected in San Francisco Bay, which is an estuarine environment.

The existence of indigenous marine bacteriophages was finally demonstrated by Spencer (47), who reported in 1955 the isolation of a phage strain lytic for marine luminous bacteria from a seawater sample taken 10 miles off the Scottish coast in the North Sea. This bacteriophage would not lyse its host in a medium containing 0.5% NaCl, but would do so at 2–3% NaCl concentration. Lysis also took place in a seawater medium when incubated at 0°C.

This first report on marine bacteriophages did not generate appreciable interest in these viruses. The next publication on this topic appeared in 1960, again by Spencer (48), and until 1970 only three more reports on isolations of unquestionable marine phage–host systems were published (7,26,52). Although the number of publications dealing with marine phage has increased considerably, our knowledge of these members of microbial communities in the seas and oceans is still very limited.

5.2. DEFINITION OF MARINE BACTERIOPHAGES

Owing to the obligate parasitic nature of viruses, any definition of marine bacteriophages should take into account the phage hosts. In other words, phage–host systems (PHS) are to be defined as marine ones. It is important that both parts of a PHS depend on factors typical of marine environments, though to varying degrees. The phage must depend on such factors not only for adsorption to, infection of, and reproduction within the host cell, but also for their stability in the free state.

Based on what is generally accepted as setting marine bacteria apart from terrestrial species, Spencer (47) proposed to define marine bacteriophages accordingly: their reproduction and stability should depend on salinities and temperatures usually encountered in their habitats. Furthermore, their heat tolerance should be lower than that observed with terrestrial bacteriophages, which generally are not severely harmed by exposure to 60°C or higher for short periods of time (10–30 min).

Spencer's (47) first isolate, which was inactivated within 15 min at 45°C, fits perfectly into this definition. However, with additional phage strains at hand, he found out that bacteriophages of marine origin differ greatly in their heat tolerance (49). One of his phage strains was not inactivated at all during 60 min at 65°C. He also observed that the sensitivity to organic solvents is an equally unreliable means to distinguish marine from terrestrial phage. Physiologically and ecologically more important are his findings concerning the impact of the ionic environment on the stability and infectivity of the phage tested (49).

Presently there is no more accurate definition of marine bacteriophages than Spencer's, but greater importance than before is attached to salinity and ecologically relevant temperature than to tolerance to elevated temperature or organic solvents.

5.3. OCCURRENCE OF MARINE BACTERIOPHAGES

From the available information it can be concluded that marine bacteriophages are present in marine waters and sediments that support bacterial life. Because, however, specific phage strains generally seem to occur at very low concentrations, their detection depends on methods that, if available, might command unjustifiable effort. The long-lasting lack of convenient methods for the isolation of marine PHS from environments inhabited by sparse bacterial populations certainly is one reason for lack of knowledge in this area.

5.3.1. Numbers of Marine Bacteriophages

Almost no information exists on the number of particles of specific phage strains in seawater, and none at all with regard to marine sediments. Such information can be obtained only by directly plating the sample or, less accurately, by the most-probable number method. Several authors have attempted unsuccessfully to isolate phages by direct methods, and others, although occasionally successful, did not publish their findings.

Spencer (48) detected four phage strains in concentrations ranging from 1 to 100 plaque-forming units (PFU) per 10 mL of North Sea water collected 10 miles off the Scottish coast. In seawater sampled near Helgoland (North Sea) Moebus (unpublished) observed between 1 and 3 PFU/mL with five unidentified bacterial isolates derived from water samples taken previously. In sharp contrast, Ahrens (1) in the brackish Kiel Bight (Baltic Sea) found up to 25,500 and 36,500 PFU/mL with two host strains of *Agrobacterium stellatum*, A18 and A22, respectively.

5.3.2. Distribution of Marine Bacteriophages

By far the greater part of information on the presence of bacteriophages in marine environments is based on isolation via enrichment cultures. These are used in two modes. Most commonly a sample, which may remain untreated before use or may be treated in different ways to reduce the number of naturally occurring bacteria, is inoculated with a large number of cells of the presumptive host and incubated. The host bacteria preferentially are isolated from the same sample, which therefore must be stored for some time at low temperature. The other mode, developed by Hidaka (21) and modified by Moebus (37), basically employs samples enriched with nutrients only and incubated immediately after collection. Bacteria used to detect phage are isolated meanwhile from platings of the respective sample performed prior to the start of the enrichment culture. The latter mode until now was used only with oceanic water samples.

5.3.2.1. Isolation from Marine Waters. In addition to the reports of Spencer (47,48) and Ahrens (1) a few more are available that, in some cases, describe PHS of questionable marine nature. Carlucci and Pramer (6) enriched for phage in seawater collected 200 m offshore at New Jersey, using *Serratia marinorubra* as host. Because the lytic activity of the phage was found not to depend on the presence of seawater in the medium they probably cannot be accepted as marine ones. Stevenson and Albright (50) isolated a PHS from water taken in the intertidal zone in British Columbia. The host grew well between 5 and 30°C and needed marine salts "in excess of 0.25%," which is a farily low concentration. The respective requirements of the phage were not reported. Espejo and Canelo (14) isolated a bacteriophage strain from water collected 1 mile off the coast of Chile in a polluted bay. The growth requirements of this lipid-containing phage were not determined; its host depended on temperatures below 30°C and on a NaCl concentration similar to that of seawater (15).

Kakimoto and Nagatomi (28) derived six PHS from Kinko Bay (Japan). According to their salt requirements the hosts were classified as marine (four strains) or halophile. All of them were more temperature sensitive than their phage, which survived at 60°C for a nonreported period of time. Hidaka (18,20) described 20 PHS isolated from samples taken in Kagoshima Bay or south of Kyushu at 50 m depth. Most of the phage survived when kept for 30 min at 50°C, and only four of them were completely inactivated by this treatment (19,20). Three of the latter strains were also highly sensitive to chloroform. To obtain these phage about 200 bacterial isolates were screened for phage sensitivity.

Among 629 bacterial isolates from water and sediment collected in the Arabian Sea by Chaina (7) 10 were found to be sensitive to phage present in the samples. The actual source of the PHS (water or sediment) was not revealed. The phage remained viable for 30 min at 50°C but were killed at 60°C after more than 2–5 min.

Hidaka et al. (23) sampled water at 14 stations from various depths around the Ryukyu Islands and determined the numbers of heterotrophic bacteria per milliliter and that of isolates sensitive to phage present in the respective sample. In water sampled at 50 m depth, between 10 and 50% of the bacteria were phage sensitive. Although the numbers of heterotrophs in samples from 100 m depth were more or less similar to those in 50-m samples, considerably fewer phage-sensitive isolates were found at 100 m depth. Samples from 300 m depth yielded lower numbers of heterotrophs, and only a few PHS were observed. The number of different bacteriophage strains found per 250 mL of seawater ranged from 1 to 16, 1 to 7, and 1 to 3 for samples taken at 50, 100, and 300 m depth, respectively.

Finally, there are two reports dealing with PHS isolation from open ocean environments. Hidaka (21) sampled water from five depths between 1 and 300 m at 13 stations in the southwestern Pacific. Of 576 bacterial isolates 72 were found to be susceptible to bacteriophages. PHS were present at all stations down to 50 m. With water from 50 m depth somewhat greater diversity in PHS was observed than with samples taken from 1–3 m below the surface. In seven samples taken from 200 m depth only one PHS was found, whereas none of seven samples from 300 m depth was positive.

Moebus (37) collected water from 6 m below the surface at 48 stations between the European continental shelf and the Sargasso Sea. PHS were found in 42 samples, their numbers positively correlated with that of colony-forming units (CFU) present in the respective sample. With 213 among 931 bacterial isolates 258 bacteriophage strains were detected. Detailed studies regarding the bacteriophage sensitivity patterns of the Atlantic Ocean bacteria revealed that 326 of the 733 strains tested were susceptible to one or several of the 258 bacteriophage strains. Of the 326 phage-sensitive bacteria 250 differed from each other by their phage sensitivity patterns, whereas 241 of the 258 bacteriophage strains differed by their host ranges (39). Almost all host bacteria were classified as marine ones (40).

5.3.2.2. Isolation from Marine Sediments.

Hidaka (20) isolated a PHS from Kagoshima Bay mud (depth 230 m), which later was described in detail by Hidaka and Shirahama (24). The temperate phage was inactivated by about 99% within 30 min at 45°C. Wiebe and Liston (52) described a PHS isolated from North Pacific Ocean sediment retrieved from 825 m depth. The phage was stable for 24 hr at 45°C but was completely inactivated within 10 min at 50°C. Johnson (26) isolated a phage sensitive to 50°C from Indian Ocean sediment taken at a depth greater than 3000 m.

Zachary (54) examined mud samples from salt marshes, collected in the intertidal zone at various sites along the eastern and southern coast of the United States, for the presence of bacteriophages active against *Vibrio* (*Beneckea*) *natriegens*. Phage with diverse morphology were found in all samples in which the salinity was 8‰ or higher. Two phage strains were investigated in detail (55,56) and were found to differ considerably in temperature sensitivity (55).

Finally, the PHS isolated by Keynan et al. (30) must be mentioned, although the actual source of the phage is not known. It was derived from an enrichment culture set up with mud collected at Woods Hole, Massachusetts, a dead squid, and a large inoculum of the host. The phage activity was decreased by 50% following exposure to 62°C for 18 min.

5.3.2.3. *Additional Information on Marine Bacteriophages.*   Recently, bacteriophage-like particles have been observed by electron microscopy in bacterioplankton preparations. Johnson and Sieburth (27, see also 44) recovered phage particles, after concentrating bacteria 2000–5000 times from estuarine (Narragansett Bay, Rhode Island), shelf, and open ocean water. These phage particles were most abundant in the estuarine samples. Torrella and Morita (51), from their observations made with samples from Yaquina Bay, Oregon, estimated that the actual number of virions ranged from 10^3 to more than 10^4/mL, depending on the sampling site. Observations of this kind do not provide information on the nature of the phage. They argue, however, for more intensive investigations of the role of bacteriophages in microbial communities.

There is considerable interest in bacteria that spoil seafood (fish, shellfish, crustaceans) or even cause illness in humans, and phage serve as a valuable tool to classify such bacteria by phage typing. Such investigations are mainly concerned with the physiology and ecology of the respective host bacteria, and valuable information on phage is gained as a by-product.

Delisle and Levin (10–13) reported the isolation of phage active against psychrophilic pseudomonads and the characteristics of a few selected PHS. Although no seawater media were needed to propagate host bacteria and phage, there are indications of differences among the phage populations depending upon their source: Boston harbor, spoiled fish fillets, and municipal sewage. During a study with *Pseudomonas putrefaciens* (11) including 4 terrestrial and 23 "marine" strains (the latter derived from haddock fillet), 40 strains of bacteriophages lytic for 14 "marine" bacterial isolates were obtained from the aforementioned sources. Only one of the "marine" pseudomonads was susceptible to phage from sewage. On the other hand, no phage active against the four terrestrial strains of *P. putrefaciens* were detected from spoiled fish fillets. Whether these differences reflect the involvement of indigenous marine phages, however, remains dubious.

Much attention has been paid during the last 20–30 years to *Vibrio parahaemolyticus,* a food-poisoning pathogenic bacterium, which occurs abundantly in nutrient-rich inshore waters and sediment as well as in animals living in such environments. As shown by Baross and Liston (2), *V. parahaemolyticus* depends on the presence of NaCl in the medium but grows well at both 37 and 40°C and therefore is not regarded as an indigenous marine species. Bacteriophages active against it were isolated mainly from estuarine water and sediment, but also from diverse animals and even from the feces of a food-poisoned patient (3,9,25,31,32,41,45). Their temperature tolerance varies greatly. Koga et al. (32) found some of their phage to be inactivated (within 30 min) at 45°C, but others only at 55 or 60°C. Strain v6 of Nakanishi et al. (41), isolated from a lysogenic *V. parahaemolyticus,* even remained unaffected at 60°C for 80 min.

Finally, it should be stressed that nonmarine bacteriophages increasingly attract interest as indicators of pollution in inshore environments by enteric microorganisms (see Chapter 8).

5.4. ECOLOGICAL OBSERVATIONS ON MARINE PHAGE–HOST SYSTEMS

There is great interest in the ecological role of bacteriophages as related to their impact on bacterial populations by either reducing, via lysis, the number of sensitive cells or by transferring genetic material from one cell to another. No information concerning these aspects is available with regard to marine phage. From the review of the relevant literature as presented above this lack of knowledge is easily understandable: owing to the generally very low concentration in marine environments of phage particles attacking specific bacterial hosts it seems impossible to trace the development of PHS in their natural habitat.

Even under favorable conditions unequivocal information is difficult to obtain. Thus Ahrens (1) determined the seasonal fluctuations in the numbers of agrobacteria and their phage in the Kiel Bight. While during the first half of the year relatively low numbers of colony-forming units (CFU) and PFU were found per milliliter, a substantial increase in CFU per milliliter was observed in August followed by pronounced fluctuations during September through November. At the same time a burst occurred in the number of PFU (up to 25,500 PFU/mL, mean about 16,000 PFU/mL), which lasted for several weeks in August and September. In the second half of September a short-lived burst in both the CFU and PFU counts was observed.

One could argue that in the absence of the phage the CFU peak observed in August would have lasted for some time, or that a more pronounced increase in CFU numbers toward the CFU peak in late September would have been found, or that the latter CFU peak was short-lived owing to the large number of PFU (36,500 PFU/mL). However correct this argument may be, there is no proof to substantiate it. The only definite conclusion is that an increase in the concentration of sensitive bacteria was a precondition for the growth of the bacteriophage population. This is supported by the fact that Ahrens (1) always encountered lower numbers of phage than of sensitive bacteria.

Based on the results of investigations performed with nonmarine PHS in chemostat cultures (8,35,42) it can be safely assumed that careful studies under the much more complicated natural conditions are unlikely to produce results that can be correctly interpreted with regard to the impact of bacteriophages on indigenous bacterial populations. In chemostat cultures it was observed that, although the composition of the populations of both the phage and bacteria may drastically change in the course of the experiment by the occurrence and enrichment of mutants, an equilibrium between phage and the initially inoculated type of sensitive bacteria was finally attained. This equilibrium is characterized by numbers of bacteria and phage oscillating at different levels, with the numbers of phage exceeding that of bacteria. The latter observation is in contrast to the findings of Ahrens (1) stated above, and therefore it can be concluded that the fluctuations in CFU counts observed by Ahrens were only in part caused by phage infection.

It is, of course, a fact that infection of sensitive bacteria by virulent phage results in the elimination of the infected cells and in the production of progeny virions, and there can be no doubt that this also happens in marine environments. However, beyond this generality, the questions regarding the ecology of PHS in marine waters and sediments remain unanswered, questions such as the following:

How long do free phage particles survive under natural conditions?

How metabolically active must a cell be to become infected by adsorbed phage and produce new virions?

How many progeny virions are produced under natural conditions?

To what extent are naturally occurring events reflected by results obtained under laboratory conditions?

As the following compilation of present knowledge will show, very little is known to provide answers to such questions.

5.4.1. The Ionic Environment

By definition marine bacteriophages depend on salinities characteristic of their natural habitat. In most investigations dealing with PHS from marine sources the ionic requirements were met via the use of media prepared with natural or artificial seawater. In some cases (46,47) sea salts were replaced by NaCl.

5.4.1.1. Ionic Environment and Lytic Activity of Bacteriophage.
Working with two phage strains Spencer (49) found that the ability to lyse their hosts depended on the presence of Mg^{2+} in addition to Na^+. In one case the proportion between both ions was critical; in the other it was not. With the PHS isolated by Wiebe and Liston (52) lysis took place in seawater broth at 10 and 22°C, but only at 10°C in a medium containing 0.05 M Mg^{2+} in addition to 0.085 M NaCl (the latter being the lowest concentration at which the host would grow). Poor lysis occurred when Mg^{2+} was replaced by 0.01 M Ca^{2+}. Based on these experiments one cannot ascertain which of the various steps (adsorption, infection, replication) leading to bacterial lysis by phage was affected by the ionic conditions.

5.4.1.2. Ionic Environment and Adsorption of Phage onto Cells.
Several investigators have reported the adsorption rates of phage (22,24,26,28) but the influence of ions on adsorption was investigated only in two studies. The phage isolated by Keynan et al. (30) adsorbed almost equally well (>90%) within 10 min in seawater, distilled water, 3% NaCl, and 3% NaCl plus 15 mM Mg^{2+} + Ca^{2+}, as well as with 10^{-3} M EDTA added to the last mentioned medium. Zachary (55) tested two phage strains in a four-salts nutrient medium with varying NaCl concentrations ranging from 0.06 to 0.41 M NaCl. This medium, besides NaCl and organic nutrients, was composed of 3.8 mM KCl, 0.018 M $MgSO_4 \cdot 7H_2O$, and 3.8 mM $CaCl_2 \cdot 2H_2O$. Phage nt-6 adsorbed poorly (less than 30%) during 20 min at 27°C under all conditions tested, whereas phage nt-1 showed a strong dependence on NaCl concentration. Only 8% phage adsorbed at 0.06 M NaCl concentration in contrast to more than 80% adsorption observed at 0.16 and 0.25 M NaCl. With strain nt-1 it was also found that KCl compensated for NaCl when added to media containing only 0.06 or 0.08 M NaCl to raise their osmolarity to that of four-salts medium containing 0.16 M NaCl.

5.4.1.3. Ionic Environment and Bacteriophage Replication.
Zachary (55) also investigated the influence of NaCl concentration on phage production employing the four-salts nutrient medium in one-step growth experiments. With phage nt-1 an increase in burst size from 12 to more than 500 PFU/cell was observed with NaCl concentration increasing from 0.06 M to between 0.25 and 0.41 M NaCl. The latent period decreased from 90 min at 0.06 M NaCl to 45 min at 0.25 M NaCl and increased again to 50 min at higher NaCl concentrations. With phage nt-6 the maximal burst size of about 600 PFU/cell was observed at NaCl concentrations of 0.13 and 0.16 M, whereas the lowest burst size of 300 and 311 PFU/cell was found with 0.06 M as well as with 0.33 and 0.41 M NaCl. The latent period of this strain decreased from 90 min at 0.06 M NaCl to 55 min at 0.16 M and increased again to 60 min at the higher NaCl concentrations tested. With both phage it was found that KCl could partially compensate for NaCl, both in regard to burst size and latent period.

5.4.1.4. Ionic Environment and Survival of Marine Bacteriophage.

The phage described by Keynan et al. (30) was stable in natural and artificial seawater for at least 24 hr at 4°C. In solutions of 3, 2, and 1% NaCl it was inactivated by about 70, 80, and 90%, respectively, under otherwise identical conditions. Addition of 15 mM Mg^{2+} + Ca^{2+} to the NaCl solutions resulted in a decrease in inactivation to 35, 35, and 86%, respectively. In distilled water and in solutions containing 1 or 15 mM Mg^{2+} + Ca^{2+}, however, less than 1% phage survived after 2 hr at 4°C. The protective function of Mg^{2+} and Ca^{2+}, therefore, depended on sufficiently high Na$^+$ concentration.

Quite different observations were made by Spencer (49), although his experiments were performed at 40°C. Phage P/14 remained stable at this temperature for at least 10 min when suspended in seawater, but it was inactivated to less than 1% after this time of exposure in distilled water and NaCl solutions of between 0.5% and 2.75% NaCl. In 3% NaCl solution inactivation to about 0.1% occurred within 2 min. In a solution containing 0.495% MgCl$_2$ and 2.75% NaCl, more than 10% of the phage survived after 10 min, and in the presence of 0.495% or 0.99% of MgCl$_2$ alone inactivation also was much slower than in NaCl solutions.

Hidaka (19) tested five phage strains for infectivity after 24 hr at 5°C in distilled water, 0.5% and 3% NaCl solutions, ⅙ and full-strength artificial seawater (ASW), as well as in seawater broth (SWB). To the first five media 0.05% polypeptone and 0.01% yeast extract were added and the pH of all media was adjusted to 7.6–7.8. The five phage strains survived by 100% in SWB and ASW. In the other four media one strain was completely unaffected and three strains lost some of their activity, with the most pronounced effect observed in 0.5% NaCl solutions. Only one phage, 06N-58P, was totally inactivated in distilled water and both NaCl solutions, and some inactivation was found in ⅙ ASW. This strain is the only RNA-containing marine phage reported so far (22).

Experiments lasting for up to 120 days were performed by Zachary (55), who tested two phage strains for their survival in untreated, autoclaved, and filter-sterilized estuarine water (EW, salinity 14‰), pond water, distilled water, and two versions of his four-salts nutrient medium containing 0.06 or 0.16 M NaCl. Phage nt-1 survived at 20°C for 120 days in all these media except untreated EW (14% survival after 10 days). Phage nt-6 was also rapidly inactivated in untreated EW but, in contrast to strain nt-1, was markedly affected in distilled water (5% survival after 30 days) and pond water (45% survival after 120 days). It survived for 120 days in the four-salts media and for at least 60 days in sterilized EW.

In summary, the effects of ionic conditions on different phage strains varies greatly. Mg^{2+} or Mg^{2+} + Ca^{2+} generally exerted a positive influence on survival and lytic activity of the phage, whereas that of Na$^+$ varied considerably. Within the range of concentrations tested the effect of Na$^+$ in some cases was weak or negligible but in others it was very distinct, as observed in survival and adsorption studies. Because these observations were made with a small number of phage strains and a wide range of methods, it cannot be ascertained to what extent they really are typical of marine bacteriophages. More information is needed which should be collected, if ever possible, under experimental conditions simulating natural ones.

5.4.2. Temperature

Experiments on the influence of temperature on marine bacteriophages have mostly been performed at elevated temperatures (40°C and above) in an attempt to establish differences in regard to heat sensitivity between marine and nonmarine phage. In none of these cases were the employed temperatures ecologically relevant. As evident from the information presented in Section 5.3, most of the phage strains tested were insensitive to 50°C during 30 min, and many of them were stable at even higher temperatures.

Other information on the influence of temperature on marine bacteriophages is scant because most PHS were investigated at only one temperature. Such studies were largely done at between 20 and 25°C, but 15°C (50) and 28°C (14) were also studied.

Spencer (49), employing a qualitative spot test, observed his phage strains to cause lysis at about 1°C. At 30 or 37°C, at which temperature the respective host grew well, several of his phages were inactive. Strain P/14 lysed its host at 21°C but not at 22°C. With the PHS investigated by Wiebe and Liston (52) plaque formation was observed only between 0 and 23°C, although the host grew at temperatures up to 33°C. The optimal temperature for lysis was between 5 and 12°C. In contrast, little or no effect of temperature between 20 and 37°C was found with the phage isolated by Keynan et al. (30).

The PHS described by Johnson (26) is unique insofar as plaque formation, which was observed at 6 and 25°C but not at 30°C, depended greatly on the temperature at which the host was grown prior to use. The best lawn and plaque counts were obtained with cells grown at 30°C and plates incubated at 18°C for plaque formation.

Two phage strains tested by Zachary (56) for burst size and latent period were strongly influenced by temperature. With phage nt-1 the burst size (PFU/cell) of 520 at 27°C was reduced to 115 at 17°C, while the latent period increased from 45 min at 27°C to 170 min at 17°C. With phage nt-6 the burst size of 500 at 27°C was even reduced to 27 at 17°C, but the latent period of 55 min at 27°C increased to only 120 min at 17°C. At 10°C the latent period of both strains was longer than 10 hr.

Moebus (38) observed marked differences between plaquing efficiencies of 39 PHS incubated at 5, 15, and 25°C, respectively. These PHS, which were isolated from eastern North Atlantic water in February, produced the highest plaque counts at 15°C. With some of these PHS an inhibition reaction of the bacteria to phage was observed, generally at 25°C, as was the case with nearly 360 PHS isolated from the Atlantic. Of the PHS characterized by the inhibition reaction at 25°C, 269 were further tested by a qualitative spot test. Plaques were produced by 54 PHS at 5 and/or 15°C. With some of these PHS it was found that plaques developed at 15°C yielded only a few progeny phage which could not be propagated to produce high-titer phage stocks.

5.4.3. Hydrostatic Pressure

Of the three PHS isolated from sediments retrieved from 230 m or more below the surface (20,26,52) only that of Wiebe and Liston (52) was investigated at increased hydrostatic pressure. Initially 9×10^6 CFU/mL of the host, *Aeromonas* sp., were mixed with 3×10^2 PFU/mL and incubated at 1, 90, and 200 atm, respectively. Temperature and duration of incubation were not reported. Although the bacteria grew to 2×10^7 CFU/mL, the PFU titer increased by factors of (about) 10^4, 10^3, and 10^2, respectively. These findings demonstrate that the host bacteria were able to grow, if only slightly, and to propagate the phage at even higher hydrostatic pressure than this PHS was subjected to under natural conditions, which was 82.5 atm.

5.4.4. Anaerobiosis

The effect of anaerobiosis was studied by Zachary (56) on PHS isolated from salt-marsh sediments which, except for the upper few centimeters, are anaerobic. The host bacterium, *Vibrio natriegens,* is a facultative anaerobe and was found to propagate the phage strains nt-1 and nt-6 anaerobically. In one-step growth experiments performed in a modified four-salts nutrient medium the reproduction of phage nt-1 was similar under anaerobic and aerobic conditions. The rate of PFU produced per cell per minute of latent period was 10.0 under anaerobiosis and 10.9 under aerobiosis. This was due to an increase in burst size (from 600 to 800 PFU/cell) during a longer latent period (80 min instead of 55 min). With phage nt-6 a similar effect of anaerobiosis on the duration of the latent period was found. However, the burst size under anaerobic conditions was drastically reduced from 2700 to 110 PFU/cell, resulting in a rate of PFU/cell/min of latent period of only 1.4 instead of 54.0 under aerobic conditions.

5.4.5. Bacterial Hosts

Bacteria of various taxonomic groups were reported as hosts of bacteriophages isolated from estuarine or marine environments: *Cytophaga* (48,49), *Pseudomonas* (7,10,15,18,20,21,28,40,48), *Agrobacterium* (1), *Achromobacter* (7,20), *Vibrio/Beneckea* (7,18,20,26,28,30,40,46,48), *V. parahaemolyticus* (3,9,25,31,32,41,45), *Aeromonas* (20,21,52), *Photobacterium* (16,21,47), *Lucibacterium* (21), and *Flavobacterium* (7,18,20,40,48). The largest number of PHS have been reported for the family Vibrionaceae, and among these the genus *Vibrio* has received the greatest attention. This may in part be due to the specific interest in *V. parahaemolyticus* by several authors.

The preponderance of *Vibrio* as host of marine phage may reflect numerical dominance of these bacteria in marine environments, but it may equally well depend on the isolation procedure employed. From the available data Moebus and Nattkemper (40) calculated that at least 37% of 1382 bacterial isolates collected at 48 stations in the Atlantic (37) belonged to the *Vibrio/Aeromonas* group. However, the finding of 100% of phage-sensitive strains in this group was interpreted as an artifact caused by the type of enrichment culture employed.

5.5. EVALUATION OF PRESENT KNOWLEDGE OF MARINE BACTERIOPHAGE ECOLOGY

Although the number of reports dealing with the isolation of PHS from various estuarine and marine habitats is still low, it seems reasonable to assume that bacteriophages may occur in any environment sustaining bacterial life. The failure of attempts to detect marine bacteriophages is probably due to the lack of suitable methods, as was the case during the first half of this century when no phage could be found in offshore waters.

Meanwhile some progress has been made in regard to efficient methods enabling the isolation of numerous PHS from seawater samples containing as little as 1 CFU/mL (37). However, methods suited for meaningful investigations of the ecological importance of marine bacteriophages are still lacking. This refers to our inability to determine the actual number of host-specific virions, which, at least in most marine environments, is very low, as well as to methods for laboratory investigations.

For many years following the detection of indigenous marine bacteriophages these viruses retained the status of a curiosity. Justifiably, much attention was first paid to the question whether the few isolates collected until the early 1970s were really marine phage and to characters separating marine phage from nonmarine ones. These initial efforts confirmed the existence of truly marine PHS, but they also resulted in the knowledge that marine phage are not much different from the others. These findings may have curbed potential interest in marine phage.

Marine phage cannot be studied quantitatively in most of their habitats. This experience seemingly blocked any effort to determine PFU numbers even in environments that have the potential to support large numbers of cultivable bacteria, such as estuarine sediments. Ahrens' study (1) in Kiel Bight is the exception to the rule. All but two (1,50) isolations of marine bacteriophages reported after 1960 employed enrichment cultures. In theory, enrichment cultures should yield information on the concentration of specific phage in a sample, either by the most-probable number method or by the use of subsamples of increasing volume, but no such attempts have been reported.

Laboratory investigations of bacteriophages concerning, for example, adsorption rate or phage reproduction depend on the presence of large numbers of bacteria, at least initially, to allow rapid adsorption of phage onto cells. The bacteria needed usually are grown in media that, as compared to most marine environments, are very rich in organic nutrients, and in most cases the experiment itself is performed in media of the same or of similar composition. The information obtained by such experiments is of questionable value in the study of phage ecology. It is difficult to extrapolate the results found with PHS cultivated in enriched media to natural situations. The presently available information on marine PHS including that on their adaptability to temperature, hydrostatic pressure, or anaerobiosis is therefore of little value in answering the question, how are marine PHS maintained in nature?

Difficulties in maintaining PHS mainly rest with the phage, which, as obligate parasites, depend on their hosts. Therefore, bacteriophages must survive until they meet with suitable host cells in which they multiply in the virulent fashion. In principle, there are two possibilities for phage to survive: either as free virions or within a cell which they are able to lysogenize. Information regarding both possibilities is almost nonexistent at present.

Storage experiments employing seawater and lasting for 1 month or longer were reported only three times (1,49,55). Spencer (49) stored phage in filtered seawater at 0°C in the dark and at room temperature (20–22°C) in daylight for 1 month. The phage remained stable for most of the storage period but some inactivation was observed toward the end of it (no quantitative data were presented). Ahrens (1) found that in raw seawater of 16‰ salinity at 20°C one phage strain was inactivated within 4 weeks (from 10^5 PFU/mL to nil) and another in 2 weeks (from 2×10^6 to 1 PFU/mL).

Zachary (55) performed a comparative study of bacteriophage survival in untreated and sterilized estuarine water. The phages nt-1 and nt-6 remained stable for at least 60 days in autoclaved and filter-sterilized samples, but were reduced to 11 and 0.1% of the original titer, respectively, in untreated samples after the same time of storage.

From these findings it seems reasonable to assume that the indigenous microbial flora was involved in phage inactivation. This is supported by several other observations. Johnson (26) found the host of his phage to produce a filterable (0.22 μm) antiphage agent which was sensitive to 100°C. Other information on antiviral activity of seawater related to marine bacteria was compiled by Kapuscinski and Mitchell (29).

Besides the production of antiviral agents by bacteria there is the possibility that phage are lost during storage in raw seawater because of adsorption to nonhost cells (those that are unable to reproduce the infecting phage). With numerous marine bacteriophage strains Moebus (38) observed an inhibition reaction of the bacteria causing clear to turbid spots on bacterial lawns without phage reproduction. In liquid cultures of such phage–bacteria combinations the phage rapidly adsorbed to the cells while, depending on the initial phage titer, a portion of the cells lost the ability to form colonies.

Similar observations have also been made with nonmarine phage (5,36). The results of phage-typing experiments (38,39) and of the taxonomic evaluation of bacteria by Moebus and Nattkemper (40) indicate that the inhibition reaction to bacteriophage is fairly common among closely related marine bacteria.

Finally, in raw seawater phage also may be lost by adsorption to host cells which, for one reason or another, are metabolically inactive and remain so even when withdrawn with a sample to be plated for PFU counts. At any rate, it must be expected that the indigenous microbial flora of seawater plays a major role regarding the survivability of marine phage.

Another type of long-storage experiment, with the phage maintained in cell lysates, has been performed with numerous marine phage strains. Phage stocks are usually prepared with some kind of seawater broth. Therefore, the stability of phage in these media may not bear any relationship with phage survival under natural conditions. However, information on this topic reveals the considerable variation among marine phage.

Only Hidaka (19) reported on the stability of marine phage in stock suspensions stored at 5–8°C for 6 months. Three of five strains remained fully stable and one lost some of its activity, but the activity of the fifth strain rapidly decreased by 1 $\log_{10}$/month. Similar observations were made by me with most of the approximately 300 phage strains collected. With virulent phage, generally weak inactivation was observed for up to 7 years of storage at 8°C, whereas with temperate phage it was found that fresh stock suspensions must be produced approximately every year. However, there are exceptions, the most exceptional being a phage strain that will not survive in seawater broth for more than 3–7 days (Moebus, unpublished).

Regarding lysogeny among marine bacteria even less information is available than on longevity of marine phage. We do not know if lysogeny is as common as with nonmarine species. Also, the influence of ecologically important factors on the establishment of lysogeny or on the relative frequency of prophage entering the replicative (virulent) cycle is not known. Only one temperate marine phage has been investigated thus far (24). It was found to lysogenize between 10 and 16% of the cells and its burst size was calculated to be 25 PFU/cell in broth culture, being the lowest observed with marine phage.

The number of progeny phage particles produced by infected cells, that is, the burst size, certainly is an important factor with regard to the maintenance of PHS in nature. The burst sizes reported (1,22,24,28,30,46,55,56) range between 25 and 2700 PFU/cell, in accordance with burst sizes observed with nonmarine phage under comparable nutritive conditions. Of special interest are the results of Zachary (55,56), who determined the burst sizes of two phage strains under several different conditions of temperature, oxygen tension, NaCl concentration, and organic nutrients.

Zachary (55,56) used two media that were the same in regard to the concentration of four salts (see Section 5.4.1.2) but differed in organic nutrient content. One medium (4SN) contained per liter, in addition to the salts, 10 g Difco nutrient broth, 5 g Difco peptone, and 2.5 g Difco yeast extract. The other medium (M4SN) was modified by the addition of Tris buffer and 10 g glucose/L. Thus both media were extremely rich in organic nutrients. In 4SN medium containing 0.25 M NaCl the burst size (at 27°C) of strain nt-1 was 520 PFU/cell, and in M4SN medium of the same NaCl concentration it was 600 PFU/cell. Under identical conditions the burst sizes of phage nt-6 were 500 PFU/cell in 4SN and 2700 PFU/cell in M4SN medium. These findings, demonstrating the influence of organic nutrients on the reproduction of phage, emphasize the need for further investigations performed under more natural conditions.

Preliminary attempts of this kind were made by Moebus (unpublished). Seawater collected near Helgoland was autoclaved and used for the cultivation of four PHS that have a common host (strain H4). The bacteria in this medium grew from an initial 3×10^3 CFU/mL to a final titer of 5×10^6 CFU/mL within 1.3, 2, and 7 days at 25, 15, and 5°C, respectively. The four phage strains were tested for reproduction and survival.

Inactivation of phage in bacteria-free controls was rapid with all strains, depending mainly on incubation temperature. At least 99% of PFU were lost after 4–8 days at 5°C, the most favorable temperature for survival. With two phage strains causing slightly turbid plaques, inactivation at any of the three temperatures was too fast to be compensated for by reproduction. The other two phage strains, producing clear plaques, reached final titers of 10^7 PFU/mL at all temperatures tested when the cultures were started with 3×10^4 PFU/mL. The maximum PFU titer was attained at about the time when the bacteria entered the stationary growth phase. With 3×10^3 PFU/mL present initially, phage reproduction was observed only at 15 and 25°C. However, during most experiments it ceased after 1 day of incubation at a titer of 4–8 $\times$ 10^4 PFU/mL, followed by phage inactivation at a rate similar to that observed with the controls. At 25°C the increase in PFU titer sometimes was halted only after 1 day and, without marked reduction in PFU counts meanwhile, resumed after another 1–2 days of incubation. In such cultures the maximum titer of 10^7 PFU/mL was reached not before the eighth day of incubation. This observation indicates that reproduction of these two phages did not depend on maximum metabolic activity of the host cells, although it was much faster with exponentially growing cultures.

The rapid inactivation of the phage may have been an anomaly caused by some unknown property of the seawater sample, which was collected when the early summer phytoplankton bloom abruptly broke down (seemingly within 24 hr).

Large numbers of host bacteria *and* phage were necessary to ensure continual bacteriophage production during these experiments. Neither are usually present in marine environments. Wiggins and Alexander (53), from their findings obtained with nonmarine PHS, concluded "that bacteriophages do not affect the number and activity of bacteria in environments where the density of the host species is below the host cell threshold of about 10^4 CFU/mL.'' If correct, this conclusion should also apply to marine PHS. Then reproduction of bacteriophage in most marine waters is a rare event affecting only a minimal portion of the population of any specific host bacterium present. In sediments, the relation between bacteria and their phage may be distinctly different.

Summarizing our present knowledge on marine bacteriophages it must be stated that it is, to say the least, insufficient to answer most questions concerning the ecology of these viruses. Ecologically important factors, such as ions and temperature, have been studied with a few PHS; however, up to the present almost all experiments have been performed under conditions too different from natural ones. This is especially true with regard to the media employed, which are usually extremely rich in organic nutrients, as compared to seawater and even most types of sediment. The results obtained under such conditions at the most serve as a clue of what differences may exist between PHS in their natural environment as regards, for example, their burst sizes, but they do not provide information on the burst sizes of the respective PHS realized in their natural habitat. The same is true for adsorption rate, latent period, and other parameters. Therefore, from the ecological point of view most of the available information on marine bacteriophages is of limited value.

Admittedly marine phage pose very serious problems for their investigation in natural environments or under simulated natural conditions. In most marine environments phage seem to be present at numbers too low to be determined quantitatively. Even if this difficulty could be overcome by better methods than presently available, there would be no chance to trace the development of PHS in such habitats. Therefore, to gain insight in the ecology of marine PHS, laboratory investigations are indispensable. However, these must be performed under conditions that simulate natural ones to the greatest possible extent.

REFERENCES

1. Ahrens, R. 1971. Untersuchungen zur Verbreitung von Phagen der Gattung *Agrobacterium* in der Ostsee. Kieler Meeresforsch. *27*:102–112.

2. Baross, J., and J. Liston. 1970. Occurrence of *Vibrio parahaemolyticus* and related hemolytic vibrios in marine environments of Washington state. Appl. Microbiol. *20*:179–186.

3. Baross, J., J. Liston, and R. Y. Morita. 1978. Incidence of *Vibrio parahaemolyticus* bacteriophages and other *Vibrio* bacteriophages in marine samples. Appl. Environ. Microbiol. *36*:492–499.

4. Baross, J., J. Liston, and R. Y. Morita. 1978. Ecological relationship between *Vibrio parahaemolyticus* and agar-digesting vibrios as evidenced by bacteriophage susceptibility patterns. Appl. Environ. Microbiol. *36*:500–505.

5. Blair, J. E. and R. E. O. Williams. 1961. Phage typing of streptococci. Bull. World Health Org. *24*:771–784.

6. Carlucci, A. F., and D. Pramer. 1960. An evaluation of factors affecting the survival of *Escherichia coli* in sea water. IV. Bacteriophages. Appl. Microbiol. *8*:254–256.

7. Chaina, P. N. 1965. Some recent studies on marine bacteriophages. J. Gen. Microbiol. *41*:xxv.

8. Chao, L., B. R. Levin, and F. M. Stewart. 1977. A complex community in a simple habitat: An experimental study with bacteria and phage. Ecology *58*:369–378.

9. Colwell, R. R., T. E. Lovelace, L. Wan, T. Kaneko, T. Staley, P. K. Chen, and H. Tubiash. 1973. *Vibrio parahaemolyticus*—isolation, identification, classification, and ecology. J. Milk Food Technol. *36*:202–213.

10. Delisle, A. L., and R. E. Levin. 1969. Bacteriophages of psychrophilic pseudomonads. I. Host range of phage pools active against fish spoilage and fish-pathogenic pseudomonads. Antonie van Leeuwenhoek J. Microbiol. *35*:307–317.

11. Delisle, A. L., and R. E. Levin. 1969. Bacteriophages of psychrophilic pseudomonads. II. Host range of phage active against *Pseudomonas putrefaciens*. Antonie van Leeuwenhoek J. Microbiol. *35*:318–324.

12. Delisle, A. L., and R. E. Levin. 1972. Characteristics of three phages infectious for psychrophilic fishery isolates of *Pseudomonas putrefaciens*. Antonie van Leeuwenhoek J. Microbiol. *38*:1–8.

13. Delisle, A. L., and R. E. Levin. 1972. Effect of temperature on an obligately psychrophilic phage–host system of *Pseudomonas putrefaciens*. Antonie van Leeuwenhoek J. Microbiol. *38*:9–15.

14. Espejo, R. T., and E. S. Canelo. 1968. Properties of bacteriophage PM2: A lipid-containing bacterial virus. Virology *34*:738–747.

15. Espejo, R. T., and E. S. Canelo. 1968. Properties and characteristics of the host bacterium of bacteriophage PM2. J. Bacteriol. *95*:1887–1891.

16. Hastings, J. W., A. Keynan, and K. McCloskey. 1961. Properties of a newly isolated bacteriophage of luminescent bacteria. Biol. Bull. *121*:375.

17. d'Herelle, F. 1917. Sur un microbe invisible antagoniste des bacilles dysentériques. C. R. Acad. Sci. Paris. *165*:373–375.

18. Hidaka, T. 1971. Isolation of marine bacteriophages from sea water. Bull. Jap. Soc. Sci. Fish *37*:1199–1206.

19. Hidaka, T. 1972. On the stability of marine bacteriophages. Bull. Jap. Soc. Sci. Fish. *38*:517–523.

20. Hidaka, T. 1973. Characterization of marine bacteriophages newly isolated. Mem. Fac. Fish., Kagoshima Univ. *22*:47–61.

21. Hidaka, T. 1977. Detection and isolation of marine bacteriophage systems in the southwestern part of the Pacific Ocean. Mem. Fac. Fish., Kagoshima Univ. *26*:55–62.

22. Hidaka, T., and K. Ichida. 1976. Properties of a marine RNA-containing bacteriophage. Mem. Fac. Fish., Kagoshima Univ. *25*:77–89.

23. Hidaka, T., T. Kawaguchi, and M. Shirahama. 1979. Analytical research of microbial ecosystems in seawater around fishing ground. I. Distribution of bacteriophage systems in seawater around Ryukyu Island arc. Mem. Fac. Fish., Kagoshima Univ. *28*:47–55.

24. Hidaka, T., and T. Shirahama. 1974. Preliminary characterization of a temperate phage system isolated from marine mud. Mem. Fac. Fish., Kagoshima Univ. *23*:137–148.

25. Hidaka T., and A. Tokushige. 1978. Isolation and characterization of *Vibrio parahaemolyticus* bacteriophages in sea water. Mem. Fac. Fish., Kagoshima Univ. *27*:79–90.

26. Johnson, R. M. 1968. Characteristics of a marine *Vibrio* bacteriophage system. J. Arizona Acad. Sci. *5*:28–33.

27. Johnson, P. W., and J. McN. Sieburth. 1978. Morphology of non-cultured bacterio-

plankton from estuarine, shelf and open ocean water. Abst. Ann. Meet. Am. Soc. Microbiol., N95, p. 178.

28. Kakimoto, D., and H. Nagatomi. 1972. Study of bacteriophages in Kinko Bay. Bull. Jap. Soc. Sci. Fish. *38*:271–278.

29. Kapuscinski, R. B., and R. Mitchell. 1980. Processes controlling virus inactivation in coastal waters. Water Res. *14*:363–371.

30. Keynan, A., K. Nealson, H. Sideropoulos, and J. W. Hastings. 1974. Marine transducing bacteriophage attacking a luminous bacterium. J. Virol. *14*:333–340.

31. Koga, T., and T. Kawata. 1981. Structure of a novel bacteriophage VP3 for *Vibrio parahaemolyticus*. Microbiol. Immunol. *25*:737–740.

32. Koga, T., S. Toyoshima, and T. Kawata. 1982. Morphological varieties and host range of *Vibrio parahaemolyticus* bacteriophages isolated from seawater. Appl. Environ. Microbiol. *44*:466–470.

33. Kriss, A. E. 1963. Marine Microbiology, pp. 284–288. Oliver & Boyd, Edinburgh-London.

34. Kriss, A. E., and E. A. Rukina. 1947. Bakteriofag w more. (Bacteriophages in the sea.) Dokl. Akad. Nauk SSSR *57*:833–836.

35. Levin, B. R., F. M. Stewart, and L. Chao. 1977. Resource-limited growth, competition, and predation: A model and experimental studies with bacteria and bacteriophage. Am. Nat. *111*:3–24.

36. Maiti, M. 1978. Mode of action of bacteriophage ϕ149 on cholera and El Tor vibrios. Can. J. Microbiol. *24*:1583–1589.

37. Moebus, K. 1980. A method for the detection of bacteriophages from ocean water. Helgoländer Meeresunters. *34*:1–14.

38. Moebus, K. 1983. Lytic and inhibition responses to bacteriophages among marine bacteria, with special reference to the origin of phage–host systems. Helgoländer Meeresunters. *36*:375–391.

39. Moebus, K., and H. Nattkemper. 1981. Bacteriophage sensitivity patterns among bacteria isolated from marine waters. Helgländer Meeresunters. *34*:375–385.

40. Moebus, K., and H. Nattkemper. 1983. Taxonomic investigations of bacteriophage sensitive bacteria isolated from marine waters. Helgoländer Meeresunters. *36*:357–373.

41. Nakanishi, H., Y. Iida, K. Maeshima, T. Teramoto, Y. Hosaka, and M. Ozaki. 1966. Isolation and properties of bacteriophages of *Vibrio parahaemolyticus*. Biken J. *9*:149–157.

42. Paynter, M. J. B., and H. R. Bungay. 1969. Dynamics in coliphage infections, pp. 323–335. *In* D. Perlman (ed.), Fermentation Advances. Academic, New York.

43. Raettig, H. 1958. Bakteriophagie 1917–1956, part II. Gustav Fischer, Stuttgart.

44. Sieburth, J. McN. 1979. Sea Microbes, pp. 61–63. Oxford University Press, New York.

45. Sklarow, S. S., R. R. Colwell, G. B. Chapman, and S. F. Zane. 1973. Characteristics of a *Vibrio parahaemolyticus* bacteriophage isolated from Atlantic coast sediment. Can. J. Microbiol. *19*:1519–1520.

46. Smith, L. S., and A. P. Krueger. 1954. Characteristics of a new *Vibrio*-bacteriophage system. J. Gen. Physiol. *38*:161–168.

47. Spencer, R. 1955. A marine bacteriophage. Nature *175*:690.

48. Spencer, R. 1960. Indigenous marine bacteriophages. J. Bacteriol. *79*:614.

49. Spencer, R. 1963. Bacterial viruses in the sea, pp. 350–365. *In* C. H. Oppenheimer (ed.), Symposium on Marine Microbiology. Charles C Thomas, Springfield, Illinois.

50. Stevenson, J. H., and L. J. Albright. 1972. Isolation and partial characterization of a marine bacteriophage. Z. Allg. Mikrobiol. *12*:599–603.

51. Torrella, F., and R. Y. Morita. 1979. Evidence by electron micrographs for a high incidence of bacteriophage particles in the waters of Yaquina Bay, Oregon: ecological and
taxonomical implications. Appl. Environ. Microbiol. *37:* 774–778.

52. Wiebe, W. J., and J. Liston. 1968. Isolation and characterization of a marine bacteriophage. Mar. Biol. *1*:244–249.

53. Wiggins, B. A., and M. Alexander. 1985. Minimum bacterial density for bacteriophage
replication: Implications for significance of bacteriophages in natural ecosytems. Appl.
Environ. Microbiol. *49*:19–23.

54. Zachary, A. 1974. Isolation of bacteriophages of the marine bacterium *Beneckea natriegens* from coastal salt marshes. Appl. Microbiol. *27*:980–982.

55. Zachary, A. 1976. Physiology and ecology of bacteriophages of the marine bacterium
Beneckea natriegens: salinity. Appl. Environ. Microbiol. *31*:415–422.

56. Zachary, A. 1978. An ecology study of bacteriophages of *Vibrio natriegens*. Can. J.
Microbiol. *24*:321–324.

57. ZoBell, C. 1946. Marine Microbiology, pp. 82–83. Chronica Botanica, Waltham, Massachusetts.

6

ECOLOGY OF SOIL BACTERIOPHAGES

S. T. WILLIAMS, A. M. MORTIMER, AND LESLEY MANCHESTER

Department of Botany, University of Liverpool, PO Box 147, Liverpool L69 3BX, United Kingdom

6.1.   INTRODUCTION

Microbial ecology may be defined as "the study of the distributions, activities, and interactions of microbial organisms within their natural habitat." This necessitates the isolation of the organism, its identification, measuring its activity in both pure and mixed culture, assessing its interactions with other living cells, and determining its responses to abiotic environmental factors. These ideals are seldom met for soil microbes, despite much effort by microbial ecologists. Our knowledge of the ecology of phage in soil is still less adequate than that of their potential hosts. The reasons for this are partly due to the technical difficulties faced when studying such minute particles in the complex soil environment. However, there has also been a surprising lack of interest in soil phage ecology, despite abundant evidence that they are widespread in soil. Most information obtained has been purely qualitative, with soil being used as a convenient source of phage isolates for studies in genetics, taxonomy, and other disciplines. This situation has provoked some interesting speculation on the status of phage in soil. Thus Reanney and Marsh (42) suggested that if phage occurred in soil at 0.1% of the titer obtained in the laboratory, they would be the most numerous genetic objects in that habitat, but Reanney et al. (43) postulated that most phage genomes in nature probably exist integrated into the DNA of their host cells.

   The ecology of phage in soil clearly merits more attention. It provides a challenging subject applying and testing theoretical concepts of host–parasite interactions. In addition, phage may exert some control on the many important chemical transformations in soil, either directly by lysis of key microbes or indirectly as agents of genetic exchange between and within microbial populations. Finally, more accurate knowledge of their micro- and macro-distribution will facilitate their isolation for use as tools in the laboratory (63). This chapter aims to present an assessment of bacteriophage ecology in soil with some emphasis on *Streptomyces* phage, with which we have been most concerned.

6.2.   EVIDENCE FOR THE EXISTENCE OF PHAGE IN SOIL

Basically, two approaches have been taken to detect and isolate phage in soil. These are enrichment procedures, which demonstrate the existence of one or more phage particles in the sample, and direct counts, which estimate the number of viable phage in a given weight of soil. In both cases detection of phage depends on the lysis of selected host strains *in vitro*. There have been few, if any, convincing demonstrations of phage in soil by direct observation.

6.2.1. Enrichment Procedures

Although there are many variations in detail, all these procedures involve the incubation of broths containing a soil sample and a high concentration of potential host cells. Incubation conditions and periods are usually selected to be conducive to host growth and sometimes suitable nutrients are included in the liquid medium. After incubation, the larger particles of soil and microbial material are removed by centrifugation and then filtered or chemically sterilized supernatant is tested against the putative host in broth or agar media. This method is both simple and effective, resulting in the isolation of a wide range of phage which propagate on a variety of bacteria, some examples of which are given in Table 6.1.

6.2.2. Direct Counts

Ideally it should be possible to determine the number of selected phage particles in a known weight of soil by testing dilutions of bacteria-sterile extracts of untreated, fresh samples against one or more potential host strains *in vitro*. Although this approach clearly depends on the range and choice of host strains, it can provide quantitative ecological data that are lacking when enrichment procedures are used.

However, most of the limited number of attempts to obtain direct counts of phage in soil have met with little success. For example, Reanney and

TABLE 6.1. Examples of Phage for Soil Bacteria Isolated by Specific Enrichment

Propagation species	References
Amorphosporangium auranticolor	Wellington and Williams (59)
Arthrobacter globiformis	Casida and Liu (11)
Bacillus stearothermophilus	Reanney and Marsh (42)
Bacillus subtilis	Brodetsky and Romig (8)
Bacillus spp.	Tan and Reanney (57)
Nocardia spp.	Williams et al. (61)
Oerskovia turbata	Prauser (40)
Pseudomonas aeruginosa	Bradley (7)
Rhizobium trifolii	Barnet (3)
Rhodococcus spp.	Prauser (41)
Streptomyces spp.	Wellington and Williams (59)
Streptoverticillium spp.	Wellington and Williams (59)

Marsh (42) found it necessary to enrich soil samples with nutrients and to incubate at high temperatures to obtain counts of *Bacillus stearothermophilus* phage ranging from 2.0×10^1 to 4.0×10^7 PFU/g dry soil. A similar approach was made by Reanney and Teh (44) to estimate titers of phage of other *Bacillus* species. Casida and Liu (11) found that phage for *Arthrobacter globiformis* were rarely detected unless soil was nutritionally amended and incubated. The difficulties of obtaining direct counts from untreated soil may be a true reflection of the numbers of free phage particles in the soil, the low titers being due to any or all of a variety of factors. These include the inactivity of host microbes, adsorption of phage to soil colloids, and denaturation of free phage in the soil environment, all of which are discussed below. However, it is also possible that poor efficiency of extraction procedures may have contributed to the low counts obtained. Some support for this possibility was obtained by Lanning and Williams (23) and Williams and Lanning (62), who assessed and developed procedures for making direct counts of *Streptomyces* phage in soil.

The efficiency of procedures at each stage of extraction was assessed by determining the percentage recovery of a streptomycete phage added to sterile soils at known titers. Almost total recovery of phage was achieved by reciprocal shaking of soil suspensions in nutrient broth containing 0.1% (w/v) egg albumin at pH 8.0 and centrifugation of soil extracts at $1200 \times g$ for 15 min. Filter sterilization of soil extracts resulted, however, in losses of 33–45% of the titer obtained in unfiltered extracts, but such losses could be avoided by substituting chloroform sterilization, providing that the phage were resistant to chloroform. Recovery of phage added to three soils ranged from 34 to 87% with filter sterilization and from 93 to 100% with chloroform sterilization. Both procedures were applied to fresh samples from a range of natural soils using six potential streptomycete hosts; some of the results obtained are given in Table 6.2. Detectable numbers of phage particles occurred in each soil, the greatest being in the nutrient-rich compost. Chloroform sterilization generally gave higher counts than filter sterilization, and the importance of using soil isolates rather than type cultures as putative hosts was demonstrated. Considering that a limited number of strains from one genus were used, the counts suggest that the total number of extractable phage in soil is much greater than has been previously suggested.

6.2.3. The State of Phage in Soil

It seems reasonable to assume that phage may be involved in either virulent or lysogenic associations with their hosts in soil, but current methods do not provide any direct evidence of the state of the phage in its natural habitat. Thus Reanney et al. (43) suggested that phage counts gave no reliable indication of the concentration of extracellular phage in soil and all that can be concluded is that the potential numbers of phage per unit of soil could be very high. The detection of phage, whether by direct or enrichment meth-

TABLE 6.2. Direct Counts of Streptomycete Phage in Soils and Compost

Streptomycete host	Number of phage[a]	
	Filter sterilization	Chloroform sterilization
Compost		
Soil isolate Mx1	2917	12387
Soil isolate Mx2	3916	12008
Soil isolate Mx3	7263	22950
S. lavendulae ISP 5069	0	0
S. michiganenis ISP 5015	0	0
S. griseus ISP 5236	397	806
Arable		
Soil isolate Mx1	225	275
Soil isolate Mx2	642	1283
Soil isolate Mx3	667	1858
S. lavendulae ISP 5069	0	0
S. michiganenis ISP 5015	0	0
S. griseus ISP 5236	227	660
Garden		
Soil isolate Mx1	233	1517
Soil isolate Mx2	207	558
Soil isolate Mx3	558	2868
S. lavendulae ISP 5069	0	0
S. michiganenis ISP 5015	0	0
S. griseus ISP 5236	0	0

[a] Plaque-forming units per gram of soil (dry weight).

Source: Data modified from Lanning and Williams (23).

ods, relies upon a virulent reaction against the test strain *in vitro*. Nevertheless it is always possible that phage in natural lysogenic associations may by chance encounter an indicator culture in the laboratory. Phage that are temperate for hosts *in vitro* may be readily isolated from soil. Three phage obtained from soils in different localities produced turbid plaques with a characteristic morphology on *Streptomyces coelicolor* (13). Surviving growth from the central area of the turbid plaques was lysogenic and resistant to lysis by the homologous phage. Similar plaques provided a new temperate phage of *Streptomyces venezuelae* (54), and our own observations suggest that this is a frequent phenomenon on isolation plates. Such observations underline the potential of phages to form lysogenic associations in soil.

However, there seems to be no reason to doubt the existence of free virulent phages in soils. The majority of the isolates is virulent *in vitro*, often showing a high degree of polyvalency within a given bacterial genus (21,59). This would clearly be advantageous to a virulent phage in the natural habitat and contrasts with the often relatively restricted host range of temperate phage (e.g., 13,54). Finally, the survival capacity of free phage added to soil under various environmental conditions is generally quite high (see below).

6.3. FACTORS INFLUENCING THE STABILITY OF FREE PHAGE IN SOIL

There is evidence that most soil bacteria have only spasmodic periods of activity in micro-sites dispersed within the soil mass (see below). Therefore it is likely that virulent phage in the absence of an active susceptible host must exist for considerable periods as free virions in the soil and be subjected to environmental factors and their fluctuations.

6.3.1. Stability in Sterile and Nonsterile Soil

The effects of environmental factors ideally should be studied in soil under controlled laboratory conditions, but unfortunately much information has been derived, sometimes of necessity, solely from phage suspensions in broths. If phage stability in soil is to be assessed, sterile and/or nonsterile soil samples may be used. The former provides a more uniform reproducible system, whereas the latter is clearly a closer approximation to the natural environment. Relatively few comparisons of these systems have been made. Williams and Lanning (62) added known amounts of a streptomycete phage to soils to assess the effects of various environmental factors on its stability. Loss of stability was generally greater in nonsterile soil but the patterns of response to environmental factors were similar to those in sterile soil. Manchester (27) recorded a decrease in stability of streptomycete phage added to a sand dune soil, but there were no significant differences between sterile and nonsterile soil. Total inactivity of the phages occurred after 5 weeks.

6.3.2. Soil Colloids

Reanney and Marsh (42) suggested that the behavior of soil phage *in vitro* may be misleading, because many could be adsorbed to colloidal particles in their natural habitat. There is evidence that phage and other viruses adsorb to clays, and this may result in their protection or inactivation (14), the results of studies to date being somewhat contradictory. Although Bystricky et al. (9) indicated that the adsorption of phages to materials such as bentonite did not affect their infectivity, a study of *Arthrobacter* phage (37) showed that most phage added to solid phases having a high cation-exchange capacity could not be eluted in an infective state. It was concluded that the phage were being inactivated either by the adsorption or the elution process. Sykes and Williams (55) found that many streptomycete phage isolated from soil were adsorbed by unsubstituted or sodium treated kaolin, but fewer were adsorbed by sodium and calcium montmorillonite or by calcium- and aluminum-treated kaolin. The montmorillonites did, however, protect one of the phage during experimental manipulations. Although the majority of adsorbed phage were still infective, attempts to obtain free infectious phage by desorption were largely unsuccessful, suggesting that inactivation occurred during the elution procedure.

6.3.3. pH

There have been few studies on the effect of soil pH on phage stability, but it seems likely that extremes of pH (<4.0 or >8.0) may lead to inactivation of most soil phage. Sykes et al. (56) showed that phage isolated from nonacidic soils were active against both acidophilic and neutrophilic streptomycetes *in vitro* but none were detected in soils with a pH below 6.0, although they were known to contain acidophiles. Isolated phage were relatively stable between pH 5.5 and 9.0 in sterile soils and broths but were rapidly inactivated at lower pH levels (Table 6.3). Williams and Lanning (62) obtained a streptomycete phage from a neutral soil which remained infective at pH 3.0 in broth and sterile soil. However, it was propagated on a neutrophilic host and all attempts to infect acidophiles were unsuccessful. Similarly all attempts to isolate streptomycete phage from acidic soils failed. The possible relevance of small scale variations of pH within the soil to phage stability was demonstrated by Sykes and Williams (55). Fifty percent inactivation of phage adsorbed to kaolin occurred at pH 6.1, compared with pH 4.9 for free phage. Differences in pH (Δ pH) between that at or near to colloid surface (pH_s) and that of the bulk phase (pH_b) are well known; their relevance to

TABLE 6.3. Survival of a Streptomycete Phage in Soils of Different pH

Soil	Soil pH	Phage recovered[a]	
		0 hr	200 hr
Podzol (F1 layer)	3.7	0	0
Podzol (F2/H layer)	3.9	0	0
Podzol (A2 horizon)	4.0	0	0
Podzol (A1 horizon)	4.2	1×10^2	0
Clay soil	6.1	3×10^3	3×10^1
Garden soil	7.0	2×10^3	2×10^2
Saline soil	8.0	6×10^4	3×10^3

[a] Plaque-forming units per gram dry weight of soil. The initial concentration of phage added to soil was 1×10^6 PFU/g dry weight of soil.

Source: Adapted from Sykes et al. (56).

microbial activities in soil has been emphasized (33). Thus adsorbed phage may encounter a pH lower than that of the bulk soil and hence be more susceptible to inactivation in moderately acidic soils.

6.3.4. Temperature

Information on effect of temperature on soil phage is very limited. Most soil phage are probably inactivated at high temperatures that are unlikely to occur in their environment. Thus Williams and Lanning (62), studying the stability of a streptomycete phage added to soil incubated at temperatures ranging from 5 to 40°C, found greatest survival at low temperatures and total inactivation at 40°C. However, significantly greater tolerance for high temperatures is shown by phage for thermophilic bacteria such as *Thermoactinomyces, Micropolyspora* (22), and *Bacillus stearothermophilus* (51). Phage for the latter species were isolated by Reanney and Marsh (42) from all soils tested included one from a geothermal area. Maximum development of these phages in enriched soil culture occurred at 45°C. The thermostability of phage for various *Bacillus* species after brief treatment at 70°C was quite variable (44); surprisingly, the least heat-tolerant phage infected a strain of the thermophilic *Bacillus stearothermophilus*.

6.3.5. Other Factors

Few other environmental factors have been studied as far as phage survival in soil is concerned. Water content was found to have little effect on a streptomycete phage except for a decrease at high moisture holding capacities (62), and the addition of herbicides at practical concentrations exerted no significant effect on two phage for *Streptomyces chrysomallus* (47). The limited data available suggest that a significant proportion of free phage can remain infective for appreciable periods under the environmental conditions prevalent in many soils. As suggested by Reanney et al. (43), the packaging of DNA in phage heads may constitute a means of preserving extracellular DNA in natural environments.

6.4. FACTORS INFLUENCING THE REPLICATION OF PHAGE IN SOIL

Although the kinetics of phage–host interactions in laboratory culture is well understood, there have been few attempts to apply and test the principles elucidated *in vitro* to the soil environment. Extrapolations from the laboratory to the soil environment can be justified as the most logical means of progress given the current state of knowledge, but such projections must obviously be based on many tentative assumptions. Studies of interactions in soil necessarily compound the inadequacies of the ecological data available not only for phage but also for their hosts. It is clear that for a phage to replicate in soil it must encounter, become adsorbed to, and infect a suitable host that is metabolically active. Therefore, the distribution and frequency of sites within the soil supporting active bacterial growth are of major significance. Encounters between a phage and its host presumably occur by chance, for it can be assumed that phage particles act as passive entities in the soil solution. A major factor influencing the distribution of phage in soil is probably the location of previously lysed host bacteria. Encounters may also be facilitated by the adsorption of both phage and bacterial cells to colloidal materials. When a potentially successful encounter occurs, its subsequent development depends on the influence of environmental factors on the phage during adsorption and penetration and on the host activity after infection.

6.4.1. The Availability of Active Hosts

Attempts have been made to distinguish between oligotrophic and copiotrophic bacteria in natural environments (39). Oligotrophs predominate in nutrient-poor environments and show relatively high growth rates in culture at low concentrations of energy-yielding substrates, the latter being attributed to the bacterium's relatively low substrate saturation constant (K_s) and low maximum growth rate, μ_m (19). Copiotrophs, in contrast, require higher nutrient concentrations ($\times 100$) in culture and have relatively high K_s and μ_m values. Oligotrophs and copiotrophs may be equated with the long-known autochthonous and zymogenous groups of Winogradsky (65), respectively, or the ''K-strategists'' and ''r-strategists'' of Hirsch et al. (19). The relevance of these concepts to soil bacteria was considered by Williams (60), who concluded that although there was good evidence for the existence of copiotrophs and facultative oligotrophs in soil, there was little proof for the occurrence of strict oligotrophs.

At first sight this conclusion to some extent conflicts with the estimates of energy input to the soil and its bacterial populations, which indicate that most soil masses present oligotrophic environments for chemo-organotrophic microbes. Estimates of bacterial growth rates derived from measurements of total energy inputs suggest long generation times ranging from 18hr (38) to 1200 hr (2). However, the meaning and possibly the accuracy of such estimates are questionable. These figures probably reflect short intermittent periods of relatively rapid growth interspersed by periods of dormancy, rather than prolonged periods of extremely slow growth, although the soil environment is clearly not conducive to growth rates achieved under ideal laboratory conditions. Also the significance of temporal and spatial discontinuities of energy sources in the soil was emphasized by Williams (60). Thus, for example, sites of relatively high nutrient concentration occur during the initial stages of decomposition of fresh organic materials and on the surface of live roots. Numbers of bacteria on freshly fallen oak litter reached 2×10^{10} cells/g dry weight of organic matter (34), whereas Bowen and Rovira (6) recorded 1×10^6 bacteria/mm^3 on roots of *Lolium perenne*. Most soil bacteria produce resting cells under conditions inimical to growth; these may be easily recognized structures such as *Bacillus* endospores or *Streptomyces* arthrospores, or may involve more subtle changes in cell structure, as seen in *Arthrobacter* (16). Thus many soil bacteria seem to fall into the zymogenous category of Winogradsky (65), and as concluded by McLaren (33), it has yet to be shown that any continually active biomasses occur in soil.

If we accept these concepts of the behavior of bacteria in soil there are several consequences for their phage. The distribution of hosts is not random and opportunites for infection are sporadic in both space and time. There is also a need for phage to "mimic" the host by rapid replication in the restricted growth periods and to maintain sufficient numbers of potentially infective virions in the soil environment. The concentration of active hosts in micro-sites may well increase the chances of encounters at least in the short term. Unfortunately, there are no conclusive data on the distribution of phage within soil micro-habitats; but it may be that counts derived from the soil mass underestimate their concentration and significance in such sites. Available evidence suggests that phage can replicate and retain their stability for significant periods under most environmental conditions in soil.

6.4.2. Soil Colloids

There is evidence that phage may adsorb to clay colloids (see above), and that bacterial cells can adsorb and become coated with clay minerals and other surface-active particles (28,29). Both may influence the frequency of host–phage interactions in soil. Ruddick and Williams (48) found that streptomycete spores were strongly adsorbed to kaolin, as were phage lysing these bacteria (55), suggesting that phage and their potential hosts may be concentrated within the same microsites in soil. However, Roper and Marshall (45, 46), studying the interactions between phage and *Escherichia coli* in saline sediments, found that adsorption of montmorillonite and other colloidal materials to the bacteria protected them from attack by phage. Casida and Liu (11) concluded that phage of *Arthrobacter globiformis* indigenous in soil were present in a masked state that was not due merely to their loose physical adsorption to soil materials. The interactions between two streptomycetes and their respective phage, added to nonsterile soil and incubated at 15°C under laboratory conditions, were studied by Manchester (27). The amendment titers approximated to the predetermined concentrations of streptomycete spores and phage in the soil. Comparison of the results with those obtained for the same phage–host systems in broth culture indicated that interactions were more limited in the soil, but that infection efficiency was increased (Table 6.4). It was suggested that the latter was due to the adsorption of phage and host spores to soil particles.

TABLE 6.4.   Comparison of the Infection Efficiencies of Streptomycete Phage in Soil and Broth

Host–phage combinations	Infection efficiency[a]	
	Soil	Broth
Streptomyces Mx1 and phage ϕmx1	3.5×10^{-6}	4.5×10^{-9}
Streptomyces Mx8 and phage ϕmx8	1.9×10^{-5}	4.1×10^{-8}

[a] Infection efficiencies were derived from number of hosts infected/ (the initial number of hosts present $\times$ the initial number of phage present).

6.4.3. pH

There is evidence that pH can have a marked effect on the stability of free phage in soil (see above) and the effects of pH on the activity of soil bacteria are well documented, the majority behaving as neutrophiles *in vitro*. pH can also have differential effects on the adsorption of cells and phage to colloidal particles (e.g., 50, 48, 55). However, there have been few attempts to assess the influence of pH on the interactions between phage and bacteria in soil. Sykes et al. (56) isolated a phage from a neutral soil which attacked not only neutrophilic streptomycetes but also lysed acidophiles at pH 5.5 *in vitro*. This provided a tool to study the effect of pH on phage replication in a variety of hosts with different pH optima. Acidity was found to have differential effects on adsorption, penetration, and the length of the latent period, the latter being related to the metabolic activity of the host. The step most susceptible to acidity appeared to be during infection or early replication, whereas the ultimate intracellular replication of the phage was least influenced by the external pH. Titers of phage obtained from the various hosts growing at their optimum pH were similar.

6.4.4. Temperature

Although it appears that only extremes of temperature have a marked influence on the stability of free phage in soil, the replication cycle is clearly more sensitive because it reflects the metabolic activity of the host. Most soil bacteria behave as mesophiles in culture, with optimum growth rates between 25 and 30°C, despite the fact that temperatures in many soils, even near the surface, seldom reach these levels. Thus, for example, the annual range of temperatures at 30 cm in a soil under grass at Oxford, England was from about 2 to 17.5°C (49). Therefore, the results of experiments conducted at optimum laboratory temperatures may not be typical of the soil environment.

Seeley and Primrose (52) investigated the effect of temperature on the ecology of aquatic coliphages and recognized three physiological groups based on their efficiency of plating at different temperatures. The distribution of the three types of phage correlated closely with the temperature of the environment from which they were isolated. However, such detailed studies have not yet been made on soil phage. Reanney and Marsh (42) examined the development of phage to the thermophilic *Bacillus stearothermophilus* in enriched soil incubated at various temperatures. Maximum development occurred at 45°C with substantial numbers at 30 or 55°C. In a similar experiment conducted at 37°C, a phage-lysing *Bacillus circulans* showed lag, exponential, and plateau phases similar to those in monoculture except that the time scale was much extended (57).

The replication of streptomycete phage over a range of temperatures including those typical of soil was studied by Manchester (27). At lower temperatures the latent period increased and the burst size decreased, but the rise period showed no clear-cut trend (Table 6.5). It was also noted that phage produced fewer progeny at a typical soil temperature of 15°C. Overall it was concluded that the size of the pool of phage in soil was restricted by the prevailing temperatures.

6.5. MODELING PHAGE–HOST INTERACTIONS

6.5.1. A Population Dynamics Perspective

Intrinsically the abundance of a species is determined by the resources available for population growth. For phage, the primary resource is the presence

TABLE 6.5.   The Effect of Temperature on Parameters of Replication of Streptomycete Phage from Soil

	Host and phage								
	S. lavendulae ISP 5069 and ϕ85			*Streptomyces* sp. Mx1 and ϕmx1			*Streptomyces* sp. Mx8 and ϕmx8		
				Temperature (°C)					
Parameter	12	15	27	12	15	27	12	15	27
Latent period (min)	120	70	60	540	250	60	900	440	300
Burst size (PFU/spore)	100	224	398	19	100	148	13	66	336
Rise period (min)	80	10	142	110	100	93	170	135	200

Source: Adapted from Manchester (27).

of susceptible hosts. *A priori* therefore, populations of bacterial hosts and their phage are likely to exist in a coupled dynamic state in which bacterial populations are regulated in some manner by phage abundance, whereas phage numbers themselves are a function of host availability. Moreover, such "coexistence" may not necessarily be stable. The dynamics of this interaction in soil is by no means a straightforward example of classic "predator–prey" relationships. Four features in particular demand comment in this connection.

First, phage populations themselves may be structured, containing virions as well as perhaps both virulent and temperate classes (10). Most commonly, phage are distributed between two classes, that is, virulent or temperate forms. In the absence of reproduction, phage persisting as a virion "bank" decline in response to factors influencing their survival in the soil environment (see earlier). Increments to this bank can arise only through reproduction within hosts. For some phage types a reserve also exists in the form of prophage in lysogens.

Second, hosts may not be uniformly susceptible. In bacteria, for example, adsorption is known to differ among species, and in actinomycetes only particular stages in the life cycle, for example, germ tubes and mycelial tips, are susceptible to infection.

Third, host populations themselves may be regulated by competition for abiotic resources. The form this takes and its significance depend on both temporal and spatial resource distribution. Single-cell prokaryote populations in a homogeneous environment represent the simplest case. With a constant resource renewal rate, host populations grow at a rate determined by a unique limiting resource and equilibrium population sizes are governed by the magnitude of resource inputs. The stability of the steady-state condition and return times to equilibrium, after perturbation, are governed by the form of the (usually monotonic) relationship between growth rate and concentration of limiting resource and the magnitude of the perturbation. In hosts that show a mycelial growth form, reproduction by sporulation is often triggered by quite different subsets of resource requirements than those for mycelial growth. Pulses of spore production occur, for example, when carbon availability is low. Thus populations of discrete colonies arising from spore germination increase in size in fluctuating environments, whereas populations of mycelial tips within a clone continue to increase if the environment provides constant resource renewal. Both resource spectra are likely to occur in the soil environment at large but with different periodicities and duration. In these circumstances, the existence of any stable equilibria in host numbers (tips or colonies) is likely to be a phenomenon difficult to detect. Nevertheless, at appropriate spatial and temporal scales, resources may fluctuate in a periodic fashion and to differing amplitudes that determine overall host densities acting through density-dependent regulation.

Host generation time is the fourth feature requiring comment and is in part related to the preceding point. Whereas generation times in some hosts (such as unicells) are exceedingly short, and hence population responses to increased resource supply potentially very rapid, developmental times in filamentous forms may be of significant duration in determining response times to alterations in resource availability and hence relative host abundance.

Three of the features addressed above reflect ecological strategies in heterogeneous environments. Structured populations and phasic growth pattern with resting cells, virions, and spores are characteristic traits. Implicitly, they provide measures of population stability in the long term and have important consequences for host–parasite relationships. From the ecological perspective, the principal question of interest is what determines the abundance of phage and to what degree is there a stable association between host and phage populations. This question has been addressed in depth by a number of workers both theoretically (e.g., 26) and experimentally (12,20). Such studies of chemostat environments unambiguously demonstrated the importance of the interaction of resource concentration (particularly that which determines maximal cell density), adsorption rates of phage, burst sizes, and relative growth rates of bacteria. The apparent stability of chemostat populations of bacteria and virulent phage may also depend on refuges. The adhesion of phage to inert objects and walls in a chemostat may give protection to hosts in their vicinity.

Most theoretical studies that have contributed to our understanding of the conditions necessary for the coexistence of bacteria and phage have originated from models based on differential equations, in essence based on chemostat models (36). Their appropriateness rests on the assumption of continuous growth in a homogeneous environment. In the next section we present the application of a difference equation model to a phage–streptomycete interaction as an alternative approach to studying the conditions necessary for the coexistence of phage and their hosts in soil.

6.5.2. A Difference Equation Model for a Complex Phage–Host Association

Consider a soil environment in which resources at any one point in time are patchily distributed in space. The patchy distribution is, however, changing through time such that at each patch there is a cyclical phase of renewal and exhaustion of resources for host growth. The host displays a filamentous growth with spores developing from hyphal tips. In this environment host populations display (in effect) discrete generations; that is, lowered resource availability induces sporulation, but on a larger spatial scale all growth stages of the host are present. Colony establishment from spores, moreover, is density dependent. Both filamentous tips and young germinated spores of the host offer sites for phage adsorption. Each infectable unit, tip or germling, on lysis releases phage in known burst size which are returned to the virion bank. From this bank each virion has a defined probability of successfully adsorbing to a susceptible unit. This scenario, overall, offers a partial mimic to the association of a phage and its host in a soil environment, assuming that the association is specific.

The Model.   The roots of the model, Equations 1a and 1b, are derived from the work of Nicholson and Bailey (35) and the subsequent extensions by Hassell (17) and Beddington et al. (4).

$$H_{t+1} = s \, H_t \exp \left[r(1 - H_t/K) - aP_t) \right] \qquad (1.a)$$

$$P_{t+1} = z \, c \, H_t \left[1 - \exp \left(- aP_t \right) \right] \qquad (1.b)$$

Descriptions of the generic form of the model can be found in these papers. A discussion of application of this method is given by Manchester (27). H_t and H_{t+1} are host population densities measured as infectable units per unit soil volume over one generation. r is the intrinsic rate of natural increase of the host. The number of units present, however, is regulated in a density-dependent manner such that the soil volume displays an equilibrium density or carrying capacity, K. The exponent term in Equation 1a represents that fraction of the host population that escapes phage infection. The parameter s is included to take into account the fraction of the population that exists as spores. The term a is the infection efficiency of a phage, namely, the probability of an individual phage successfully replicating in a host unit, and P_t and P_{t+1} are the numbers of phage present at times t and $t + 1$. In Equation 1b, c is the mean burst size and z is the fraction of the virion bank that survives to the next phage generation.

Table 6.6 gives measurements derived from experimental investigations of two streptomycete–phage systems in a soil environment. These values reflect representative ranges that have been assessed in laboratory conditions. Characteristically, for both these systems host intrinsic rates of increase are high; both have the potential to increase by at least 270-fold in each generation. As May (30) has pointed out, these rates in relation to K may have important consequences on the stability of the equilibrium condition, depending on the exact nature of density-dependent regulation. Moreover, patterns of dynamic behavior are intrinsic to host populations in the absence of phage. They may be seen as gradual damped oscillations to equilibrium numbers, stable limit cycles in which the population fluctuates around the equilibrium level revisiting two or four or even more points time and time again, or wholly irregular and chaotic fluctuations. The response to a lowering of host density by phage infection must necessarily be seen against this background.

Equations 1a and 1b make it possible to simulate population trajectories for combinations of parameter values. Figure 6.1a illustrates the outcome for the set given in Table 6.6. Regardless of initial population sizes both components become extinct in a limited number of generations. Host streptomyce-

TABLE 6.6. Representative Parameter Values for Two Streptomycete–Phage Systems

Parameter	System[a]	
	MXI + φmx1	MX8 + φmx8
Infection efficiency, a, in soil	2.1×10^{-5} to 3.5×10^{-6}	1.85×10^{-5} to 9.37×10^{-5}
Carrying capacity, K (g soil)	8.5×10^{6}	6.4×10^{6}
Burst size, c, per infection	19–100	13–66
Intrinsic rate of increase, r	5.76	5.70
Virion survival[b]	0.66	0.83
Spore survival[b]	0.13	0.11

[a] Generation time was 11 days.
[b] Probability of survival over one host generation.

Source: From Manchester (27).

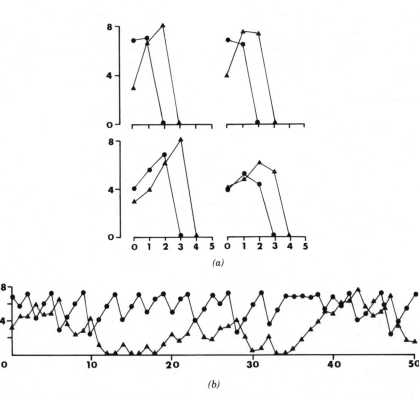

Figure 6.1. Population trajectories in the *Streptomycete* MX1 and φmx1 association as described by Equations 1a and 1b. ● streptomycete; ▲ phage. Population sizes are measured on a logarithmic scale. (*a*) Infection efficiency of 10^{-6} as measured in soil. (*b*) Infection efficiency of 3.2×10^{-8} as measured in broth.

tes cannot withstand the intensity of predation, and phage in the absence of host decline at the mortality rate z of the virion bank. It is only when the infection probability is reduced by at least two orders of magnitude that a prolonged period of coexistence of phage and host occurs (Figure 6.1b).

Deeper insight into the dynamic properties of the model may be gained by an analytical approach in which the stability of the system can be assessed. It involves an analysis of the effects of perturbations as arising from phage parasitism from equilibrium positions of host populations. Such an investigation can expose and demarcate those parameter values that result in damping of the initial perturbation in subsequent generations in comparison to those that lead to divergent oscillations in population densities. The results may be expressed in a stability diagram in which the extent of a perturbation q (magnitude of phage parasitism) from equilibrium is plotted against the host intrinsic rate of increase. Within this parameter space it is possible to define regions of stability (prolonged coexistence of phage and host) and instability (where ultimately both components become extinct) which are separated by zones in which the system oscillates at an amplitude determined by the extent of the initial perturbation. Hassell and May (18) and Elseth and Baumgardner (15) give details of the method. Figure 6.2 illustrates the stability domains for two infection efficiencies for the MX1 + ϕmx1 association. At low infection efficiency, $a = 10^{-12}$, the association can recover from a wide range of perturbations and shows considerable stability

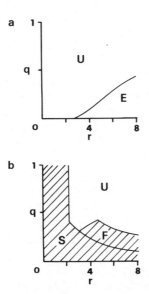

Figure 6.2. Stability domains for the interaction between *Streptomycete* MX1 and ϕmx1 as governed by Equations 1a and 1b. q measures the extent of perturbation from equilibrium, and r the intrinsic rate of natural increase. Burst size, c, is 100 and virion and spore survival are as given in Table 6.6. In (*a*) infection efficiency is 10^{-7} and in (*b*) is 10^{-12}. S = stability, F = wide fluctuations, U = unstable, E = extreme oscillations leading to rapid phage extinction.

to a wide range of perturbations but at higher efficiencies, $a = 10^{-7}$, no stable regions are apparent. These diagrams illustrate the complex nature of this dynamic system as a whole. On the one hand, host populations may fluctuate in accordance with intrinsic regulatory properties alone that may in certain circumstances be chaotic. Alternatively, a degree of phage parasitism may introduce stability to the association. Also demonstrated is the considerable importance of phage infection efficiency in affecting the outcome. Equally, the exact position of the domains in parameter space in part depends upon the relative magnitude of other parameters. It is thus the combination of parameter values that must be assessed when considering ecological strategies followed by phage persisting in soil environments.

6.6. ECOLOGICAL STRATEGIES

Bacteriophages in soil represent a class of "predators" (64) that have been paid scant attention despite the obvious parallels with other disciplines, e.g., disease epidemiology and pathology. May (31) explored the dynamic properties of hosts with discrete nonoverlapping generations whose density was regulated by a lethal parasite spreading in epidemic fashion throughout host generations before host reproductive age was attained. He concluded that such a straightforward deterministic system surprisingly exhibited chaotic dynamics.

In the case of phage and streptomycete hosts when infection efficiencies are lower than 10^{-7}, the interaction is also chaotic and inherently unstable. This provides an adequate explanation of the wide variation observed when phage counts are made from natural soils. It is also clear from Figure 6.2 that for any stable coexistence to occur, infection has to be a much rarer event, with efficiencies being lowered by at least five orders of magnitude. As probably all predators and parasites do, phage too remain between the "devil and the deep blue sea" (24). In this instance the sea is the soil environment in which phage become inactivated. The phage population in the virion bank may have a surprisingly short half-life; Manchester (27) recorded an average 9.4 days for streptomycete virions. If this is a realistic approximation for all phage, it is clear that for persistence in a soil ecosystem, reproduction within hosts is essential. Yet populations of host infectable units constitute a highly unpredictable resource for a predator, the "bedevilment" being not only wide fluctuations in host numbers, but also their transient presence in both time and space.

Theoretical and experimental studies of insect parasitoid populations and their hosts have shown that there is a range of predator and prey responses that can alter the nature of a one-sided antagonistic interaction (see 5 for an introduction). In particular the form of the responses by predators to prey density and prey aggregation have important stabilizing effects on the interaction. In the absence of a definitive description of the behavior of phage in the soil, firm conclusions are difficult to draw, but two points are salient. The first is that experimental studies suggest that the functional responses for streptomycete phage are probably of "type 2" (5,27). Thus the rate of phage parasitism decreases as streptomycete host density increases and large host populations suffer proportionally less mortality than small ones. In consequence the effect on the interaction is destabilizing. The second point derives from examination of the effect of spatial heterogeneity in the soil environment. This variation may present the host with a refuge in which total protection from infection can occur. Such a refuge may be a result of both chemical and physical properties of the soil (see above). The outcome may be that either a constant number or a constant proportion of hosts are protected. Interestingly, analytical investigations of a constant proportion refuge showed that in the streptomycete–phage system at infection efficiencies observed in soil, allocation of 80–90% of the host population to a refuge resulted in a stable equilibrium point. Given a measure of stability in the soil environment it is therefore feasible for phage and host to display long-term coexistence.

The above points impinge directly on considerations of ecological strategies followed by phage. These must be judged in the overall context of coevolution within host–parasite associations. May and Anderson (1) point out that if the fitness of parasites in any way depends on the survival of the host (or vice versa), then selection may favor host and parasites with mutually beneficial characteristics. Phage and their hosts clearly fall in this category. In a detailed assessment of bacteria and phage coevolution, Lenski and Levin (25) concluded that it was not necessary to envisage an endless arms race between host defense and parasite counterdefense as embodied in the Red Queen hypothesis (58). Furthermore, Stewart and Levins (53) have argued that phage polymorphisms stabilize phage abundance in resource-limited environments. Temperance as opposed to virulence was hypothesized to be advantageous because lysogenic phage colonies are able to husband resources in poor environments, whereas lysogeny is also an adaptation of phage to maintain populations when host densities oscillate between those necessary for phage maintenance. Detailed experimental evidence has yet to be provided but these studies clearly give a rigorous theoretical basis for further work.

6.7. CONCLUSIONS

In this review we have attempted an analytical discussion of our current understanding of the ecology of soil phage. Any attempt at synthesis is difficult because of the fragmentary nature of present knowledge and the lack of application of a clear conceptual framework. Chemostat models provide ideal theoretical and experimental tools for analysis of phage dynamics interacting with prokaryote hosts, but difference equations of the form discussed above provide an alternative method. Both suffer the criticism that they are abstractions from reality, but they provide important foci for the critical appraisal of the role of environmental factors. The models indicate the considerable importance of parameter values relative to one another. It is only with clear insight of this that the description of the biotic and abiotic mechanisms affecting the process can be pursued fruitfully. Even so, this approach is limited by its dualistic view of one host–one parasite interactions. Understanding the assembly rules by which microbial communities are governed in the soil remains a formidable but fascinating challenge. Phage probably play a far greater role in such communities than has hitherto been appreciated.

REFERENCES

1.   Anderson, R. M., and R. M. May. 1978. Regulation and stability of host-parasite population interactions. I. Regulatory processes. J. Anim. Ecol. *47*:219–249.

2.   Babiuk, L. A., and E. A. Paul. 1970. The use of fluorescein isothiocyanate in the determination of the bacterial biomass of a grassland soil. Can. J. Microbiol. *16*:57–62.

3.   Barnet, Y. M. 1972. Bacteriophages of *Rhizobium trifolii*. 1. Morphology and host range. J. Gen. Virol. *15*:1–15.

4.   Beddington, J. R., C. A. Free, and J. H. Lawton. 1975. Dynamic complexity in predator–prey models frames in difference equations. Nature *255*:58–60.

5.   Begon, M. E., and A. M. Mortimer. 1982. Population Ecology. Blackwell, Oxford, England.

6.   Bowen, G. D., and A. D. Rovira. 1973. Are modelling approaches useful in rhizosphere biology ?, pp. 443–452, *In* T. Rosswall (ed.), Modern Methods in the Study of Microbial Ecology. Swedish Natural Science Research Council, Stockholm.

7.   Bradley, D. E. 1966. The structure and infective process of a *Pseudomonas aeruginosa* bacteriophage containing RNA. J. Gen. Microbiol. *45*:83–96.

8.   Brodetsky, A. M., and W. Romig. 1966. Characterisation of *Bacillus subtilis* bacteriophages. J. Bacteriol. *90*:1655–1663.

9.   Bystricky, V., G. Stotzky, and M. Schiffenbauer. 1975. Electron microscopy of T-1 bacteriophage adsorbed to clay minerals: Application of critical point drying method. Can. J. Microbiol. *21*:1278–1282.

10. Campbell, A. M. 1961. Conditions for the existence of bacteriophage. Evolution *15*:153–165.

11. Casida, L. E., and K. C. Liu. 1974. *Arthrobacter globiformis* and its bacteriophage in soil. Appl. Microbiol. *28*:951–959.

12. Chao, L., B. R. Levin, and F. M. Stewart. 1977. A complex community in a simple habitat: An experimental study with bacteria and phage. Ecology *58*:369–378.

13. Dowding, J. E., and D. A. Hopwood. 1973. Temperate bacteriophages for *Streptomyces coelicolor* A3(2) isolated from soil. J. Gen Microbiol. *78*:349–359.

14. Duboise, S. M., B. E. Moore, C. A. Sorber, and B. P. Sagik. 1979. Viruses in soil systems, pp. 245–285. *In* H. D. Isenberg (ed.), CRC Critical Reviews in Microbiology, Vol. 7. CRC Press, Boca Raton, Florida.

15. Elseth, G. D., and K. D. Baumgardner. 1981. Population Biology. Van Nostrand, New York.

16. Gray, T. R. G. 1976. Survival of vegetative microbes in soil, pp. 237–364. *In* T. R. G. Gray and J. R. Postgate (eds.), The Survival of Vegetative Microbes. Cambridge University Press, Cambridge.

17. Hassell, M. P. 1978. The dynamics of arthropod predator–prey systems. Princeton University Press, Princeton, New Jersey.

18. Hassell, M. P., and R. M. May. 1973. Stability in insect host–parasite models. J. Anim Ecol. *42*:693–736.

19. Hirsch, P., M. Bernhard, S. G. Cohen, J. C. Ensign, H. W. Jannasch, A. L. Koch, K. C. Marshall, A. Malin, J. S. Poindexter, S. C. Rittenberg, D. C. Smith, and H. Veldkamp. 1979. Life under conditions of low nutrient concentrations: group report, pp. 357–372. *In* M. Shilo (ed.), Strategies of Microbial Life in Extreme Environments. Dahlem Konferenzen Life Sciences Research Report 13. Verlag Chemie, Weinheim, West Germany.

20. Horne, M. T. 1970. Coevolution of *Escherichia coli* and bacteriophages in chemostat culture. Science *168*:992–993.

21. Jones, D., and P. H. A. Sneath. 1970. Genetic transfer and bacterial taxonomy. Bacteriol. Rev. *34*:48–81.

22. Kurup, V. P., and R. J. Heinzen. 1978. Isolation and characterisation of actinophages of *Thermoactinomyces* and *Micropolyspora*. Can. J. Microbiol. *24*:794–797.

23. Lanning, S., and S. T. Williams. 1982. Methods for the direct isolation and enumeration of actinophages in soil. J. Gen. Microbiol. *128*:2063–2071.

24. Lawton, J. H., and S. McNeill. 1979. Between the devil and the deep blue sea: on the problem of being a herbivore, pp. 223–244. *In* Anderson, R. M., B. D. Turner, and L. R. Taylor (eds.), Population Dynamics. 20th Symp. Brit. Ecol. Soc. Blackwell, Oxford.

25. Lenski, R. E., and B. R. Levin. 1985. Constraints on the coevolution of bacteria and virulent phage: A model, some experiments and predictions for natural communities. Am. Nat. *125*:585–602.

26. Levin, B. R., F. M. Stewart, and L. Chao. 1977. Resource limited growth competition and predation: a model and some experimental studies with bacteria and bacteriophage. Am. Nat. *111*:3–24.

27. Manchester, L. 1986. Modelling the interaction of *Streptomycetes* and their phage. Ph.D. thesis, pp 289. University of Liverpool, Liverpool, England.

28. Marshall, K. C. 1968. Interaction between colloidal montmorillonite and cells of *Rhizobium* spp. with different ionogenic surfaces. Biochim. Biophys. Acta *156*:179–186.

29. Marshall, K. C. 1969. Studies by microelectro-phoretic and microscopic techniques of the sorption of illite and montmorillonite to rhizobia. J. Gen. Microbiol. *56*:301–306.

30. May, R. M. 1975. Biological populations obeying difference equations: Stable points, stable cycles and chaos. J. Theor. Biol. *49*:511–524.

31. May, R. M. 1985. 1985. Regulation of populations with non-overlapping generations by microparasites: a purely chaotic system. Am. Nat. *125*:573–584.

32. McLaren, A. D. 1975. Soil as a system of humus and clay immobilized enzymes. Chem. Scripta *8*:97–99.

33. McLaren, A. D. 1973. A need for counting microorganisms in soil mineral cycles. Environ. Lett. *5*:143–154.

34. Mindermann, G., and L. Daniels. 1967. Colonisation of newly fallen leaves by microorganisms, pp. 3–9. *In* O. Graft and J. E. Satchell (eds.), Progress in Soil Biology. North-Holland, Amsterdam.

35. Nicholson, A. J., and V. A. Bailey. 1935. The balance of animal populations. Proc. Zool. Soc. London *3*:551–598.

36. Noack, D. 1986. A regulatory model for steady state conditions in populations of lysogenic bacteria. J. Theor. Biol. *18*:1–18.

37. Ostle, A. G., and J. G. Holt. 1979. Elution and inactivation of bacteriophages on soil and cation-exchange resin. Appl. Environ. Microbiol. *38*:59–65.

38. Parinkina, O. M. 1973. Determination of bacterial growth rates in tundra soils, pp. 303–309. *In* T. Rosswall (ed.), Modern Methods in the Study of Microbial Ecology. Swedish Natural Science Research Council, Stockholm.

39. Poindexter, J. S. 1981. The caulobacters: Ubiquitous unusual bacteria. Microbiol. Rev. *45*:123–179.

40. Prauser, H. 1976. Host–phage relationships in nocardioform organisms, pp. 266–284. *In* M. Goodfellow, G. H. Brownell, and J. A. Serrano (eds.), The Biology of The Nocardiae. Academic, London.

41. Prauser, H. 1984. Phage host ranges in the classification and identification of Gram-positive branched and related bacteria, pp. 617–633. *In* L. Ortiz-Ortiz, L. F. Bojalil, and V. Yakoleff (eds.), Biological, Biochemical and Biomedical Aspects of Actinomycetes. Academic, London.

42. Reanney, D. C., and S. C. N. Marsh. 1973. The ecology of viruses attacking *Bacillus stearothermophilus* in soil. Soil Biol. Biochem. *5*:399–408.

43. Reanney, D. C., P. C. Gowland, and J. H. Slater. 1983. Genetic interactions among microbial communities, pp. 379–422. *In* J. H. Slater, R. Whittenbury, and J. W. T. Wimpenny (eds.), Microbes in their Natural Environments. Cambridge University Press, Cambridge.

44. Reanney, D. C., and C. K. Teh. 1976. Mapping pathways of possible phage-mediated genetic interchange among soil bacilli. Soil Biol. Biochem. *8*:305–311.

45. Roper, M. M., and K. C. Marshall. 1974. Modification of the interaction betweeen *Escherichia coli* and bacterio-phage in saline sediment. Microbial Ecol. *1*:1–13.

46. Roper, M. M., and K. C. Marshall. 1978. Effect of clay particle size on clay–*Escherichia coli*–bacteriophage interactions. J. Gen. Microbiol *106*:187–189.

47. Roslycky, E. B. 1982. Influence of selected herbicides on phages of some soil bacteria. Can. J. Soil Sci. *62*:217–220.

48. Ruddick, S. M., and S. T. Williams. 1972. Studies on the ecology of actinomycetes in soil. V. Some factors influencing the dispersal and adsorption of spores in soil. Soil Biol. Biochem. *4*:93–103.

49. Russell, E. W. 1973. Soil conditions and plant growth, 849 pp. Longman, London.

50. Santoro, T., and G. Stozky. 1967. Effect of electrolyte composition and pH on the particle size distribution of microorganisms and clay minerals as determined by electrical sensing zone method. Arch. Biochem. Biophys. *122*:664–669.

51. Saunders, G. F., and L. L. Campbell. 1966. Characterization of a thermophilic bacteriophage for *Bacillus stearothermophilus*. J. Bacteriol. 91: 340–348.

52. Seeley, N. D., and S. B. Primrose. 1980. The effect of temperature on the ecology of aquatic bacteriophages. J. Gen. Virol. *46*:87–95.

53. Stewart, F. M., and B. R. Levin. 1984. The population biology of bacterial viruses: Why be temperate. Theor. Pop. Biol. *26*:93–117.

54. Stuttard, C., and M. Dwyer. 1981. A new temperate phage of *Streptomyces venezuelae*: Morphology, DNA molecular weight and host range of SW2. Can. J. Microbiol. *2*:496–499.

55. Sykes, I. K., and S. T. Williams. 1978. Interactions of actinophage and clays. J. Gen. Microbiol. *108*:97–102.

56. Sykes, I. K., S. Lanning, and S. T. Williams. 1981. The effect of pH on soil actinophage. J. Gen. Microbiol. *122*:271–280.

57. Tan, J. S. H., and D. C. Reanney. 1976. Interactions between bacteriophages and bacteria in soil. Soil Biol. Biochem. *8*:145–150.

58. Van Valen, L. 1973. A new evolutionary law. Evol. Theory *1*:1–30.

59. Wellington, E. M. H., and S. T. Williams. 1981. Host ranges of phages isolated to *Streptomyces* and other genera. Zentbl. Bakt. Mikrobiol. Hyg. I Abt. 11 Suppl.:93–98.

60. Williams, S. T. 1985. Oligotrophy in soil: Fact or fiction?, pp. 81–110. *In* M. M. Fletcher and G. D. Floodgate (eds.), Bacteria in their Natural Environment. Academic, Orlando.

61. Williams, S. T., E. M. H. Wellington, and L. S. Tipler. 1980. The taxonomic implications of the reactions of representative *Nocardia* strains to actinophage. J. Gen. Microbiol. *119*:173–178.

62. Williams, S. T., and S. Lanning. 1984. Studies of the ecology of streptomycete phage in soil, pp. 473–483. *In* L. Ortiz-Ortiz, L. F. Bojalil, and V. Yakoleff (eds.), Biological, Biochemical and Biomedical Aspects of Actinomycetes. Academic, London.

63. Williams, S. T., M. Goodfellow, and J. C. Vickers. 1984. New microbes from old habitats?, pp. 219–256. *In* D. P. Kelly and N. G. Carr (eds.), The Microbe 1984: Part II Prokaryotes and Eukaryotes. Cambridge University Press, Cambridge.

64. Williamson, M. 1971. The Analysis of Biological Populations. Edward Arnold, London.

65. Winogradksy, S. 1924. Sur la microflore autochthone de la terre arable. C. R. Hebd. Seanc. Acad. Sci., Paris *178*:1236–1239.

7

FATE OF BACTERIOPHAGES IN WATER AND WASTEWATER TREATMENT PLANTS

GABRIEL BITTON

Department of Environmental Engineering Sciences, University of Florida, Gainesville, Florida 32611

7.1. INTRODUCTION

Phage were discovered 70 years ago by d'Herelle (33) and are known to be obligate parasites of prokaryotic and possibly eukaryotic microorganisms. Coliphages have been considered in sanitary microbiology as potential indicators of water quality as well as for monitoring the efficiency of water and wastewater treatment plants (52). The prime consideration is whether phage can indicate the presence or absence of human viruses, the concentration and assay of which are relatively expensive and time-consuming. This topic is discussed in more detail in Chapter 8.

Phage are associated with fecal material, suggesting that the animal intestine is a natural habitat for some of these microorganisms (21). An extensive study was carried out to determine the phage loads in fecal samples from healthy individuals as well as from patients with internal and leukemic diseases (21). It was found that about half of the fecal samples from healthy individuals were positive for phage and about 2% of the samples had phage concentrations about 10^5 plaque-forming units (PFU) per gram of feces. As regards the patients, 75% of the samples were positive for phage and 14% of the samples exceeded 10^5 PFU/g. It also seems that chemotherapy tends to reduce the phage numbers in feces to less than 10 PFU/g (21). Osawa et al. (45) examined the feces of humans and of a wide variety of animals in Japan and found that the percentage of phage-positive samples was fairly high (74–92%) for mammals and birds in a zoo as well as for pigs from a slaughterhouse. However, for humans and some domestic animals (horses, cows, fowls) the percentage of positive fecal samples was relatively low (10–30%).

Because phage are naturally present in animal feces it is evident that one can expect their presence in raw sewage entering wastewater treatment plants and, eventually, water treatment plants. The goal of this chapter is to document the fate of these microorganisms following wastewater and water treatment operations.

7.2. FATE OF PHAGE IN WASTEWATER TREATMENT PLANTS

7.2.1. Occurrence of Phage in Domestic Wastewater

Domestic sewage harbors a wide range of phage strains, the detection of which is carried out by using a variety of host bacteria. Phage concentration in raw domestic wastewater is on the order of 10^5–10^7 PFU/L (6,30,31,34). However, it seems that these levels represent only a relatively small portion of the actual number of phage in sewage. Using electron microscopy (EM), Ewert and Paynter (20) demonstrated that the plaque assay detected only 3.8% of the EM count. The total EM count in sewage varied between 1.6×10^8 and 1.2×10^{10} phage particles/L.

Some studies have focused on F^+-specific RNA phage in sewage because these phage have been advocated as indicators of fecal pollution (30,31). This suggestion is based on the knowledge that the F pili are not produced below 30°C, thus preventing these phage from multiplying in environmental waters. Methods have been developed to assay specifically for these phage (30,31). RNA phage levels in raw sewage vary between 10^4 and 10^7 PFU/L, using F^+ *Escherichia coli* or *Salmonella typhimurium* strains (18,30).

Phage, like human viruses (32,60), have a great affinity for sewage solids of biological and mineral origin. In seeded studies, this association was demonstrated for T2, T7, f2, and poliovirus type 1. Furthermore, the solids-associated viruses, with the exception of RNA phage f2, were still infective in the adsorbed state (42,53). It was found that 1.0–24% of phage in sewage effluent were solids-associated. In this study, freely suspended phage were defined as those passing through a serum-treated microporous filter with a pore size of 0.45 μm (24).

We now examine the fate of phage following various steps of sewage treatment. These steps include primary treatment, biological treatment (activated sludge, trickling filters, oxidation ponds), and finally tertiary treatment.

7.2.2. Phage Removal via Primary Treatment

Primary treatment is a physical process that includes screening, grit removal, and primary sedimentation. The detention time in the sedimentation basin is approximately 2 hr. Virus reduction by primary treatment is generally inefficient and ranges between 0 and 30% (7). Phage numbers in primary effluents range between 0.2 and 3.3 × 10^6/L (9,34). In a seeded experiment, using phage f2, Sherman (56) found that the mean phage reduction by primary treatment was 37.1% (27.8–54.2%) and 32.2% (13.4–69.4%) in 2.5 MGD (million gallons/day) and 1.5 MGD plants, respectively. The phage removal ability of primary treatment is therefore relatively low and erratic. From the data of Ignazzitto et al. (34), who used 10 strains of *E. coli* for phage assay, we estimated that the percent removal of indigenous phage ranged between 0 and 83%.

7.2.3. Phage Removal by the Activated Sludge Process

The activated sludge process essentially consists of the production of microbial biomass from soluble organic compounds. The bacterial cells subsequently undergo flocculation and the resulting flocs settle down in the sedimentation tank as sludge material. It is therefore apparent that one of the main mechanisms of phage removal by the activated sludge process is the adsorption of phage to and/or embedding within the mixed liquor flocs. Another means of phage removal is the possible inactivation by sewage bacteria or ingestion by protozoa. It has been shown that protozoa in aeration tanks ingest both bacteria and the phage attached to them (61). Most of these mechanisms were shown to control human virus removal in activated sludge systems (7).

Some investigators have studied the removal efficiency of activated sludge systems seeded with the RNA coliphage f2 in order to draw conclusions regarding the behavior of human enteric viruses in this process (4,43). In a step aeration activated sludge plant in Maryland the mean removal efficiency of f2 was very low and amounted to only 11% (43). Balluz et al. (4) reported that the removal efficiency of an activated sludge plant in England was 79.6%, and this may have been due to the relatively poor association of added f2 to the activated sludge solids. Indeed, under their experimental conditions, f2 behaved differently from poliovirus type 1 (3) with regard to association with solids and percent removal. The authors concluded that f2 was unsuitable for modeling the behavior of enteric viruses. These studies suggest that seeded experiments do not completely simulate the behavior of indigenous coliphages which, as mentioned above, occur as embedded particles within the activated sludge flocs. Thus it appears that indigenous coliphages would be more useful in evaluating the performance of the activated sludge process. Their numbers in an activated sludge in Texas were determined in the influent and in the settling basin effluent (19). The removal efficiency of this system was estimated at approximately 96%. Safferman and Morris (51) reported a similar removal efficiency in a multistage pilot plant. Another study showed a 99% removal of indigenous phage as well as a decay rate similar to that of poliovirus type 1 (26). A recent study of phage removal by the activated sludge process at the Austin sewage treatment plant showed that phage reduction varied between 96.0 and 97.3%. The average removal of enteroviruses was 92% (20a). Similarly, from the data of Bitton et al. (9), it was estimated that coliphage removal by the University of Florida activated sludge process was approximately 98%. Ayres (2) conducted a weekly sampling of *E. coli* and coliphage for a 1-year period. It was concluded that coliphages were efficiently removed following passage through the activated sludge process and were better indicators of enterovirus removal than *E. coli*. These results suggest that indigenous phage can serve as models for enterovirus inactivation in activated sludge. It was recently reported, however, that enteroviruses numbers were correlated only with coliphages that formed plaques greater than 3 mm in diameter (20a). The topic of phage suitability as indicator microorganisms is discussed further in Chapter 8 of this book.

It is thus concluded that the activated sludge process is generally efficient in removing phage, and the prime mechanism of removal is their embedding within sludge flocs and transfer to sludge following settling in the sedimentation tank.

7.2.4. Phage Removal by Trickling Filters

The trickling filter process consists of spraying sewage on a bed packed with gravel, rock, or plastic materials. The spraying of sewage on the filter material results in the formation of a microbial film that coats the filter material and helps in the degradation of organics with subsequent BOD (biochemical oxygen demand) removal. This treatment system is easier to operate than activated sludge and is widely used by the industry. However, enterovirus removal is usually low and inconsistent (7,8).

Coliphages occur in relatively large concentrations in trickling filter effluents. Kott et al. (38) reported that their levels in trickling filter effluents range between 1.8×10^5 and 4.6×10^6 PFU/100 mL. In another study (39), the levels of *E. coli* B phage in Haifa trickling filter effluents ranged between 0.8×10^6 PFU/100 mL and 2.3×10^6/100 mL. Coliphages displayed few fluctuations in numbers whereas enterovirus numbers showed larger fluctuations during the year. The authors concluded that coliphages could serve as good indicators of enteroviruses in sewage effluents. However, the percent removal of phage by the trickling filter process was erratic and was 40–50% during the winter and spring seasons but was more than 90% during the summer and fall seasons (38). In trickling filter plants seeded with bacterial phage f2, the percent reduction was approximately 30% (56). The percent removal by the filter per se (i.e., settling not included) was less than 20%. From the data of Durham and Wolf (19) we estimated that the percent reduction of phage in a Dallas trickling filter plant was 49%. The mechanism of virus and phage removal by trickling filters deserves further investigation.

7.2.5. Phage Removal by Oxidation Ponds

Oxidation ponds offer an inexpensive means of biological treatment of wastewater. The treatment of sewage is the result of biological processes carried out mainly by the activity of bacteria and algae.

Typically, phage levels in oxidation pond effluents vary between 10^5 and 10^7 PFU/L (5,6,39,50). The study of Pretorius (50) showed that coliphages were reduced at a lower rate (56.7 and 74.7%) than fecal coliforms (94 and 95.4%) in two oxidation ponds with a detention time of 20 days each. Baylet et al. (5), using a series of three oxidation lagoons with a total detention time of 65 days, reported that the percent removal of indigenous phage was 94%. The removal was even greater following treatment in a polishing pond which served as a tertiary treatment of wastewater in Leucate, France. Walker et al. (64) studied the removal of seeded phage, active against *Salmonella durham,* in an oxidation pond with a detention time of 25.2 days. The percent removal was between 80 and 90%. The phage was detected in the pond effluent on the fifth day after seeding despite the theoretical 25.2-day detention time in the pond. Short circuiting is often responsible for the relatively poor performance of oxidation ponds.

Physical factors (temperature, solar radiation, adsorption to solids) and probably biological factors are responsible for the removal/inactivation of phage in oxidation ponds. The latter would be expected to perform best under warm and sunny climates. For example, in Taber, Alberta, no phage removal was observed following passage of wastewater through an oxidation lagoon. This may be explained by the fact that the lagoon was under 1 m of ice (6). Phage may also adsorb to suspended solids and be transferred to the pond sediments where they may survive for relatively long periods. In the seeded experiment of Walker et al. (64) phage survived in water and sediments for 143 days. Phage survival was also higher than that of *E. coli* and *Salmonella*.

7.2.6. Fate of Phage Following Disinfection of Water and Wastewater

Phage (seeded or indigenous phage) have traditionally been considered as indicators for the effect of disinfectants (chlorine, iodine, ozone) on bacterial and viral pathogens in water and wastewater as well as for the study of mechanisms of inactivation of viruses.

Chlorine is the most popular and least expensive disinfectant for water and wastewater effluents despite the production of chlorinated hydrocarbons, some of which are carcinogenic to laboratory animals and possibly to man. RNA phage f2 is often used in disinfection experiments to study the efficacy of various chlorine species or the mechanism of inactivation. This phage, when suspended in an activated sludge effluent, was found to be more resistant to combined chlorine than poliovirus type 3 (17). Kott et al. (38,39) also found that f2 and MS-2 phage were more resistant to chlorination than poliovirus type 1. Longley et al. (40) seeded the effluents of secondary clarifiers with f2 and found that the phage was inactivated less than coliform bacteria in the presence of 17 mg chlorine/L. Less than 1 $\log_{10}$ inactivation was observed after 15 min contact time whereas for coliforms more than 4 $\log_{10}$ inactivation was obtained within the same time period. The authors also demonstrated that chlorine gas is superior to liquid chlorine in f2 inactivation. Other studies have also shown the superior resistance of f2 following chlorination (54,57). Lothrop and Sproul (41) showed that 28 mg/L of combined chlorine residual was required to obtain 99.99% reduction of bacterial phage T2. However, the latter was more sensitive to chlorine than poliovirus type 1 but was more resistant than coliforms. Kott et al. (39) observed that in the presence of 20 mg/L chlorine and following a 2-hr contact time, the decline of indigenous coliphage and enteroviruses was 48% and more than 99.9%, respectively. It also appears that chlorine dioxide is superior to chlorine as regards the inactivation of indigenous coliphages in wastewater effluents (40).

At final concentration of 30 mg/L in wastewater effluents, iodine had a higher inactivating effect on f2 at high pH than at lower pH (17).

Ozone is a powerful oxidant used for the disinfection of water and wastewater. It also substantially reduces the levels of color, cyanides, and other chemicals. Ozone is generally a stronger disinfectant than chlorine with respect to phage inactivation (40). Pavoni et al. (48) studied the inactivation of f2 in a secondary effluent from a package treatment plant. A complete inactivation of the phage was obtained within 5 min in the presence of 15 mg/L ozone. In distilled water it took only 15 sec to obtain complete inactivation. Similarly, in ozone demand-free water, a 7-$\log_{10}$ f2 inactivation was reached within 5 sec in the presence of 0.8 mg/L (37). T2 inactivation in ozone-free water was 99.99–99.999% in the presence of 0.2 mg/L ozone (35).

The mechanism of phage inactivation by disinfectants has been investigated using mainly phage f2. It was shown that chlorine inactivates the RNA core in intact viruses. Following chlorine addition phage were still able to adsorb to their host cells. However, the site of iodine action is the protein coat and more specifically the amino acids tyrosine and histidine (17,44). Pavoni et al. (48,49) postulated that the mechanism of phage destruction by ozone was the oxidation of the whole phage particle. More recently, Kim et al. (37) examined the mode of inactivation, using f2 labeled with ^{3}H-uridine. Following ozonation, f2 adsorption to host cells was reduced as a result of disintegration of the protein coat followed by liberation of RNA.

We have mentioned earlier that a portion of coliphages in wastewater are associated with solids. This may lead to their protection from the inactivating effect of disinfectants. Stagg et al. (59) have shown that MS-2 adsorbed to bentonite was more resistant to hypochlorous acid than freely suspended phage. In a subsequent field study, it was shown that solids-associated indigenous coliphages were less affected by chlorination than freely suspended phage. Similar results were reported for enteroviruses.

7.2.7. Tertiary Treatment and Wastewater Reclamation

Tertiary treatment is a series of steps designed to further remove trace organics, color, turbidity, nitrogen, phosphorus, heavy metals, and microbial pathogens and parasites from secondary effluents. These steps include coagulation–sedimentation with alum, lime and/or polyelectrolytes, sand filtration, activated carbon adsorption, and final chlorination. Most of these steps are described in more detail in Section 7.3.

Little work has been done concerning phage removal and/or inactivation during wastewater reclamation, using some of the steps mentioned above. Wolf et al. (66) studied the fate of f2 and poliovirus type 1 (P1) following alum treatment, mixed-media filtration, and high lime treatment of activated sludge effluents. Alum coagulation followed by mixed-media filtration removed 99.985% of f2. The efficiency of the coagulation process depended on the Al–P ratio. At a ratio of 7–1, the percent removal was 99.845%, whereas at a ratio of 0.441–1 the percent phage reduction was only 46%. High lime treatment (391 mg/L as CaO + 18.6 mg/L of $FeCl_3$) was the most efficient in inactivating totally both P1 and f2.

Wastewater renovation has been extensively investigated in South Africa. Virus, phage, and bacterial indicators removal was studied at the Stander plant (built for research purpose) in Pretoria and at the Windhoek large-scale reclamation plant in Namibia. At the latter plant, the treatment sequence consists of lime treatment, quality equalization tank with a mean residence time of 10 hr, sand filtration, chlorination, activated carbon, and final chlorination. The most efficient treatment steps were lime treatment with 99.96% phage reduction and breakpoint chlorination with 100% phage reduction (27). At the Stander plant, the effect of pH during lime treatment on phage reduction was investigated (28). A significant phage reduction (99.95%) was obtained at pH 11.2, whereas at pH 9.6 only 57% reduction was observed.

7.3. FATE OF PHAGE IN WATER TREATMENT PLANTS

There are two types of water treatment plants: conventional filter plants and softening plants. In filter plants, the treatment sequence generally consists of the following steps: prechlorination, coagulation–sedimentation, sand filtration, and final chlorination or ozonation. In softening plants the treatment sequence consists of a softening step (addition of lime and sodium carbonate), sedimentation, filtration, and final disinfection. We now review the phage removal potential of these various steps in water treatment plants.

7.3.1. Phage Removal by Chemical Coagulation

Chemical coagulation is an important treatment process for color and turbidity removal from water. It involves the destabilization of colloidal particles, resulting in the formation of aggregates that are large enough to settle out rapidly in the clarification basin. The most commonly used coagulants are alum, ferric chloride, and ferric sulfate. Most of the data on phage removal by coagulation were obtained from studies with seeded phage. At concentration of 40–50 mg/L, coagulation with alum removed 98 and 99% of T4 and MS-2, respectively (13). A 98% reduction was observed for T2 upon addition of alum at 40 mg/L (63). More than 99% f2 removal was obtained in the presence of 15–25 mg/L (55,67). Ferric chloride is also an efficient coagulant for phage and enterovirus removal. At 40–50 mg/L it removes more than 99% of f2 (55,67). However, indigenous phage removal, using this coagulant, was less efficient (92.6%) than for added phage T4 (99.7%) in a pilot water treatment plant (29).

Studies have been undertaken on the use of coagulant aids, mainly polyelectrolytes, to improve the coagulation and subsequent removal of enteroviruses and phage. It was reported that polyelectrolytes have little value as primary coagulants and do not increase f2 removal when used as coagulant aids (55,67). Similarly, Thorup et al. (63) did not show any improvement of T2 removal by alum coagulation in the presence of cationic, or nonionic polyelectrolytes. Chaudhuri and Engelbrecht (13) demonstrated that these chemicals were effective as coagulant aids for T4 but not for T2.

7.3.2. Phage Removal by Sand Filtration

Sand filtration always follows the coagulation–clarification step in water filtration plants. Sand filters operate at flow rates varying between 2 and 4 gpm/ft^2 and are routinely backwashed to clean the sand (7,8).

Sand filters remove viruses via adsorption and sieving processes. Because sand is essentially a poor adsorbent of viruses, phage removal is often due to the filtering out of particles adsorbed to solids or embedded within flocs. Phage removal by rapid sand filtration is often low and inconsistent. For example, Guy et al. (29) reported that T4 removal varied between 0 and 87%. Phage reduction was enhanced when the sand filter had been in operation for a few days. The removal of indigenous phage by rapid sand filtration was only 37.5%. Flow rate is an important parameter in a sand filter operation. Ninety seven percent of T4 were removed at a flow rate of 0.2 gpm/ft^2, whereas at 2gpm/ft^2 T4 removal was reduced to 40% (25).

7.3.3. Phage Removal by Diatomaceous Earth Filtration

Diatomaceous earth is composed of remains of siliceous shells of diatoms. It is commonly used in water treatment plants and in swimming pools. Most of the data on virus removal by diatomaceous earth filtration have been generated from studies where seeded phages were used. Diatomaceous earth is basically a poor virus adsorbent. In batch studies, less than 30% of MS-2 was adsorbed to this material (1,12). The removal efficiency may be enhanced by coating the diatomite filter with cationic polyelectrolytes or with aluminum and iron salts (1,11,12). For example, it was shown that a diatomite filter coated with 0.2–0.4 mg/g cationic polyelectrolyte (C-31) removed more than 99% of MS-2. However, the removal efficiency of diatomite may also be affected by other factors such as pH, filtration rate, organic matter, and turbidity.

7.3.4. Phage Removal by Activated Carbon

Activated carbon is derived from bituminous coal, wood, and lignite. It is used in wastewater and water treatment plants to remove odors, taste, and organic substances. It removes nitrogenous compounds and thus makes the disinfection process more efficient. It is also efficient in the removal of trihalomethane precursors. Activated carbon offers a relatively large surface area on the order of 400–1000 m^2/g (8). However, it was found that much of this surface area is not utilized for virus adsorption. Cookson and North (16) reported that the maximum surface area utilized for T4 adsorption was 18% of the total surface area. Activated carbon is a porous material but phage and virus adsorption occurs exclusively on the external surface of the coal. Cookson and his collaborators (14–16) have extensively studied the adsorption of phage T4 to activated carbon. The adsorption process obeys the Langmuir isotherm, is reversible, and is controlled by the pH and the ionic strength of the suspending medium. Furthermore, organic materials readily compete with phage for the adsorption sites on activated carbon (16,58). In the presence of sewage effluents, T2 adsorption to activated carbon varied between 29 and 75% in batch experiments and 7 and 32% in continuous flow studies (58). Poliovirus 1 adsorption was also inhibited in the presence of soluble organics (23). With regard to the adsorption mechanism, there is an electrostatic attraction between carboxyl groups on the carbon surface and amino groups on the phage surface (14,15). Some of these findings have been confirmed by other studies on phage adsorption to bituminous coal (46,47). In batch tests, the maximum adsorption of T4 to bituminous coal was less than 60% (47). In a pilot treatment plant, T4 removal by activated carbon was 76.8% when the carbon column was not backwashed and was only 21.4–53.3% upon backwashing (29). It is thus concluded that activated carbon is generally a poor adsorbent of phage and viruses. This is mainly due to the competitive adsorption of organics onto activated carbon. Thus activated carbon should not be considered primarily for virus removal in water treatment plants.

7.3.5. Phage Removal and Inactivation by Water Softening

Water softening aims at removing carbonate and noncarbonate hardness due to calcium and magnesium compounds. Lime (calcium hydroxide) softening removes only the carbonate hardness. Noncarbonate hardness necessitates the use of lime in combination with sodium carbonate. This is called the lime–soda ash process. This process results in the formation of precipitates of calcium carbonate and magnesium hydroxide (7,8). Phage inactivation/removal via water softening may be due to inactivation by the high pH generated as a result of lime addition and or to the adsorption of phage particles to the precipitates.

We have seen previously (see wastewater reclamation plants) that lime treatment effectively destroys phage as a result of the generation of high pH. Thayer and Sproul (62) have studied the extent of inactivation of bacteriophage T2 following water softening. T2 inactivation in excess of 99.9% was obtained at pH 10.8. Straight lime softening of water containing 300 mg/L of hardness resulted in only less than 60% removal of the phage. However, more than 99.999% inactivation was observed when magnesium hydroxide was the only precipitate formed. Wentworth et al. (65) confirmed these data by showing that poliovirus type 1 was also effectively removed by the excess lime–soda ash process. The higher removal efficiency observed with magnesium hydroxide precipitates may be explained by the findings that magnesium hydroxide precipitates are positively charged whereas calcium carbonate precipitates are negatively charged (10). Thus phage would adsorb more efficiently to magnesium hydroxide than to calcium carbonate precipitates.

It thus appears that water softening processes, namely, the lime–soda ash process, would be efficient in phage inactivation and removal.

7.3.6. Extent of Phage Removal in Water Treatment Plants

Limited data are available on the extent of phage reduction in water treatment plants. Because phage are being considered as potential indicators of viral pollution, their removal was compared to that of enteroviruses, rotaviruses, and bacterial indicators in water treatment plants (22,36,60a). Studies were undertaken in a 205 MGD water treatment plant in Mexico (22,36). The treatment sequence comprised flocculation with alum and a polyelectrolyte (Catfloc or superfloc), clarification, sand filtration, and final chlorination. During the dry season, phage removal after clarification was 63.6% and the overall reduction in the finished chlorinated water was 100%. During the rainy season, the overall removal was lower (91.4%). Clarification and sand filtration removed 65.7 and 93.6% of the phage, respectively. It was shown that total coliforms, fecal coliforms, fecal streptococci, and coliphages were better removed than enteroviruses or rotaviruses (36). Similar results were obtained by Gerba et al. (22). However, phage reduction was significant (99.9%) following all treatment steps (flocculation–clarification, sand filtration, and final chlorination). The most efficient step for rotavirus reduction was chlorination (Table 7.1). The indigenous phage were found in the finished water at concentrations varying from 0 to 133 PFU/L.

A recent survey of a water treatment plant in Michigan (60a) showed that enteroviruses were better correlated with coliphage than with the traditional bacterial indicators. However, the data available are too limited to propose that coliphages could serve as indicators for enteric viruses in water treatment plants.

TABLE 7.1. Percent Removal of Viruses and Microbial Indicators in a Water Treatment Plant[a,b]

Treatment[b]	Turbidity	TPC	TC	FC	FS	Coli-phages	Entero-viruses	Rota-viruses
After clarification	48	82.2	98.7	98.5	96.4	99.9	35	23.5
After filtration	68	95.7	97.9	98.5	98.7	99.9	68	0
Finished water	77	98	99.5	98.3	99.5	99.9	81	81

[a] TPC = total plate count; TC = total coliform; FC = fecal coliform; FS = fecal streptococci.
[b] The treatment sequence was the following: (1) coagulation; (2) sedimentation; (3) sand filtration; (4) chlorination.

Source: Adapted from Gerba et al. (22)

7.4. CONCLUSIONS

From our discussion on the fate of phage in water and wastewater treatment plants we may draw the following conclusions:

1. In wastewater treatment plants, the activated sludge process is the most efficient in phage removal.

2. Phage may serve as potential indicators for the removal efficiency of wastewater treatment plants with regard to viruses.

3. In wastewater reclamation plants, lime treatment and chlorination are the most efficient steps in phage inactivation.

4. In water filter plants, coagulation–sedimentation is one of the most efficient steps for phage removal.

5. In water softening plants, the softening process (i.e., addition of lime and sodium carbonate) is effective in phage inactivation and removal.

6. Owing to the limited available data we cannot at present consider phage as reliable viral indicators in water treatment plants. More research efforts should be expanded in this area.

ACKNOWLEDGMENTS

This work was supported in part by funds from the Engineering and Industrial Experiment Station, University of Florida, Gainesville, Florida.

REFERENCES

1. Amirhor, P., and R. S. Engelbrecht. 1974. Virus removal by polyelectrolyte-aided filtration. J. Am. Water Works Assoc. *67*:187–192.

2. Ayres, P. A. 1977. Coliphages in sewage and the marine environment, pp. 275–298. *In* F. A. Skinner and J. M. Shewan (eds.), Aquatic Microbiology. Academic, New York.

3. Balluz, S. A., H. H. Jones, and M. Butler. 1977. The persistence of poliovirus in activated sludge treatment. J. Hyg. *78*:165–173.

4. Balluz, S. A., M. Butler, and H. H. Jones. 1978. The behaviour of f2 coliphage in activated sludge treatment. J. Hyg. *80*:237–243.

5. Baylet, R., F. Sinegre, F. Sauze, and M. Gervais. 1980. Lagunage et virologie des eaux usees: Evolution de la charge en coliphages. Tech. Eau Assainissement *378*:37–41.

6. Bell, R. G., 1976. The limitation of the ratio of fecal coliform to total coliphage as a water pollution index. Water Res. *10*:745–748.

7. Bitton, G. 1980. Introduction to Environmental Virology. Wiley-Interscience, New York, 326 pp.

8. Bitton, G. 1980. Adsorption of viruses to surfaces: Technological and ecological implications, pp. 331–374. *In* G. Bitton and K. C. Marshall (eds.), Adsorption of Microorganisms to Surfaces. Wiley-Interscience, New York.

9. Bitton, G., L. T. Chang, S. R. Farrah, and K. Clifford. 1981. Recovery of coliphages from wastewater and polluted lake water by the magnetite-organic flocculation. Appl. Environ. Microbiol. *41*:93–97.

10. Black, A. P., and R. F. Christman. 1961. Electrophoretic studies of sludge particles produced in lime-soda softening. J. Am. Water Works Assoc. *53*:737–741.

11. Brown, T. S., J. F. Malina, Jr., and B. D. Moore. 1974. Virus removal by diatomaceous earth filtration. Part 2. J. Am. Water Works Assoc. *66*:735–738.

12. Chaudhuri, M., P. Amirhor, and R. S. Engelbrecht. 1974. Virus removal by diatomaceous earth filtration. J. Environ. Eng. Div. *100*:937–941.

13. Chaudhuri, M., and R. S. Engelbrecht. 1970. Removal of viruses from water by chemical coagulation and flocculation. J. Am. Water Works Assoc. *62*:563–567.

14. Cookson, J. T. 1967. Adsorption of *Escherichia coli* bacteriophage T4 on activated carbon as a diffusion limited process. Environ. Sci. Technol. *1*:157–160.

15. Cookson, J. T. 1969. Mechanism of virus adsorption on activated carbon. J. Am. Water Works Assoc. *61*:52–56.

16. Cookson, J. T., and W. J. North. 1967. Adsorption of viruses on activated carbon: Equilibria and kinetics of the attachment of *E.coli* bacteriophage T4 on activated carbon. Environ. Sci. Technol. *1*:46–52.

17. Cramer, W. N., K. Kawata, and C. W. Cruse. 1976. Chlorination and iodination of poliovirus and f2. J. Water Poll. Contr. Fed. *48*:61–76.

18. Dhillon, T. S., Y. S. Chan, S. M. Sun, and W. S. Chau. 1970. Distribution of coliphages in Hong Kong sewage. Appl. Microbiol. *20*:187–191.

19. Durham, D., and H. W. Wolf. 1973. Wastewater chlorination: Panacea or placebo? Water Sewage Works *120*:67–69.

20. Ewert, D. L., and M. J. B. Paynter. 1980. Enumeration of bacteriophages and host bacteria in sewage and the activated sludge treatment process. Appl. Environ. Microbiol. *39*:576–583.

20a. Funderburg, S. V., and C. A. Sorber. 1985. Coliphages as indicators of enteric viruses in activated sludge. Water Res. *19*:547–555.

21. Furuse, K., S. Osawa, J. Kawashiro, R. Tanaka, A. Ozawa, S. Sawamura, Y. Yanagawa, T. Nagao, and I. Watanabe. 1983. Bacteriophage distribution in human faeces: Continuous survey of heatlhy subjects and patients with internal and leukaemic diseases. J. Gen. Virol. *64*:2039–2043.

22. Gerba, C. P., J. B. Rose, G. A. Toranzos, S. N. Singh, L. M. Kelley, B. Keswick, and H. L. DuPont. 1984. Virus removal during conventional water treatment. Final Report, Grant #CR80933), U.S. Environmental Protection Agency, Cincinnati, Ohio.

23. Gerba, C. P., M. D. Sobsey, C. Wallis, and J. L. Melnick. 1975. Adsorption of poliovirus onto activated carbon in wastewater. Environ. Sci. Technol. *9*:727–731.

24. Gerba, C. P., C. H. Stagg, and M. G. Abadie. 1978. Characterization of sewage solids-associated viruses and behavior in natural waters. Water Res. *12*:805–812.

25. Gilcreas, F. W., and S. M. Kelly. 1955. Relation of coliform organism test to enteric virus pollution. J. Amer. Water Works Assoc. *47*:683–687.

26. Glass, J. S., and R. T. O'Brien. 1980. Enterovirus and coliphage inactivation during activated sludge treatment. Water Res. *14*:877–882.

27. Grabow, W. O. K., B. W. Bateman, and J. S. Burger. 1978a. Microbiological quality indicators for routine monitoring of wastewater reclamation plants. Prog. Water. Technol. *10*:317–327.

28. Grabow, W. O. K., I. G. Middendorf, and N. C. Basson. 1978b. Role of lime treatment in the removal of bacteria, enteric viruses and coliphages in a wastewater reclamation plant. Appl. Environ. Microbiol. *35*:663–669.

29. Guy, M. D., J. D. McIver, and M. J. Lewis. 1977. The removal of virus by a pilot treatment plant. Water Res. *11*:421–428.

30. Havelaar, A. H., and W. M. Hogeboom. 1984. A method for the enumeration of male-specific bacteriophages in sewage. J. Appl. Bacteriol. *56*:439–447.

31. Havelaar, A. H., W. M. Hogeboom, and R. Pot. 1984. F specific RNA bacteriophages in sewage: Methodology and occurrence. Water Sci. Technol. *17*:645–655.

32. Hejkal, T. W., F. M. Wellings, A. L. Lewis, and P. A. LaRock. 1981. Distribution of viruses associated with particles in wastewater. Appl. Environ. Microbiol. *41*:628–634.

33. d'Herelle, F. 1926. The Bacteriophage and its Behavior (English translation by G. H. Smith). Williams and Wilkins, Baltimore, Maryland.

34. Ignazzitto, G., L. Volterra, F. A. Aulicino, and A. M. D'Angelo. 1980. Coliphages as indicators in a treatment plant. Water, Air, Soil Poll. *13*:391–398.

35. Katzenelson, E., B. Lketter, H. Schechter, and H. I. Shuval. 1974. Inactivation of viruses and bacteria by ozone. pp. 409–421. *In* A. J. Rubin (ed)., Chemistry of Water Supply, Treatment and Distribution. Ann Arbor Science, Ann Arbor, Michigan.

36. Keswick, B. H., C. P. Gerba, H. L. DuPont, and J. B. Rose. 1984. Detection of enteric viruses in treated drinking water. Appl. Environ. Microbiol. *47*:1290–1294.

37. Kim, C. M., D. M. Gentile, and O. J. Sproul. 1980. Mechanism of ozone inactivation of bacteriophage f2. Appl. Environ. Microbiol. *39*:210–218.

38. Kott, Y., N. Roze, S. Sperber, and N. Betzer. 1974. Bacteriophages as viral pollution indicators. Water Res. *8*:165–171.

39. Kott, Y., H. Ben-Ari, and L. Vinokur. 1978. Coliphage survival as viral indicator in various wastewater effluents. Prog. Water Technol. *10*:337–346.

40. Longley, K. E., V. P. Olivieri, C. W. Cruse, and K. Kawata. 1974. Enhancement of terminal disinfection of a wastewater treatment system. pp. 166–179. *In* J. F. Malina, Jr., and B. P. Sagik (eds), Survival in Water and Wastewater Systems. University of Texas, Austin, Texas.

41. Lothrop, T. L., and O. J. Sproul. 1969. High level inactivation of viruses in wastewater by chlorination. J. Water Poll. Contr. Fed. *41*:567–575.

42. Moore, B. E., B. P. Sagik, and J. F. Malina, Jr. 1975. Viral association with suspended solids. Water Res. *9*:197–203.

43. Naparsted, J., K. Kawata, V. P. Olivieri, and V. R. Sherman. 1976. Virus removal in an activated sludge plant. Water Sewage Works *123*:R16–R20.

44. Olivieri, V. P., C. W. Cruse, Y. C. Hsu, A. C. Griffiths, and K. Kawata. 1975. The comparative mode of action of chlorine, bromine, and iodine on f2 bacterial virus, pp. 145–153. *In* J. D. Johnson (ed.), Disinfection of Water and Wastewater, Ann Arbor Science, Ann Arbor, Michigan.

45. Osawa, S., K. Furuse, and I. Watanabe. 1981. Distribution of ribonucleic acid coliphages in animals. Appl. Environ. Microbiol. *41*:164–168.

46. Oza, P. P., and M. Chaudhuri. 1976. Virus-coal sorption interaction. J. Environ. Eng. Div. *102*:1255–1262.

47. Oza, P. P., N. Sriramulu, and M. Chaudhuri. 1973. Preliminary investigation on the use of coal for removing viruses from water, pp. 195–201. Proc. Symp. Env. Poll., Central Pub. Hlth. Eng. Res. Inst., Nagpur, India, Jan. 17–19, 1973.

48. Pavoni, J. L., and M. E. Tittlebaum. 1974. Virus inactivation in secondary wastewater treatment plant effluent using ozone. pp. 180–189. J. F. Malina, Jr., and B. P. Sagik (eds.), Virus Survival in Water and Wastewater Systems. University of Texas, Austin, Texas.

49. Pavoni, J. L., M. E. Tittlebaum, H. T. Spencer, M. Fleishmann, C. Nebel, and R. Gottschling. 1972. Virus removal from wastewater using ozone. Water Sewage Works *119*:59–62.

50. Pretorius, W. A. 1962. Some observations on the role of coliphages in the number of Escherichia coli in oxidation ponds. J. Hyg. *60*:279–281.

51. Safferman, R., and M. Morris. 1976. Assessment of virus removal by a multistage activated sludge process. Water Res. *10*:413–417.

52. Scarpino, P. V. 1978. Bacteriophage indicators, pp. 201–227. *In* G. Berg (ed.), Indicators of Viruses in Water and Food. Ann Arbor Science, Ann Arbor, Michigan.

53. Schaub, S. A., and B. P. Sagik. 1975. Association of enteroviruses with natural and artificially introduced colloidal solids in water and infectivity of solid-associated virus. Appl. Environ. Microbiol. *30*:212–222.

54. Shah, P. C., and J. McCamish. 1972. Relative resistance of poliovirus 1 and coliphages f2 and T2 in water. Appl. Microbiol. *24*:658–659.

55. Shelton, S. P., and W. A. Drewry. 1973. Tests of coagulants for the reduction of viruses, turbidity and chemical oxygen demand. J. Amer. Water Works Assoc. *65*:627–635.

56. Sherman, V.R., K. Kawata, V.P. Olivieri, and J.D. Naparstek, 1975. Virus removals in trickling filter plants. Water Sewage Works. *122*: R36–R44.

57. Snead, M. C., V. P. Olivieri, K. Kawata, and C. W. Kruse. 1980. The effectiveness of chlorine residuals in inactivation of bacteria and viruses introduced by post-treatment contamination. Water Res. *14*:403–408.

58. Sproul, O. J., M. Warner, L. R. LaRochelle, and D. R. Brunner. 1969. Virus removal by adsorption in wastewater treatment processes, pp. 541–547. Adv. Water Poll. Res., 14th Intern. Conf. Water Poll. Res., Prague.

59. Stagg, C. E., C. Wallis, and C. H. Ward. 1977. Inactivation of clay-associated bacteriophage MS-2 by chlorine. Appl. Environ. Microbiol. *33*:385–391.

60. Stagg, C. E., C. Wallis, and C. P. Gerba. 1978. Chlorination of solids-associated coliphages. Prog. Water Technol. *10*:381–387.

60a. Stetler, R. A. 1984. Coliphages as indicators of enteroviruses. Appl. Environ. Microbiol. *48*:668–670.

61. Sturdza, S. A., and M. Russo-Pendelesco. 1967. Recherches ecologiques phago-bacteriennes dans le milieu exterieur. Arch. Roum. Pathol. Exp. Microbiol. *26*:125–154.

62. Thayer, S. E., and O. J. Sproul. 1966. Virus inactivation in water softening precipitation processes. J. Am. Water Works Assoc. *58*:1063–1074.

63. Thorup, R. T., F. P. Nixon, D. F. Wentworth, and O. J. Sproul. 1970. Virus removal by coagulation with polyelectrolytes. J. Am. Water Works Assoc. *62*:97–101.

64. Walker, J., H. Leclerc, and J. M. Foliguet. 1980. Etude experimentale de l'elimination des bacteriophages en bassin de lagunage. Can. J. Microbiol. *26*:27–32.

65. Wentworth, D. F., R. T. Thorup, and O. J. Sproul. 1968. Poliovirus inactivation in water softening precipitation processes. J. Am. Water Works Assoc. *60*:939–946.

66. Wolf, H. W., R. S. Safferman, A. R. Mixson, and C. E. Stringer. 1974. Virus inactivation during tertiary treatment. pp. 145–157. *In* J. F. Malina, Jr., and B. P. Sagik (eds.), Virus Survival in Water and Wastewater Systems. University of Texas, Austin, Texas.

67. York, D. W., and W. A. Drewry. 1974. Virus removal by chemical coagulation. J. Am. Water Works Assoc. *66*:711–716.

8

PHAGE AS INDICATORS OF FECAL POLLUTION

CHARLES P. GERBA

Departments of Microbiology and Immunology and Nutrition and Food Science, University of Arizona, Tucson, Arizona, 85721

8.1. INTRODUCTION

There are many potential applications of bacteriophages as indicators (Table 8.1). These include their use as indicators of sewage contamination, efficiency of water and wastewater treatment, and survival of enteric viruses and bacteria in the environment. The use of bacteriophages as indicators of the presence and behavior of enteric bacteria and animal viruses has always been attractive because of the ease of detection and low cost associated with phage assays. In addition, they can be quantified in environmental samples within 24 hr as compared to days or weeks for enteric viruses. Coliphages have been the most commonly used in this context although other bacteriophages and cyanophages have also been studied (37; Chapter 10). Much of the justification for the study of coliphage behavior in nature has been to gain

197

TABLE 8.1   Potential Applications of Bacteriophage as Indicators

As indicators of
  Feces
  Domestic sewage
  Pathogens (bacterial and viral)
  Host organism (i.e., fecal coliforms)
  Water and wastewater treatment efficiency
  Environmental fate of enteric viruses
  Water movement in surface waters and groundwaters

insight into the fate of human pathogenic enteric viruses. As a result, more is probably known about the ecology of coliphages than of any other bacteriophage group.

The potential of fecally contaminated water to transmit disease is well established. If sufficiently sensitive methods are used, a variety of pathogenic organisms can be regularly detected in sewage, although the numbers and types vary greatly upon the current health status of the contributing community. Direct detection and enumeration of pathogenic bacteria and viruses require specialized techniques and facilities. In addition, such techniques are costly and time consuming. To overcome these difficulties, early water bacteriologists developed the concept of bacterial indicators of fecal pollution. Thus the coliform group was adapted in 1914 by the United States Public Health Service as an indicator of fecal pollution of drinking water. The rates of waterborne bacterial disease outbreaks in the United States are low today primarily because of the faithful implementation of coliform standards to assess the sanitary quality of water. However, it has become increasingly apparent that they do not always indicate the presence of pathogenic viruses and protozoa (24,33,34). Thus coliphages have been suggested as potential indicators that would be more reflective of the presence of enteric viruses because of their similar structure and persistence in the environment (27,29,40,42). Attributes of an ideal indicator have been discussed by several authors (2,4,14) and are listed in Table 8.2. No single group of organisms has been found that fulfills all these criteria. However, total and fecal coliform bacteria fulfill most of them and have been successfully used in the past as effective indicators of bacterial pollution derived from fecally contaminated sources.

8.2. PHAGE AS INDICATORS OF FECAL CONTAMINATION

The ubiquity of coliphages in the feces of man and other warm-blooded animals, in sewage, and in sewage-polluted waters suggests that coliphages could reliably indicate fecal contamination of the environment. With the

TABLE 8.2 Criteria for an Ideal Indicator Microorganism

1. The indicator organism should always be present when pathogens are present and absent when pathogens are absent.
2. The persistence in the environment of the indicator and pathogen should be similar.
3. The density of the indicator organism should have some direct relationship to the density of pathogen(s).
4. Preferably, the indicator should be present in the source of pollution at levels in excess of the pathogen concentration.
5. The indicator should be at least as resistant to disinfectants as pathogens.
6. The indicator should be nonpathogenic and easily quantifiable.
7. The test for the indicator organism should be simple, rapid, and economical and should be applicable to all types of water.

Source: Modified from Refs. 2, 4, and 14.

possible exception of the spore-forming *Clostridium perfringens,* coliphages probably exhibit the greatest resistance to environmental stresses, and hence longest survival, of the bacteria and viruses excreted in the feces. They are thus potentially useful for detecting remote sources of fecal contamination. The natural environment of coliphear would appear to be that of its host, *Escherichia coli,* which is found in the intestinal tract of all warm-blooded animals. The occurrence of coliphage in feces appears to be highly variable (Table 8.3). Coliphages are more commonly isolated from the feces of lower animals and occur in greater concentrations than in man (6,12,31). Few quantitative studies have been conducted to explain these differences.

TABLE 8.3 Reported Concentrations of Coliphages in Various Sources

Source	Typical Concentration	% Positive	Reference
Feces			
Human	$0-10^6/g$	12–75	Dhillon et al. (6)
Animal	$0-10^7/g$	10–92	
Gray water[a]	$0->10^{10}/L$	12	Havalaar (18)
Septic tanks	$10^3-10^6/L$	100	Lupo (30)
Raw sewage	$10^5-10^7/L$	100	Scarpino (36)
Primary effluent	$0.2-10^6/L$	100	Bitton (3)
Secondary effluent (activated sludge)	$10^1-10^4/L$	100	Scarpino (36)
Oxidation pond effluents	$10^3-10^7/L$	100	Scarpino (36)
Slaughterhouse (pigs and calves)	$10^2-10^5/L$	100	Havelaar (18)

[a] From various household sources.

The microflora of the gut, diet, and physiological state of the animal could potentially effect the concentrations of coliphage in feces. For example, about half of the fecal samples from healthy individuals were found positive for phage in one study, whereas about 75% of the fecal specimens from ill individuals were positive (12). Osawa et al. (31) found a higher percentage of stools from zoo animals to be positive for phage than from domestic farm animals. The absence of phage in stools and/or their occurrence at varying concentrations may only be a reflection of the limitations of the specificity of the bacterial host and not necessarily the absence of phage *per se* in each individual. It would appear from the studies conducted to date that the use of phage for direct detection of fecal contamination would be limited because the ability to detect coliphage consistently has not been demonstrated.

In contrast to feces, coliphages appear to be consistently present in domestic raw sewage and have been reported to occur in concentrations ranging from 10^5 to 10^7 PFU/L (Table 8.3). Because of the limited host range for most coliphages, the actual numbers are probably much greater (9). For example, Ignazzitto et al (23) found highest recoveries from wastewater with *E. coli* B12, *E. coli* C, and *E. coli* K-12 HfrH as host strains. When *E. coli* B was used as the host strain, as suggested by Kott (26), plaque counts obtained were five to six times lower. There does not appear to be a large seasonal variation in the numbers of coliphage in raw sewage, although this topic does not appear to have been studied in detail (29). Coliphage densities in septic tanks appear to be lower than in raw domestic sewage (Table 8.3.). Lupo (30) reported coliphage densities ranging from 10^2 to 10^6 PFU/L. The geometric mean density of 8.3×10^4 PFU/L was about two to three orders of magnitude less than that reported in raw sewage. The mean total coliform density (4.6×10^6/100 mL) was only slightly less than that of raw domestic sewage. In a well-operated sewage treatment facility using activated sludge, substantial removal of coliphage may take place (11) and coliphage concentrations are substantially reduced (Table 8.3).

Coliphages have been detected in a wide variety of natural waters including lakes, streams, rivers, wetlands, and groundwater (1,13,22,27,38). Kott (27) observed concentrations of coliphage ranging from 2 to 920 PFU/100 mL in various creeks and streams in Israel. In a study of the Hudson River, which at the time received untreated human wastes, mean concentrations of coliphage were found to range from 0 to 58 PFU/100 mL (22).

Guelin (15) was the first to advocate the use of bacteriophages as indicators of enteric bacteria and viruses. His results suggested that the number of coliphages in fresh and marine waters were related to the density of coliforms. He also studied the occurrence of bacteriophages of a number of other enteric bacteria including *Salmonella typhi* and *Clostridium welchii*. Bacteriophages against both bacteria were found to vary directly with the degree of sewage pollution and they survived longer in seawater than their host (16,17).

Dienert (7) also supported the contention that bacteriophages were useful indicators of the degree of pollution of water by enteric bacteria. He cited the recovery of bacteriophages active against *Escherichia coli* and *Salmonella* from polluted surface waters but not from unpolluted ones.

Cornelson et al (5) used the host organisms *E. coli* Brx and *Salmonella typhi* ViA to study the occurrence of bacteriophages in 14 wells from three villages in Romania. The level of bacteriophages that were infective for *E. coli* Brx was higher in summer (385 bacteriophages/L) than in winter (10 bacteriophages/L) and increased after periods of rainfall. *E. coli* bacteriophages were found in 115 (82%) of 140 samples collected in the summer, 50 (57%) of 70 collected in the autumn and spring, and 50 (40%) of 126 collected in the winter. Wells that were more liable to contamination by surface waters owing to poor construction showed greater numbers of bacteriophages specific for *E. coli* and *S. typhi*. The authors also suggested that some degree of correlation existed between the frequency of *S. typhi* phage and the local incidence of typhoid fever. In another study dealing with the persistence of bacteriophages in groundwater, Etrillard and Lambert (8) reported that a well 40 m deep possessed the chemical characteristics of polluted water, but no bacterial pathogens or indicator bacteria (*E. coli*) could be detected. However, they were able to find bacteriophages active against typhoid bacteria in the well water.

Feigin (10) studied the sanitary significance of detecting bacteriophages in open reservoirs. He found a distinct relationship between the frequency of dysentery bacilli isolation and the bacteriophage present in the water. Both depended on the extent of pollution of the water with excreta and were easy to detect in small numbers. He considered bathing beaches satisfactory if no bacteriophages were present in 20 mL of water.

Zieminski (48) compared the levels of *E. coli* in river water over the period of 1 year to levels of bacteriophages infective for *Salmonella typhi, Shigella flexneri,* and *Salmonella paratyphi* B. He could not find any correlation between the *E. coli* titer and the titers of any of the bacteriophages. He concluded that the determination of bacteriophage levels was not a useful method for the detection of enteric bacterial pathogens.

The work of several other investigators has also suggested a relationship between the presence of coliform bacteria and coliphages in sewage and sewage-polluted waters (29,43,47). In these studies, it was noted that the ratio of coliphages to coliform bacteria varied greatly with the water under study. Bell (1) argued that the widely varying coliform to coliphage ratios he observed in raw sewage, lagoon effluent, and river water would make it difficult to use coliphage as a reliable index of fecal pollution. He observed that environmental factors such as temperature, turbidity, and sewage treatment processes cause considerable variation in the coliform to coliphage ratio. A study of the fluctuations of coliform determinations by MPN and coliphages in the Hudson River also failed to demonstrate a relationship between the two groups of organisms (22).

On the basis of temperature ranges for growth, Seeley and Primrose (39), divided coliphages into three classes: LT (low-temperature, 15–30°C), MT-(mid-temperature, 15–45°C), and HT (high-temperature, 25–45°C). The temperature dependence of growth appears to be a genotypically stable trait. HT and MT phage are predominantly found in feces or fecally polluted waters, whereas LT phage are predominant in unpolluted waters. Multiplication of LT phage but not of HT phage at 22°C has been demonstrated by Parry et al. (32) in river water to which *E. coli* cells were added in the laboratory. Havelaar (18) suggested that the detection of HT phage in water samples could be considered as a potential indicator of fecal or sewage contamination, whereas the presence of LT phage is unrelated to extraneous pollution. Because MT phage are found in both polluted and unpolluted waters, and because they are able to multiply at 37°C, it cannot be concluded that their presence indicates fecal pollution. A higher temperature of incubation could be useful for this purpose.

Havelaar (18) has studied in detail the use of F-specific RNA phage as indicators of fecal contamination. He seldom found these phage in human feces and suggested that they originated in the sewage environment. He thus speculated that such phage might be used as an index of sewage pollution rather than fecal pollution. The reason why F-specific phage are found in greater numbers in wastewater is not known. The host *E. coli* does not produce F pili below 30°C, so F-specific phage infection does not occur. Havelaar speculated that multiplication could occur by infection of *E. coli* cells with preformed F pili entering the sewage environment from feces.

In summary, although the use of bacteriophages as indicators of fecal contamination appears attractive, more thorough studies involving statistical comparison with other microbial and chemical indicators are needed before they could be seriously considered as meaningful indicators of fecal contamination. It is doubtful that bacteriophages will ever be substituted for such traditional microorganisms as coliform and fecal coliform bacteria as indicators of fecal contamination. However, they could potentially be useful as indicators of the source of the fecal contamination or its history. Clearly more information is needed about their ecology.

8.3. PHAGE AS INDICATORS OF ENTERIC BACTERIA

Unfortunately, in none of the previously quoted studies was a statistical analysis of the data conducted. Kenard and Valentine (25) were the first to assess analytically if a predictive relationship existed between indicator bacteria and their coliphages. Fecal coliform, total coliform, and total coliphage numbers were determined in 150 samples of sewage and river water over a 2-year period. Various linear, logarithmic, and polynomial relationships were examined. A logarithmic plot of the data indicated a high degree of correlation between coliphages and fecal coliforms. Phage counts were also related to the number of total coliforms, precipitation, and river depths, The ratio of coliphages to fecal coliforms was about 0.7–1, regardless of the contamination level, whereas the coliphage to total coliform ratio varied between 1 and 25. Using similar techniques, Wentsel et al. (45) expanded upon the original study of surface waters, treated sewage, and treated drinking water. More than 600 surface water samples were collected in this study. Correlation coefficients were 0.69 for fecal coliforms versus coliphages and 0.62 for total coliforms versus coliphages. These correlations were significant at the 99.9% level, indicating a high probability of a relationship between fecal or total coliforms and coliphages in natural waters (Figure 8.1). A difference in the relationship between fecal coliforms versus coliphage was noted between this study and that of Kenard and Valentine (25) (Figure 8.2). The slopes of the ratios between the two groups of organisms were approximately the same; however, the intercepts were different, indicating a difference in test sensitivity. The difference in sensitivity was due to the size of sample assayed for coliphage between the two studies (17.4 versus 50 mL). In comparison to natural waters, coliphages could be frequently isolated from sand-filtered and potable drinking water in the absence of coliform bacteria (45). This is probably due to the greater resistance of coliphages to removal and inactivation than coliform bacteria have. The investigators ad-

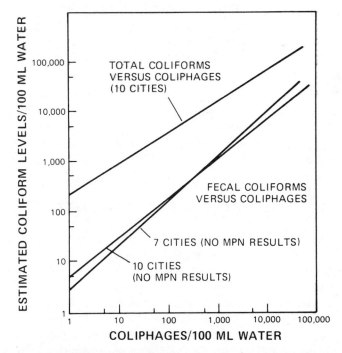

Figure 8.1 Coliform–coliphage relationships in natural waters collected from streams, rivers, lakes, and reservoirs. MPN, most probable number. Reproduced with permission from the American Society for Microbiology (45).

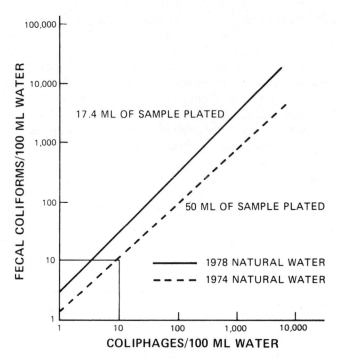

Figure 8.2 Comparison of the relationship of fecal coliforms to coliphages from two field test programs with natural water sample (streams, lakes, and reservoirs). Reproduced with permission from the American Society for Microbiology (45).

vocated the method as a rapid method for coliform detection, for results could be obtained in 6 hr instead of 24–48 hr. Although the results suggest some useful relationship between coliphage and enteric indicator bacteria in natural surface waters, coliphages are not useful indicators of enteric bacteria in treated waters.

8.4. PHAGE AS INDICATORS OF ENTERIC VIRUSES

Coliphages have received considerable interest over the last few decades as indicators of human enteric viruses (35,37) in water. The bases for proposing coliphages as indicators of enteric viruses have been detailed by Kott (27) and are listed in Table 8.4. The use of coliphages as indicators for enteric viruses is attractive because of the low cost and rapidity of phage assay. Although a relationship between the presence of coliforms and coliphages appears to exist in sewage and sewage-polluted waters, there is some conflicting information relating to the usefulness of coliphages as viral indicators. Most of the studies comparing the presence of coliphages and viruses in water have been empirical and only recently has a significant amount of data

TABLE 8.4 Bases of Proposing Coliphages as Indicators of Human Enteric Viruses

1. Coliphages are found in abundance in wastewater and polluted water.
2. The populations of coliphages exceed those of enteric viruses.
3. Coliphages are incapable of reproduction outside of the host organism.
4. Coliphages can be isolated and counted by simple methods.
5. The time interval between sampling and final result is shorter than that for enteric viruses.
6. Many coliphages are more resistant to inactivation by adverse environments and disinfection than enteroviruses.

Source: Ref. 27.

been collected for statistical analysis (11,34). These data suggest that some relationship may exist between these two groups of viruses, but clearly more extensive field studies are necessary to assess its general application.

Laboratory studies with individual coliphages such as T-even phage, MS-2, f2, etc., have demonstrated that many of these phage survive longer in natural waters and are more resistant to common disinfectants such as chlorine than are many enteroviruses (29,46). Kott et al. (28) found that coliphages T1 and T7 were inactivated at similar rates in marine waters. A rough estimate of enterovirus pollution in seawater was made on the basis of coliphage concentrations. Thus less than 10/100 mL of seawater indicated slight and recent pollution; 100/100 mL indicated recent, medium-sized pollution; and 1000/100 mL indicated recent, heavy pollution. Actual correlations between coliphage densities and enteroviruses were not made. Utilization of comparative die-off and resistance information should not be used to judge enterovirus levels in natural or other water. However, studies with individual phage do not prove the usefulness of the coliphage indicator concept because many coliphages are inactivated more readily than enteroviruses under the same conditions (46). In addition, one cannot depend on the same phage being consistently present in the environment.

Few field studies have actually been conducted comparing the relative occurrence of coliphages and enteric viruses. This is in part due to the lack of readily available methods for the concentration of enteric viruses from water until the early 1970s. Kott et al. (29) were the first to study the ratios of enteric viruses to coliphages and indicator bacteria in various types of water; for example, heavily polluted streams, creeks, groundwater, and tapwater. Ratios of coliphages to human enteric viruses were highly variable depending on the type of water and the season. Ratios of coliphages to viruses in surface waters ranged from as low as 1–1 to as high as 1000–1, whereas in wastewater the ratio was fairly constant at 1–10^5. In trickling plant effluent, the ratios varied from 1–10^5 in the spring to 1–10^3 in the winter. Enteroviruses were isolated on two out of eight occasions when coliphages were detected. However, coliform bacteria were also observed in these same samples and only 1-L volumes were tested for enteroviruses. Thus the presence of coliform bacteria had indicated that the water was already of poor quality. Also, to be useful, coliphage must be capable of indicating the presence of human enteric viruses in 10–1000 L of water because this is the range in which enteric viruses are believed to pose a health hazard (41).

Vaughn and Metcalf (44) were the next to conduct comparative studies on the occurrence of coliphages and enteric viruses in sewage and estuarine waters. At one municipal sewage treatment plant, coliphages were recovered from 11 of 25 samples examined, whereas enteroviruses were recovered from only three of these same effluents. In sewage-polluted marine waters, coliphages were recovered more frequently than enteric viruses. However, the rather inefficient gauze pad method was used for enteric virus concentration and this certainly would contribute to a low isolation rate of enteric viruses. The authors also observed significant changes in the coliphage population during the 3-year study. During the first year, more than three times as many phage were recovered in the bacterial host *E. coli* ATCC 11303 than in the host *E. coli* 9637, and almost six times as many phage were recovered in ATCC 11303 as in *E. coli* UNH. In the 2 years that followed, a shift occurred that favored the host *E. coli* ATCC 9637, but at the end of 3 years, the total numbers of phage recovered in *E. coli* ATCC 9637 and ATCC 11303 were similar. Greater annual fluctuations occurred in the frequency of the phage that infected *E. coli* B (ATCC 11303). From their study, Vaughn and Metcalf (44) concluded that coliphages were not suited as enteric virus indicators. They cited such findings as (*a*) coliphages were consistently present in raw sewage samples that yielded inconsistent enterovirus isolations, (*b*) treated sewage effluents were coliphage-positive but enterovirus-negative, and (*c*) many (63%) enterovirus isolations occurred without phage isolation. Many of these results could be explained by the limited volumes assayed by the agar overlay technique used for phage detection as compared to the most-probable number technique used by Kott (26), which was capable of detecting as few as two coliphages/100 mL.

In attempting to evaluate the use of coliphage as an indicator of activated sludge treatment, Funderburg and Sorber (11) found a statistically significant correlation between coliphage giving rise to plaques greater than 3 mm in diameter in the effluent discharges. Total coliphage concentrations and those producing plaques smaller than 3 mm could not be correlated with enteric virus concentrations.

Given major differences in the origin and ecology of enteric viruses and coliphages, it is doubtful that coliphages could successfully be used in all situations as indicators. However, they may potentially be useful as indicators of water and wastewater treatment in some situations, but a great deal of additional work on a solid statistical basis must be performed to justify the use of phage for this purpose.

8.5. CONCLUSIONS

Although the use of phage as indicators of fecal pollution offers many attractive features, their ecology is significantly different from pathogenic enteric bacteria and viruses as to place serious constraints on this application. Various studies suggest their potential application, but more extensive studies would have to be conducted placing the information on a firm statistical basis. In addition, water of varying quality having received different treatments would have to be evaluated. Perhaps through the selective use of phage groups, such as F-specific hosts (19,20,21) or those of certain plaque size (11), this could be achieved. However, the costs of this would be high and success perhaps limited to only certain situations.

REFERENCES

1. Bell, R. G. 1976. The limitation of the ratio of fecal coliforms to total coliphage as a water pollution index. Water Res. *10*:745–748.
2. Berg, G. 1978. The indicator system, pp. 1–13. *In* G. Berg (ed.), Indicator of Viruses in Water and Food. Ann Arbor Science, Ann Arbor, Michigan.
3. Bitton, G. 1987. Fate of bacteriophages in water and wastewater treatment plants. *In* S. M. Goyal, C. P. Gerba, and G. Bitton (eds.), Phage Ecology. Wiley, New York.
4. Cabelli, V. 1979. Evaluation of recreational water quality, the EPA approach, pp. 14-1 to 19-23. In A. James and L. Evison (eds.) Biological Indicators of Water Quality. Wiley, New York.
5. Cornelson, D. A., I. Sechter, G. Zamfir, V. Avram, H. Tudoranu, and H. Feller. 1957. Igiena (Bucharest). *6*:317–321.
6. Dhillon, T. S., E. K. S. Dhillon, H. C. Chau, W. K. Li, and H. C. Tsang. 1976. Studies on bacteriophage distribution: virulent and temperate bacteriophage content of mammalian feces. Appl. Environ. Microbiol. *32*:68–74.
7. Dienert, F. 1944. Hygiene of bathing places. Bull. Acad. Med. *128*:660–665.
8. Etrillard, P., and M. Lambert. 1936. Water Pollution Abstr. 9. No. 977, p. 292. As cited by P. Scarpino, Ref. 35.
9. Ewert, D. L., and M. J. B. Paynter. 1980. Enumeration of bacteriophages and host bacteria in sewage and the activated-sludge treatment process. Appl. Environ. Microbiol. *39*:576–583.
10. Feigin, T. S. 1963. On the sanitary indicative significance of detection of bacteriophage lysing dysentery bacilli in the water of open reservoirs. Zh. Mikrobiol. Epidemiol. Immunobiol. *40*:132–137.
11. Funderburg, S. W., and C. A. Sorber. 1985. Coliphages as indicators of enteric viruses in activated sludge. Water Res. *19*:547–555.
12. Furuse, K., S. Osawa, J. Kawashiro, R. Tanaka, Z. Ozawa, S. Sawamura, Y. Yanagawa, T. Nagao, and I. Watanabe. 1983. Bacteriophage distribution in human faeces: continuous survey of healthy subjects and patients with internal and leukemic diseases. J. Gen. Virol. *64*:2039–2043.
13. Gerba, C. P. 1985. Personal observations.

14. Goyal, S. M. 1983. Indicators of Viruses pp. 211–230. *In* G. Berg (ed.), Viral Pollution of the Environment. CRC Press, Boca Raton, Florida.

15. Guelin, A. 1948. Etude quantitative de bacteriophage de la mer. Ann. Inst. Pasteur. *74*:104–112.

16. Guelin, A. 1948. Etude des bacteriophages typhiques. VI—Dans les eaux. Ann. Inst. Pasteur. *79*:447–453.

17. Guelin, A. 1950. Sur las presence du bacteriophage perfringens dans les eaux et son role dans l'epuration des eaux stagnantes. Ann. Inst. Pasteur. *79*:447–453.

18. Havelaar, A. H. 1986. F-specific RNA Bacteriophages as Model Viruses in Water Treatment Processes. Ph.D. Dissertation, University of Utrecht, The Netherlands.

19. Havelaar, A. H., and W. M. Hogeboom. 1983. Factors affecting the enumeration of coliphages in sewage and sewage-polluted waters. Antonie Van Leeuwenhock J. Microbiol. *49*:387–397.

20. Havelaar, A. H., and W. M. Hogeboom. 1984. A method for the enumeration of male-specific bacteriophages in sewage. J. Appl. Bacteriol. *56*:439–447.

21. Havelaar, A. J., W. M. Hogeboom, and R. Pot. 1984. F-specific RNA bacteirophages in sewage: methodology and occurrence. Water Sci. Technol. *17*:645–655.

22. Hilton, M. C., and G. Stotzky. 1973. Use of coliphages as indicators of water pollution. Can J. Microbiol. *19*:747–751.

23. Ignazzitto, G., L. Volterra, F. A. Avlicino, and A. M. D'Angelo. 1980. Coliphages as indicators in a treatment plant. Water, Air and Soil Pollution. *13*:391–398.

24. Jakubowski, W., and J. C. Hoff. 1979. Waterborne Transmission of Giardiasis. U.S. Environmental Protection Agency. Publication No. EPA-600/9-79-001, Cincinnati, Ohio.

25. Kenard, R. P., and R. S. Valentine. 1974. Rapid determination of the presence of enteric bacteria in water. Appl. Microbiol. *27*:484–487.

26. Kott, Y. 1966. Estimation of low numbers of *Escherichia coli* bacteriophage by use of the most probable number method. Appl. Microbiol. *14*:141–144.

27. Kott, Y. 1981. Viruses and bacteriophages. Sci. Total Environ. *18*:13–23.

28. Kott, Y., B. H. Ari, and N. Buras. 1969. The fate of viruses in a marine environment, pp. 823–829. *In* S. H. Jenkins (ed.), Proc. 4th International Conf. Water Poll. Res. Pergamon, New York.

29. Kott, Y., N. Roze, S. Sperber, and N. Betzer. 1974. Bacteriophages as viral pollution indicators. Water Res. *8*:165–171.

30. Lupo, L. B. 1979. Bacteriophage as Indicators of Fecal Pollution. M. S. Thesis, University of Rhode Island.

31. Osawa, S., K. Furuse, and I. Watanabe. 1981. Distribution of ribonucleic acid coliphages in animals. Appl. Environ. Microbiol. *41*:164–168.

32. Parry, O. T., J. A. Whitehead, and L. T. Dowling. 1981. Temperature sensitive coliphage in the environment, pp. 277–280. *In* M. Goddard and M. Butler (eds.), Viruses and Wastewater Treatment. Pergamon, New York.

33. Payment, P., M. Trudel, and R. Plante. 1985. Elimination of viruses and indicator bacteria at each step of treatment during preparation of drinking water at seven water treatment plants. Appl. Environ. Microbiol. *49*:1418–1428.

34. Rose, J. B., C. P. Gerba, S. N. Singh, G. A. Toranzos, and B. Keswick. 1986. Isolation of entero- and rotaviruses from a drinking water treatment facility. J. Am. Water Works Assoc. *78*:56–61.

35. Scarpino, P. V. 1971. Bacterial and viral analysis of water and waste water, pp. 639–761. *In* L. L. Ciaccio (ed.), Water and Water Pollution Handbook, Vol. 2. Dekker, New York.

36. Scarpino, P. V. 1975. Human enteric viruses and bacteriophages as indicators of sewage

pollution, pp. 49–59. *In* A. L. H. Gameson (ed.), Discharge of Sewage from Sea Outfalls. Pergamon, New York.

37. Scarpino, P. V. 1978. Bacteriophage indicator, pp. 201–227. *In* G. Berg (ed.), Indicators of Viruses in Water and Food, Ann Arbor Science, Ann Arbor, Michigan.

38. Scheuerman, P. R., S. R. Farrah, and G. Bitton. 1987. Reduction of microbial indicators and viruses in a cypress strand. Water Sci. Tech. *19*:539–546.

39. Seeley, N. D., and S. B. Primrose. 1980. The effect of temperature on the ecology of aquatic bacteriophages. J. Gen. Virol. *46*:87–95.

40. Simkova, A., and J. Cervenka. 1981. Coliphages as ecological indicators of enteroviruses in various water systems. Bull. World Health Org. *59*:611–618.

41. Sproul, O. J. 1983. Public health, financial and practical considerations of virological monitoring and quality limits. Water Sci. Technol. *15*:33–41.

42. Stetler, R. E. 1984. Coliphages as indicators of enteroviruses. Appl. Environ. Microbiol. *48*:668–670.

43. Suner, J., and J. Pinol. 1966. Coliform bacteriophages and marine water contamination. Adv. Water Pollut. Res. *3*:105–118.

44. Vaughn, J. M., and T. G. Metcalf. 1975. Coliphages as indicators of enteric viruses in shellfish raising estuarine waters. Water Res. *9*:613–616.

45. Wentsel, R. S., P. E. O'Neill, and J. F. Kitchens. 1982. Evaluation of coliphage detection as a rapid indicator of water quality. Appl. Environ. Microbiol. *43*:430–434.

46. Yates, M. V., and C. P. Gerba. 1985. Virus persistence in groundwater. Appl. Environ. Microbiol. *49*:778–781.

47. Zaiss, U. 1981. Dispersal and fate of coliphages in the River Saar. Zbl. Bakt. Hyg. B. *174*:160–173.

48. Zieminski, S. 1958. Comparative studies on the presence of the *E. coli* group and bacteriophages of some pathogenic intestinal bacteria in river water. Roczn. Panst. Zabl. Hig. *9*:225–231.

9

BACTERIOPHAGES OF INDUSTRIAL IMPORTANCE

MARY ELLEN SANDERS

Biotechnology Group, Miles Laboratories, Inc., P. O. Box 932, Elkhart, Indiana 46515

9.1. INTRODUCTION

The industrial fermentation environment, where productivity depends largely on the growth and metabolic activity of bacterial starter strains, can provide an opportunity for bacteriophage to proliferate and thereby interfere with starter culture functioning. Phage infection of a commercial fermentation process can result in substantial economic setbacks from product loss, spoilage of raw materials, and nonproductive operation costs. The extent of economic damage caused by phage-inflicted fermentations is difficult to assess. Confidentiality of production records and failure to confirm the causes of slow or failed fermentations leave estimates of financial losses to speculation. However, bacteriophage contamination of dairy (56), casein (150,151), wine (132), acetone–butanol (97), sake (53), and glutamic acid (41) fermentations has been reported (Table 9.1). It is important to note, however, that although lytic bacteriophages have been isolated for many commercially used bacteria, confirmed reports of abnormal fermentation due to phage infection have not always been published. This appears to be the case with

TABLE 9.1 Incidences of Commercial Phage Infections

Product	Infected bacterium	Refs.
Cheddar, cottage cheese	*Streptococcus lactis*	167
	Streptococcus cremoris	
Yogurt	*Lactobacillus bulgaricus*	104
	Streptococcus thermophilus	1,108
Swiss cheese	*S. thermophilus*	23
	Lactobacillus lactis	2
	Lactobacillus helveticus	2
Blue cheese	*Leuconostoc cremoris*	128
Italian cheese	*L. bulgaricus*	2,108
	S. thermophilus	1,108
Viili	*S. cremoris*	121
	S. lactis subsp. *diacetylactis*	121
	Leuconostoc cremoris	121
Yakult	*Lactobacillus casei*	125
Casein	*S. cremoris*	150,151
Natto	*Bacillus subtilis* var. *natto*	170
Wine	*Leuconostoc oenos*	22,132
Sake	*Leuconostoc mesenteroides*	53
Polymyxin	*Bacillus polymyxa*	96
Colistin	*Bacillus colistrium*	96
Bacitracin	*Bacillus licheniformis*	96
L-Glutamic acid	*Brevibacterium lactofermentum*	41
	Microbacterium ammoniaphilum	41
Acetone–butanol	*Clostridium acetobutylicum*	97
	Clostridium saccharoperbutylacetonicum	40

Lactobacillus plantarum used in sausage fermentations (95) and actinomycetes used in antibiotic production (98). Obviously, criteria other than the existence of homologous lytic phage must be met for phage to have industrial impact.

In order for an inhibitory phage infection to occur, the phage must (*a*) have a means of contaminating the fermentation, (*b*) exhibit replication rates during the fermentation that are rapid enough to affect the culture growth, and (*c*) affect the dominant or a unique component of the starter culture. These criteria are met in several commercial fermentations, making the means of identification and control of phage contamination an important concern. This chapter reviews bacteriophage contamination of industrial fermentations. Because much is reported on the incidences and dynamics of phage infections in dairy fermentations, this chapter draws heavily on examples from the dairy industry, from which much can be learned. Reviews of phage of industrial significance include those of Hongo et al. (41), Rudolf (113), Ogata (96), and, specifically for the dairy industry, Klaenhammer (56), Teuber and Lembke (149), Huggins (42), and Davies and Gasson (21).

9.2. IDENTIFICATION OF PHAGE PROBLEMS IN INDUSTRIAL FERMENTATIONS

The occurrence of an abnormal industrial fermentation generates immediate concern, but not necessarily immediate understanding of the cause. Bacteriophage infection is only one of many possible detriments to microbial activity. Contamination of the fermentation medium with chemical inhibitors or antibiotics, improper environment control (temperature, pH, dissolved oxygen, nutrient fed), or genetic variation within the starter culture can all lead to ineffective fermentations. Nonphage inhibition in dairy fermentations is discussed by Roginski et al. (111,112). Confirmation of the role of bacteriophage in fermentation failures requires differentiation from these nonphage inhibitions. Strategies employed by the dairy industry for detection of phage infection are discussed by Huggins (42), Cox (19), and Hull (47).

Phage infection of industrial cultures may be accompanied by the decrease or cessation of end-product formation, culture lysis, shifts in strain dominance of mixed-strain cultures, decreased fermentation efficiency, or defective end products. Furthermore, especially in food fermentations, the development of improper starter culture may allow the growth of food pathogens. The specific symptoms of phage infection depend on the affected fermentation. A pattern frequently seen in phage-infected milk fermentations is a gradual decrease in the rate of lactic acid production in successive vats using the same starter strains (161). Presumably this corresponds to a gradual buildup of lytic phage in the cheesemaking environment, and a concomitant destruction of a portion of the culture. Eventually, the phage sensitive strain is totally inhibited, and the fermentation fails.

Confirmation of phage inhibition of fermentative activity requires the specific detection of the infective phage. Rapid screening for phage-directed inhibition, such as spot tests (45) or one-cycle activity tests (15,32,34), are not necessarily effective at differentiating between inhibitions caused by chemical inhibitors and those caused by phage. However, these screening methods are rapid, inexpensive, and not labor intensive, and they do serve the purpose of repeating, on a laboratory scale, the inhibition witnessed during production. Unlike inhibitory chemicals, phage provided with the proper host can direct their own replication. Therefore, when a lytic phage population is diluted and plated, as with the soft agar overlay method described by Adams (3), individual plaques result. Furthermore, the self-replicating nature of phage can be shown by increased inhibition with successive passages through an activity test, indicating the buildup of a population, instead of the dilution of a chemical agent. The differentiation of phage from bacteriocin inhibitions is described by Tagg et al. (145).

Some nonmicrobiological methods of phage detection have also been used. Predominant among these, but too sophisticated for routine industrial use, is electron microscopy. This technique confirms the presence and morphology of phage particles, and is especially helpful when suitable indicator hosts are not available, as is the case with many lysates of induced temperate phage (43,57,147). An enzyme-linked immunosorbent assay was proposed to detect phage in whey (71). Antibodies coupled with alkaline phosphatase were capable of detecting phages down to 9.8×10^6 PFU/mL and were specific for immunologically related phages. Impedance, which is inversely correlated with bacterial growth, can also be used to detect bacterial inhibition. Waes and Bossuyt (162) reported that 10^5 lytic phage per milliliter could be detected within 2 hr via their inhibition of impedance decrease in pre-rennet milk. This test, however, was not specific for phage, but for any inhibitor of bacterial growth.

In many cases, attempts were made to study the phage in more depth at a laboratory scale. This entails purification of the phage, identification of an indicator strain, and description of appropriate assay conditions for the phage–host pair under study. Frequently, inference from previously isolated phage attacking the same type of bacteria leads to the definition of adequate assay conditions. However, in some cases, the phage may require unique treatment for successful cultivation in the laboratory environment. For example, attempts to describe phage for thermophilic lactic acid bacteria were hampered owing to failure to detect plaque formation (109,133) and led to a study to optimize plating efficiencies (133). Parameters such as temperature, pH, aeration, medium composition, and possible strain or phage interactions should be considered in describing a detection/enumeration system for industrial phage. Attention must be paid to how laboratory-scale parameters differ from those at the production level, where the phage–host interaction is known to be effective. Use of the production medium or production growth scheme may be inconvenient at the laboratory bench, but may be essential to detect a useful degree of lytic phage development. Factors affecting phage development are discussed more fully in Section 9.4.1.

9.3. INCIDENCES OF PHAGE INFECTION OF INDUSTRIAL PROCESSES

9.3.1. Dairy

Of all fermentation industries, the dairy industry appears to be most burdened with abnormal fermentations caused by phage. Examination of the fermentation process reveals four critical points that render dairy fermentations especially susceptible to phage infection.

1. The fermentation is conducted in open vats, inviting contamination of the process by phage harbored by the air, milk, utensils, equipment, and personnel.

2. The fermentation medium, milk, is pasteurized but not sterilized. This heat treatment effectively eliminates the natural starter bacteria present in the milk, forcing the fermentation to rely exclusively on added starter. The choice of starter does increase the cheesemakers' control over the fermentation quality and time because strains with proper flavor characteristics, acid-producing activity, and compatibility can be selected. But it also limits the number of phage-unrelated strains that are relied upon to execute the fermentation. Should lytic phage develop against the dominant strain(s), no wild flora remain to aid the fermentation.

3. Cheesemaking protocols vary with cheese type, but all employ one or more of a variety of mixing steps, including starter addition, and stirring during cooking, washing, and milling (159). These steps effectively disperse phage throughout the product, encouraging secondary infections.

4. With modernization of the cheese industry, rapid turnover of cheese vat is demanded. This increases the degree of cell growth per day and decreases the manufacturer's tolerance for delayed acid production. It has been calculated that up to 10^{18} bacterial cells may be produced daily in individual cheeseplants (68). The opportunity for phage proliferation in cell growth of this magnitude is apparent.

Phage for species of group N streptococci (56), Leuconostoc (21), Lactobacilli (135), and *Streptococcus thermophilus* (1,23) have been isolated and implicated in improper fermentation of a variety of fermented milk products. Yet a survey of the literature suggests that phage infection of the streptococcal components of dairy starter strains is more frequently observed. This may reflect (*a*) more prevalent use of these cultures; (*b*) a genetic or physiological predisposition of the streptococci to phage infection; or (*c*) the existence of phage against the lactic streptococci that are more destructive than phage against other bacteria. Although phage were first implicated in slow acid production in dairy fermentations in 1936 (167), phage infection of cheddar, cottage, Italian, Swiss, and blue cheeses and yogurt manufacturing processes remains a production concern (Table 9.1).

9.3.2. Sausage

The inhibition of commercial sausage fermentations by phage has not been reported to date. However, occasional failure of sausage fermentations by commercial starter bacteria has led to recognition of the potential for phage involvement in inhibition of meat cultures. The fermented meat industry has begun to rely on producing sausage with commercial cultures rather than natural flora, to the extent that greater than 50% of fermented sausage is thought to be produced with commercial cultures (130). This shift in process technology may be accompanied by an increased vulnerability to phage inhibition, depending on the selection criteria and compositions of the commercial cultures and on the sausage production technology. However, the limited research that has been published in this area does not indicate that immediate concern is warranted.

Laboratory-scale sausage fermentations resulted in delayed acid production when deliberate phage contamination of a homologous strain was employed (95,157). The fermentations recovered, however, and acid production, presumably by phage-resistant mutants, was normal by the end of the 11-day fermentation (95). The delay in acid production did not appear to pose a microbiological safety or spoilage risk, but no organoleptic evaluations of the finished products were reported. The detailed examination of *Lactobacillus plantarum* phage fri isolated from meat starter showed it to have a very narrow host range, but a large burst size of 200 (156). Until commercial reports indicate otherwise, it is safe to assume that sausage fermentation is not seriously threatened by phage attack, perhaps owing to the long fermentation time and diverse wild flora present in the raw materials.

9.3.3. Casein

Phage infection of *Streptococcus cremoris* and *S. lactis* strains used in casein manufacture has been reported from New Zealand (150,151). Although the phage infection scenario is similar to that found with these bacteria in cheesemaking the problem is more easily tolerated because the requirements for casein manufacture are less stringent than those for cheesemaking. Because the manufacture of casein requires only acid production by the starter, there is no concern about subtle strain contributions to the flavor or texture of the final product. Therefore, slow acid production can be offset by the direct addition of lactic acid (150). Recently, a defined starter system for casein manufacture was described (35).

9.3.4. Wine

The metabolism of the dicarboxylic malic acid to monocarboxylic lactic acid is a favorable bioconversion for decreasing the acidity of certain wines (65). *Leuconostoc oenos* is known to perform this decarboxylation. Sozzi et al. (132) briefly reported that an abnormal malolactic fermentation was found to be caused by bacteriophage against the predominant strain of *Leuconostoc oenos* in the wine. The extent and reversibility of this delayed decarboxylation was not discussed. However, their report does suggest a potential threat of bacteriophage to the wine industry.

9.3.5. Glutamic Acid

Phage infection of bacteria used for L-glutamate production has been well documented by the Japanese industry (41,96). An interesting aspect of phage infection for this industry is that a diversity of bacteria appear to be affected, including *Brevibacterium lactofermentum, Microbacterium ammoniaphilum,* and *Corynebacterium* (41), even though single strains are used for individual fermentations. Symptoms of phage infection include abnormal bacterial growth, delays in fermentation, decrease in yield and turbidity, and finally cell lysis.

9.3.6. Acetone–Butanol

The production of acetone and butanol by fermenting clostridia has been affected by bacteriophage attack (97). One particular strain, *Clostridium saccharoperbutylacetonicium* N1, desired for its high butanol-producing ability, was particularly plagued by phage infections (96). The replacement of this strain by phage-resistant mutants did not eliminate the phage problem, with replacement strains being required up to 12 times in 1 year. As is common to most phage-mediated abnormal fermentations, the most destructive clostridial phages were high multiplicity phages with burst sizes as high as 500 (96). The current production of acetone–butanol does not use a *Clostridium*-based fermentation. However, if this process technology is revived, phage susceptibility of the production strains may be an important characteristic.

9.4. FACTORS AFFECTING PHAGE DEVELOPMENT

9.4.1. External Factors

The expressed resistance of a bacterial strain to phage is determined by the inherent characteristics of both the phage and the host strain, as well as the environmental conditions during the interaction. Therefore, the description of a strain as "phage resistant" is meaningless unless reference is made both to the phage with which the strain was challenged and to the conditions of the assay. Conditions affecting phage development include temperature, ionic environment, pH, nutrient concentration, and growth phase of the host cell. This section examines these factors and how they relate to phage detection, efficiency of development, and host range determinations. It must be remembered, however, that information gained in a controlled laboratory setting, with defined strains and phage, may not always be directly applicable in the field, where the nature of phage contamination cannot be dictated. Studies concerned with external factors affecting development of phage of the lactic streptococci are emphasized in this section because of the shortage of information dealing with other commercially important bacteria.

The requirement for divalent cations for phage development has been established for many phage. For the lactic streptococcal phage, where phage infection occurs predominantly in a calcium-rich medium e.g., milk, Ca^{2+} is required for phage growth (17). Potter and Nelson (106) found that calcium concentrations in the range of 0.3–2 mM allowed lactic phage development. The cation requirement has been shown to be important for *E. coli* phage λ and T4 for adsorption (31) but for some stage beyond adsorption in the lactic streptococci (103,106). Media formulated to chelate divalent cations (83,105) can be effective in inhibiting the development of phage, without negatively affecting bacterial growth. Beyond the specific requirement for divalent cations, media that promote host strain growth generally also enable phage development. However, the growth optima for host and phage do not always coincide.

The ability of phage to plaque on lactic streptococcal strains has been found to be influenced by a variety of factors. The growth phase of the host strain can affect the efficiency of plating (EOP) of both homologous and heterologous phage. However, the use of stationary phase cells has been suggested for some phage–host systems (104,106) and logarithmic cells for others (85). Therefore, optimal growth phase requirements for specific systems should be determined, because no generalization can be made. However, it should be noted that an increase in the level of dead cells or injured cells could result in phage adsorption to a cell that would not support its replication, thereby effectively decreasing the observed degree of plaque-forming units.

The pH, as regulated by lactose concentration in M17 medium, was found to alter the EOP of heterologous lactic streptococcal phage–host systems by several log cycles (148). Using 1% lactose in M17 medium, the pH dropped to 5.1 after 18 hr at 22°C, and the EOP was 2 log cycles greater than the same assay conducted with cells grown in M17 medium containing the normal lactose concentration of 0.5% (final pH 5.7). These results are somewhat contradictory to those reported by Cherry and Watson (12), who found that at pH less than 5.0, lysis of an *S. lactis* strain by its homologous host did not occur. Therefore, optimal pH conditions apparently vary with the phage–host system under investigation. In addition, Terzaghi and Terzaghi (148) found that use of sugars other than lactose, which permitted the attainment of comparable pH levels after growth, did not completely produce the same increased EOP effect. Therefore, the effect seemed somewhat sugar-dependent. It was also noted that the EOP of heterologous phage–host interactions was more variable with respect to assay conditions than were homologous interactions.

The effect of temperature on phage development has also been studied. Hunter (49) reported on the temperature effects on phage development in the lactic streptococci. His results were the first to indicate that bacterial growth and homologous phage multiplication in the lactic streptococci followed different temperature dependencies. A number of similar studies have shown that increased phage assay temperature (from 30 to 37 or 40°C) enhanced the development of certain lactic streptococcal phage (48,58,82,114,116). In some cases, this may be due to a decrease in latent period (54,171), an increased burst size (58), or decreased adsorption capacity (117). However, Terzaghi and Terzaghi (148) found that increasing the assay temperature from 22 to 34°C decreased the EOP for certain heterologous phage–host crosses. Mullan et al. (90) reported that certain homologous phage–host interactions were sensitive to growth at 38–40°C. Fourteen phage were found to be unable to lyse their normal host strains when grown at 38 and/or 40°C, even though the host strain grew well, implying the presence of a temperature-sensitive target in phage development. However, Sozzi et al. (134) examined 19 strains of mesophilic lactic streptococci and showed only three to be temperature sensitive, a much lower frequency than reported by Mullan et al. (90). The minimum, maximum, and optimum bacterial growth temperatures closely paralleled phage development temperatures for all but three strains tested, and it was concluded that bacteriophages multiply whenever the bacteria grow. This conclusion, however, appears to be an oversimplification in view of several other reports (48,49,90,105). The studies of Mullan et al. (90) indicated that growth at high temperatures can inhibit the development of homologous phage for lactic streptococci, but these results, too, cannot be accurately applied to all lactic streptococcal phage. In general, it appears that high-efficiency, homologous, phage–host crosses may be hindered or unaffected by growth at elevated temperatures, but phage development of low-efficiency heterologous crosses may be significantly enhanced by heat shock (105,114) or elevated growth temperatures (105).

9.4.2. Biological Factors

The relationship established between a phage and its host dictates the degree of the threat imposed by the phage to cell survival. The phage–host relationship can be lytic, lysogenic, or somewhere in between.

The lytic cycle of phage infection, first described by Twort in 1915 (139), is the most destructive of phage–host interactions. The cell is lysed, releasing many viable phage progeny. The development potential for lytic phage is immense; burst sizes can range from 3 to 500 per infected cell. With a burst size of 200 and a latent period of 40 min, phage have the potential to increase in number from 1 to 1.6×10^9 in 120 min. The destructive potential for a lytic phage of this type is obvious. Slower developing phage do exist and are proportionately less threatening because of lower replication rates.

The lysogenic state, most completely understood for *E. coli* phage λ (37), is known to exist for many lactic streptococci (43,60,74,81,110) and lactobacilli (126,140). Lysogeny enables the phage to coexist with the host cell. Although there is a finite energy demand on the cell for maintenance of the lysogen, and lysogenic conversion (7) may alter the cell's phenotype, the lysogenized bacterial cell is able to grow and metabolize essentially as usual (31). However, the threat of the temperate phage to industrial cultures is significant. The lysogenic strain may serve as a source of lytic phage in the fermentation environment. This may occur by mutation of induced phage particles from lysogenic to lytic; or in the presence of multiple starter strains, a temperate phage for one strain may prove to be lytic for another.

The transition of phage from temperate to lytic has long been suspected as a source of destructive phage in the dairy industry (69). However, documentation of the transition of a phage from lysogenic to lytic has proved to be difficult to obtain. Hosts sensitive to lysis by temperate phage lysates were relatively rare (36,57,69), even though intact phage particles were demonstrated via electron microscopy. Furthermore, Jarvis (51) showed no significant homology of DNA from 36 strains of lactic streptococci to DNA from 11 lytic phage. Also, no homology was detected between DNA of 25 lytic phage of various morphological types and DNA from three temperate phage, although the three temperate phage appeared related. Therefore, this lack of DNA similarity between temperate and lytic phage of the mesophilic lactic streptococci and of the lytic phage and host genome provides significant evidence against lysogenic phage as a source of lytic phage against the mesophilic lactic streptococci.

However, Shimizu-Kadota et al. (126), investigating the cause of lysis of *Lactobacillus casei* S-1 during the production of lactic acid beverages, were successful at demonstrating the derivation of a destructive virulent phage from a temperate phage harbored by the starter strain. Furthermore, the use of a prophage-cured derivative of S-1 prevented subsequent fermentation failures due to phage (125). The transition from temperate to lytic phage appeared to be due to the insertion of a segment of host DNA into the phage genome (123,127). Otherwise, no difference between the temperate and lytic phage genomes was observed. This region was subsequently identified as a transposon, 1SL*1*, the first identified for a *Lactobacillus* (124). These experiments conclusively showed that, at least with lactobacilli, prophage can give rise to virulent phage. The involvement of a transposable genetic element provides a genetic mechanism for the conversion of lysogenic to virulent phage. Further evidence for the relatedness of temperate and virulent phage of the lactobacilli was provided by Mata et al. (80). They showed that 2 temperate and 11 lytic phage crossreacted with antisera and contained homologous DNA.

Whereas the lysogenic state persists because of the incorporation of the phage DNA into the bacterial genome, loose associations between lytic phage and host bacteria can also be established. Phenotypic resistance can be established in a genotypically phage-sensitive culture, providing a means of long-term perpetuation of a lytic phage along with a sensitive host strain (8). For example, in the case of phage T7 infection of *Shigella dysenteriae* (59), lysis of the first round of infected bacteria results in the release of an enzyme that interferes with the functioning of the phage receptors in the uninfected bacteria. Low levels of lytic phage are produced, but the culture is not totally lysed. Transfer to fresh medium dilutes out the enzyme, and cells again become phenotypically sensitive and susceptible to infection. In this way, a culture can serve as a self-perpetuating source of lytic phage but still establish a degree of resistance to the phage. In the case of the filamentous, single-stranded DNA phage, the phage is virulent but not lytic. Viable phage particles leak out of the cell, causing growth inhibition, but not lysis, of the host strain (79). Again, these phage could be carried for a long time in a culture of susceptible host cells. Subculture in the presence of antiphage serum can readily distinguish nonlysogenic from lysogenic phage–host interactions. From an empirical perspective, the danger of the lysogenic or pseudolysogenic status of commercial strains rests with the possibility that the phage released from these interactions may potentially serve as lytic phage for other strains during use. Therefore, any steps the starter culture technologist can take to eliminate the source of any phage particle, lytic, lysogenic, or pseudolysogenic, would be judiciously taken.

9.5. COPING WITH PHAGE INFECTIONS

Any program dealing with the problem of phage infection of industrial strains must be concerned with both the phage and the strains. It is the integration of knowledge about these two biological entities that will give the fermentation technologist hope in controlling phage inhibition of starter strains. The control of phage of industrial significance has been discussed in reviews by Klaenhammer (56), Wigley (168), Lawrence (66,67), and Huggins (42) for the dairy industry, and by Hongo et al. (41) for the amino acid industry. These reviews should be consulted for supplementary information.

Strategies for controlling industrial phage problems have focused on:

1. Decreasing contamination of fermentation processes from external sources of phage.
2. Development of starter strains resistant to prevalent phage.

3. Addition of chemical agents that specifically inhibit phage but not the production strains.

9.5.1. Prevention of Phage Contamination

An important factor in the control of phage infection of industrial strains is the blocking of phage invasion routes. Here a distinction can be made between food fermentations, which are conducted with little concern for asepsis, and most other industrial bioconversions, which use enclosed fermentation vessels, sterile substrates, and filtered air, and are otherwise protected from external contamination. As we have seen, however, both types of fermentations are susceptible to phage attack. This section discusses sources of phage contamination and precautions helpful in preventing such contamination.

Good fermentation facility design and consistent, sanitary operation conditions can do much to decrease external phage contamination. For example, phage infection of the glutamic acid fermentation was especially concentrated during plant improvement processes (96). Although phage levels in the immediate environment may not always correlate with phage infection during fermentation, phage invasion into the fermentation tank is generally preceded by increased levels of phage in the immediate environment (41). Therefore, sanitation of equipment and pipelines, prevention of cross-contamination of seed tanks, and protection of fermentation vessels from air, fermentation by-products, or personnel are critical in minimizing phage invasion.

For aseptic fermentations, critical points in preventing phage invasion are seed-culture preparation and air filtration. Bulk starter production, tank design, and production practices to minimize external phage contamination were reviewed by Wigley (168). In aerobic fermentations, it is especially important to provide a source of air that is properly filtered to prevent air inlets from becoming a significant source of phage contamination. In nonaseptic fermentations, minimizing external sources of phage is even more critical, because the possible number of invasion routes is increased. In this case, attention must be focused more keenly on plant sanitation and flow patterns through the plant. The flow of air, raw materials, fermentation by-products, and even personnel should follow the flow of the product. In a well-designed plant, stages of production should be physically separated. A reverse in this flow enables cross-contamination from the phage-laden final stages to the initial phage-susceptible stages of the fermentation.

Conscientious manufacturing practices can do much to decrease the contamination of fermentation vessels with external sources of phage. However, phage may still invade a fermentation through lysogenic or phage-carrying cultures. This emphasizes the importance of a strain selection program that monitors the phage-producing characteristics of potential fermentation strains.

9.5.2. Phage-Resistant Mutants

The most common response to a phage-inhibited commercial strain is to attempt to isolate a phage-resistant mutant of that strain. This approach has been used for phage problems encountered in the glutamic acid (39), acetone–butanol (96), casein (151), and cheese (56) industries. Mutant selection is a direct approach to solving phage inhibition problems, but in practice, obtaining phage-resistant mutants that retain the high performance capabilities of the optimized production strains is difficult. Furthermore, little control over the types of contaminating phage in the production facilities can be exerted. Therefore, although the phage resistance pattern of a strain can be rapidly determined in the laboratory, the fate of the strain once released into the production setting is difficult to predict. For these reasons, although phage resistant mutants have contributed to controlling phage infections, they have not provided the final solution to industrial phage problems.

In fermentation industries affected by phage, the following criteria have been recently described for the development of a strain selection program (42,46,72,96,150). First, well-defined single or multiple strain starter cultures should be used. In food fermentations, where undefined mixed or multicomponent cultures are used repeatedly, diverse populations of phages are often isolated from the environment and fermentation vessel (33). Therefore, finding strains resistant to the indigenous phage populations can be difficult (72). Minimizing the number of available host strains decreases the types of contaminating phage (146). Second, the production facility (air, equipment, fermentation by-products and products, and seed cultures) should be monitored routinely for phage inhibitory to production strains. Third, any strains exhibiting a high degree of sensitivity to indigenous phage should be removed from production. In many cases, strains that support only slowly replicating phage owing to long latent periods or small burst size can be used productively (42) and need not be removed. Fourth, a phage-sensitive strain must be replaced by a strain resistant to prevalent phage populations. The resistant strain may be an unrelated strain equally capable of effecting the fermentation, or a phage-resistant mutant of the sensitive strain. Phage-resistant mutants can be derived by challenge of the sensitive strain with one or more inhibiting phages. Phage banks containing stocks of phage contaminating previous fermentations are useful for this purpose. As with any mutation program, the clever part of this scheme lies in the selection of the phage-resistant mutants retaining desired functionality. Metabolic and physiological properties have been shown to accompany mutations to phage resistance (72) rendering a large portion of the phage-resistant mutants nonfunctional. Plating methods that can select desirable fermentative traits among mutants are very useful. To differentiate between fast and slow lactic acid-producing lactic streptococci, a milk-based medium was developed by Huggins and Sandine (44) and has proved useful in the selection of phage-resistant mutants with suitable activity (42). Further characterization of phage-resistant strains with respect to important functional criteria (activity, efficiency, compatibility) may be necessary. Finally, these strains may be used or stored as backup strains for use as needed.

In the dairy industry, alternative means of controlling the phage sensitivity of the starter strains are used. In the United States, many cheesemaking facilities automatically rotate commercially obtained phage-unrelated strains for successive production runs (42). This eliminates the need for the labor-intensive responsibility of monitoring phage levels when using one defined culture, yet adds the variability to production because several different cultures are used. For recent discussions of strain rotation programs, see Huggins (42) and Klaenhammer (56). Phage-resistant cultures have also been prepared for casein (151) and cheese (136) fermentations by the continuous addition of phage-contaminated whey into the bulk culture. The bulk culture that emerges is resistant to the phage of current threat in the plant. However, continuous addition of destructive phage may alter the nature of the culture, affecting the repeatability and quality of the fermentation end product (136). The inconsistent acid production that accompanies culture produced in this manner is generally not tolerable in modern production facilities.

9.5.3. Antiphage Agents

The usefulness of antiphage agents to control lytic phage development is difficult to determine. Most available information describes the use of antiphage compounds on a laboratory, not a production, scale. One notable exception is the use of phosphates to prevent phage development by the chelation of Ca^{2+} and Mg^{2+} (30). Phosphated media have been tested at the laboratory and pilot plant scales and are now widely used for the preparation of phage-free bulk starters in the dairy industry. To be useful, an antiphage agent must:

1. Protect commercially important strains from phage attack, without affecting strain growth and end-product formation.
2. Not adulterate the final product, or have an adverse affect on product quality (e.g., purification characteristics, flavor, texture).
3. Be cost-effective.
4. Have broad-spectrum activity against phage.

Some antiphage agents compatible with commercial microorganisms and with demonstrated efficacy on a laboratory scale include ascorbic acid (92,93), basic amino acids (94), D-glucosamine (50), terpenes, (169), cation chelators (30,100,120,122), Fe^{2+} (160), surface-active agents (99), and spermine (27).

Ascorbic acid was shown to inactivate phage of *E. coli, B. subtilis,* and *Lactobacillus casei* (91,93). In the most dramatic case, free *L. casei* phage J1 incubated with 10 mM ascorbic acid at pH 7.3 at 37°C in 0.02 M Tris buffer irreversibly decreased phage titers by 2.5 log cycles within 3 min (91). However, inactivation kinetics in nutrient broth were markedly slower, and inactivation in fermentation media was not attempted. Evidence obtained from *in vitro* studies indicated that the mode of action of ascorbic acid inactivation of phage was the induction of single-strand breaks in the phage DNA. Further studies are necessary to determine the utility of ascorbic acid for inhibition of commercial phage infections. Certainly the nontoxic nature of ascorbic acid makes it an attractive possibility for food industry applications.

D-Glucosamine (50) was also shown to be toxic to phage of *Lactobacillus casei*. The concentration for maximum phagocidal activity was high, 0.1 M, and required aeration and lengthy incubation (75 min for 3 log cycle reduction). Similarly, for basic amino acids, concentrations for maximum phage inhibitions were 0.7 M for arginine and 0.5 M for histidine (94). Concentrations of thiol reducing agents required for a 2 log cycle reduction in *L. casei* phage were 5 mM/15 min for glutathione, 30 mM/60 min for 2-mercaptoethanol, and 1 mM/30 min for dithiothreitol (92). Applicability of D-glucosamine, basic amino acids, and thiol-reducing agents to phage inactivation under industrial conditions is questionable, considering required concentrations, time, and reaction conditions. Testing of these compounds under fermentation conditions is essential for evaluation of their usefulness.

Findings by Wolf et al. (169) do indicate that antiphage agents can be industrially significant. These workers evaluated terpenes obtained from black pepper oil, cinnamon flower oil, cardamom oil, linalyl acetate, cinnamic aldehyde, safrole, carvone, and *cis/trans*-citral and found that pretreatment of strains of *S. cremoris*, *S. lactis*, *S. diacetylactis*, and *L. bulgaricus* for 2–12 hr with terpenes protected the strains from homologous phage attack. Concentrations of terpenes required were below those showing any toxic effect on the bacterial strains, a negative effect on flavor, or harmful effects to human consumers. No target site or mechanism of protection were discussed.

The most widely used antiphage agent in the dairy industry is phosphate. Phosphates in immediately (30) or slowly (120) soluble forms are specifically added to bulk starter media and serve the purpose of chelating divalent cations specifically required by phage for development (17). This specific target is also widely conserved among phage of the lactic acid bacteria, and therefore this is an effective antiphage chemical. However, some non-calcium-requiring phage have been isolated (131) and therefore phosphates do not necessarily offer complete protection. Phosphates are added to the starter medium to prevent the buildup of phage prior to inoculation of the cheese milk; they are not added during the fermentation. Therefore, phage attack during the fermentation is not prevented by this measure. Commercially available phosphated media have been shown to: vary in effectiveness (119), adversely effect the activity of some starters (70), and add to the cost of starter preparation. For these reasons, some cheesemakers prefer to prevent phage buildup in bulk starters by focusing on control of external phage contamination by scrupulous sanitation, aseptic technique, and the use of well-designed bulk starter tanks (69,119).

Spermine was shown to stop *S. lactis* phage development irreversibly at concentrations as low as 5×10^{-4} *M* (27). Lysis was prevented if spermine was added up to 21 min after phage infection. Therefore, spermine did not act at the infection/injection point of phage development. Yet spermine-treated cells lysed by enzymatic or physical means did not release any intact phage particles or phage components. Therefore, spermine appeared to prevent phage from synthesizing new phage material. Furthermore, cells treated with spermine maintained resistance to homologous phage, indicating that the antiphage activity directed a permanent change in the bacterium. The usefulness of this compound in the protection of strains during fermentation or for the isolation of phage-resistant mutants has not been explored, to my knowledge, on an industrial scale.

The role of anti-phage agents in control of commercial phage inhibition of production strains has been explored by many investigators. Many different chemical agents have been tested but only a few showed antiphage activity, and of these even fewer seem applicable to the fermentation processes. Perhaps more in-depth understanding of the developmental process of industrially significant phage could lead to the identification of compounds with activity against specific target sites common to a variety of phage types. Finally, the identification of specific antiphage agents may provide research tools for more in-depth studies of industrially significant phage.

9.6. GENERAL BACTERIAL INTERFERENCE WITH PHAGE PROLIFERATION

As pointed out previously, the simple isolation of phage-resistant mutants is not always a suitable solution to phage-sensitive starter cultures. However, a more directed approach to the development of phage-resistant strains is difficult, because the genetics, physiology, and biochemistry of phage defense mechanisms are so poorly understood, especially in industrially significant bacteria. Certain strain-directed mechanisms that interfere with phage have, however, been described (26).

Phage defense mechanisms that protect the cell against the development of lytic phage could theoretically act at target sites all along the lytic cycle for phage development: adsorption, injection of DNA, survival of DNA, phage-directed protein synthesis, phage DNA replication, assembly, and cell lysis. Figure 9.1 simplistically illustrates the lytic cycle of phage infection and describes some possible mechanisms for specific antiphage activities that could theoretically be exhibited by bacterial strains. In some cases these activities have been shown experimentally, whereas others remain speculative. However, increased demand for knowledge of defense mechanisms

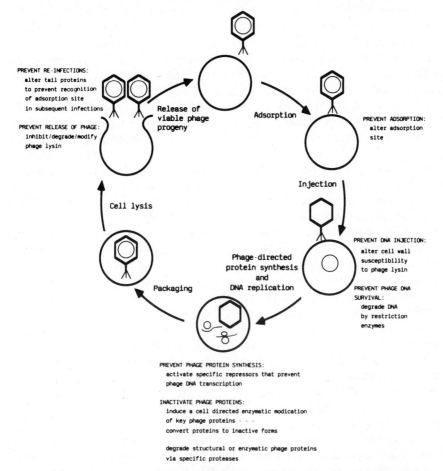

PREVENT RE-INFECTIONS:
alter tail proteins
to prevent recognition
of adsorption site
in subsequent infections

PREVENT RELEASE OF PHAGE:
inhibit/degrade/modify
phage lysin

Release of
viable phage
progeny

Adsorption

PREVENT ADSORPTION:
alter adsorption
site

Injection

PREVENT DNA INJECTION:
alter cell wall
susceptibility
to phage lysin

PREVENT PHAGE DNA
SURVIVAL:
degrade DNA
by restriction
enzymes

Cell lysis

Phage-directed
protein synthesis
and
DNA replication

Packaging

PREVENT PHAGE PROTEIN SYNTHESIS:
activate specific repressors that prevent
phage DNA transcription

INACTIVATE PHAGE PROTEINS:
induce a cell directed enzymatic modication
of key phage proteins · · ·
convert proteins to inactive forms

degrade structural or enzymatic phage proteins
via specific proteases

Figure 9.1 Fact and fantasy of bacterial interference with phage development. Described are some potential mechanisms whereby a cell could defend itself against phage. In some cases, these mechanisms have been experimentally verified (e.g., restriction/modification systems, alteration of cellular adsorption sites). In most cases, with industrially important phage–host crosses, however, reports suggest an inhibition of phage development, but mechanistic studies have not been pursued.

against phage may stimulate research activities to differentiate between the fact and fantasy of possible defense mechanisms.

A common route to the acquisition of phage resistance is for the host strain to undergo a genetic change that results in an alteration of the phage receptor site on the cell surface. The cell no longer adsorbs the phage and thereby blocks any phage development. This genetic change may occur via simple mutation (24), lysogenic conversion (158), or acquisition of certain plasmids (101).

Another form of protection that is used by cells against foreign, invading DNA is the operation of restriction/modification systems. These systems, characterized by host-dependent replication of phage, operate in a variety of bacterial genera (7,84). Restriction/modification systems function by the cooperation of two enzymes, a restriction endonuclease that recognizes a specific nucleotide sequence in invading DNA and if it is unlabeled, hydrolyzes it; and a modification methylase that chemically labels the host DNA at the specific recognition sequence, to render it immune to intrinsic restriction.

Restriction/modification is not fully effective in protecting the cell from the invasion of foreign DNA. Certain environmental factors can influence the efficiency of restriction/modification systems, such as growth phase of the cells (75), composition of the growth medium (75), and heat shock (159). Furthermore, restriction efficiency on invading DNA can be affected by the number of recognition sites on the DNA molecule, the *in vivo* number of molecules of the enzymes harbored by the cell, the relative *in vivo* activities of the restriction endonuclease as compared to the modification methylase, and also by the multiplicity of infection of invading DNA molecules. As a defense mechanism against phage, therefore, restriction/modification systems are usually characterized by a low plating efficiency of 10^{-3}–10^{-5} of the phage on the host cells (6). However, the rare phage that escapes restriction is modified by the cellular enzymes, rendering the phage immune to the cell's restriction activity upon reinfection. These modified phage are capable of maximal plating efficiency on the originally restrictive host cell. This mechanism of the generation of modified phage can undermine the effectiveness of restriction/modification systems as a defense against phage, assuming (*a*) sufficiently high titers of challenging phage, and (*b*) repeated exposure of the host strain to modified phage populations.

The capacity of classical restriction/modification systems to act on any unmodified, invading, double-stranded DNA that possesses the appropriate recognition specificities renders them functional against a broad range of phage. This trait makes them attractive as general mechanisms for defense against phage. A recent patent by Hershburger and Rosteck (38) claims the use of cloned restriction/modification systems for use in preventing phage infection of recombinant DNA-containing bacterial cultures.

However, the utility of restriction/modification systems may be somewhat limited owing to their inherent leakiness, which results in the release of modified phage. In addition, some phage have been shown to develop mechanisms to overcome classical restriction (63,64,140). The gene that controls this function in T3 and T7 phage, *ocr,* has been mapped within gene 0.3 region of T3 and T7 DNA. The *ocr+* gene product exerts its protective effect by turning off the cellular restriction enzyme so that it can not act on the unmodified T3 and T7 DNA. The T3 gene 0.3 codes for an *S*-adenosylmethionine (SAM) hydrolyzing activity, presenting an attractive mechanism of action, because SAM is required by the Type I restriction enzymes involved. But phage mutants deficient in SAM hydrolysis still express the *ocr* function. Furthermore, *in vitro* cleavage of phage T3 and T7 DNA has been demonstrated by the restriction enzymes that fail to act *in vivo*. Therefore, the mechanism of action of the *ocr* gene product is unclear. However, the evolution of this trait in T3 and T7 emphasizes the ability of phage to counteract defenses evolved by the hosts.

A novel mechanism for the functioning of host-controlled modification of viruses has been described (61,62). Passage of phage T7 through *E. coli* strains B (harbors classical restriction/modification system) or O (does not harbor classical restriction/modification system) results in the host-controlled modification of phage T7. However, the mechanism of modification was not via a classical modification methylase, for the DNA of unmodified phage never entered the restricting host cells. It was found that this "nonclassical" restriction/modification system restricted phage by establishing a barrier to adsorption and modified phage by structurally altering their tails so they adsorbed more efficiently. Plaque development differed by up to 6 log cycles when modified phage were compared to unmodified phage. Differences in adsorption were host-controlled and therefore reversible. The biochemical basis for this modification is not known. However, it is likely to result from specific modifications(for example, acetylation or phosphorylation) of the phage tail protein. The implications of the identification of this"nonclassical" restriction/modification system is that host-controlled replication identified *in vivo* is not necessarily a function of classical restriction endonuclease and modification methylase activity. This nonclassical restriction/modification is suggested in some data presented by Keogh (54) working with the lactic streptococci. She reported that 73% of phage c1 adsorbed to strain C1. However, only 5% adsorbed to C1 when phage c1 first replicated through strain DR7. These data demonstrate host-dependent adsorption capacity and indicate that a system similar to that in T3 and T7 phage may occur in the lactic streptococci.

Another mechanism primarily described in species of *Escherichia, Salmonella,* and *Shigella,* whereby bacterial cells prevent the development of phage, is abortive infection. In most cases abortive infection is characterized by failure to produce phage progeny, even though adsorption, phage DNA injection, and early phage protein synthesis are normal (27). Infection usually results in the degradation of host DNA, thereby killing the host cell. Abortive infection frequently occurs in strains harboring certain extrachromosomal elements, such as the lysogenic phage λ and P22, and by plasmids such as the F factor, R factor, and colicinogenic plasmids (27). Unique properties characterize the abortive infection expressed by each of these elements. However, the general mechanism of abortive infection appears to be a cell membrane modification that leads to permeability dysfunctions and cell death (18,28). The expression of such a mechanism, if effective against a variety of phage, could readily minimize phage proliferation and therefore be a useful trait in a phage-threatened commercial strain. Death of the initially infected cell would be insignificant, as long as phage levels did not increase.

Cells that harbor lysogenic phage are generally protected from superinfecting phage. For example, phage P22, which normally infects *Salmonella typhimurium,* has developed several strategies for superinfection immunity and superinfection exclusion (143). The first mechanism, found in other well-studied lysogenic phage, is an immunity/repression system (10). This system acts by the expression of genes coding for two repressor proteins. These repressors prevent transcription of P22 genes necessary for lytic development and also prevent homologous superinfecting phage from lytic development. A second mechanism of protection from superinfecting phage is accomplished by a lysogenic conversion, which results in alteration of the antigenic structures of the host cell surface, thereby preventing adsorption by certain phage(158). A third mechanism acts by preventing the injection of superinfecting phage DNA (144). Although the gene for this trait is mapped, the biochemistry of the prevention of DNA uptake is not known. Finally, P22 lysogens prevent superinfection by a system similar to abortive infection (143). Phage DNA is adsorbed and injected, and early gene functions are transcribed and expressed. However, phage development is not completed. Wild-type P22 phage are immune to the system, although some heteroimmune *Salmonella* phage are susceptible to it.

The preceding discussion of defense mechanisms against phage reveals that bacteria can express diverse strategies that interfere with the destructive nature of lytic phage development. In some cases, defense against phage might not be the primary function of the system (e.g., abortive infection); or the system may be a function of a lysogenic phage and therefore is not useful to the host to prevent phage infection, only phage superinfection (e.g., superinfection exclusion). Some defense mechanisms may act against only one, or a very narrow group, of phage and therefore may not confer resistance to a wide variety of phage (e.g., prevention of adsorption, lysogenic immunity). The existence of a mechanism that would permanently prevent the proliferation of all challenging phage is doubtful, especially considering the capacity of phage to overcome a defense mechanism. Indeed, entrance of exogenous DNA into a cell can confer an advantage to the host cell under selective conditions (e.g., R factors, colicinogenic factors) or facilitate evolutionary advancement by providing a substrate for genetic recombination or exchange (107). Therefore, completely blocking the influx of infecting genomes may not be selectively advantageous. However, from the point of view of the fermentation technologist, identification of phage defense systems may provide the information necessary to select, or genetically construct, starter strains that consistently and predictably demonstrate the required level of phage resistance during commercial use.

9.7. PHAGE–HOST INTERACTIONS FOR THE LACTIC ACID BACTERIA

The description of phage-resistance mechanisms in industrially significant bacteria will require research that focuses on the biochemistry and genetics of phage, host strains, and phage development processes. Some strides made over the past 10 years with lactic acid bacteria have led investigators closer to understanding phage interactions with these commercially important bacteria.

Characterization of phage DNA has been described for phage of both lactic streptococci and lactobacilli (Table 9.2). Improved methods for phage purification and phage DNA isolation have also been reported (76,77). Phage of the lactic acid bacteria appear to harbor double stranded, linear DNA with cohesive ends enabling spontaneous circularization *in vitro* (73,141,166). The DNA of *L. casei* lysogenic phage FSW was found to be circularly permuted (127), and its conversion to a lytic phage, FSV, appeared to be mediated by transposition of a sequence from the host genome. These contributions to methodology and phage genetics help provide basic information necessary to pursue the molecular biology of phages of lactic acid bacteria.

Protein analysis of phage of the lactic acid bacteria has been limited. Stetter et al. (141) showed that *L. casei* phage PL-1 contained seven major protein bands and at least nine minor bands, with apparant molecular weights ranging from 12,000 to 140,000. In the same publication, studies on phage-controlled transcription indicated that the RNA polymerase present during phage infection was indistinguishable from the host RNA polymerase. Alatossava and Pyhtila (5) analyzed structural proteins of *Lactobacillus lactis* phage LL-H by SDS–polyacrylamide gel electrophoresis and cross-reaction with antiserum raised against intact phage particles. Later, Trautwetter et al. (155) created a partial gene bank of this phage DNA in *E. coli* and identified fragments of DNA encoding five of the seven major structural proteins. These genes were expressed in *E. coli* and proteins were detected by immunobloting. One protein was shown to produce lytic activity against the LL-H host strain.

Perhaps the best-studied phage proteins to date are the phage lysins (88,89,102,153,154). The lysin produced by the prolate ϕC2(W) was shown not to be a structural component of the phage (88). This is in contrast to a phage-bound lysin from m13 phage (102). It is, however, isolated from lysates of sensitive hosts and demonstrates lytic activity against many strains on which the phage cannot replicate. These findings contribute to our understanding of the "nascent phenomenon" (14): lysin released by phage infection of a phage-sensitive component of a culture can lead to the lysis of a phage-resistant strain in a multiple strain culture. Three other prolate phage were found to produce similar lysins, whereas no isometric phage examined produced the lysin (88). The broad host range of prolate phage (20) suggests that this lysin may play a role in the destructive nature of these phage. The lysin showed a pH optimum of 6.5–6.9, was heat labile (>47°C), and showed a temperature optimum of 37°C (89). The enzyme acted upon the N-acetylmuramyl-N-acetylglucosamine linkage in the cell wall (89). Parameters such as growth phase of the cell, carbohydrate substrate, growth temperature, lyophilization, and concentration of osmotic stabilizer affecting the lytic activity of partially purified phage lysin from ϕC2(W) on *S. lactis* C2 were studied (87). Furthermore, the isolation of lysin-insensitive variants of

TABLE 9.2 Physical Characteristics of DNA from Phage Isolated from Lactic Acid Bacteria

Phage	Host	Size[a]	G/C ratio (%)[b]	Buoyant density (g/mL)	T_m (°C)	Restriction map	Refs.
PL-1	Lactobacillus casei	25 Mdal	44	1.702	87.5	No	145
P008	Streptococcus diacetylactis	19.6 Mdal	37.5	ND	84.7	Yes	140 72
φFSW	L. casei	27 Mdal	ND	ND	ND	Yes	126
c2tl	S. lactis	22.6 Mdal	ND	ND	ND	No	56
c2t2	S. lactis	23.8 Mdal	ND	ND	ND	No	56
643	S. lactis S. cremoris	14.9 Mdal	38	1.697	85.5	Yes	76
LL-H	L. lactis	34 kb	ND	ND	ND	Yes	5,155

[a] Mdal = megadaltons; kb = kilobase pairs.
[b] No unusual bases found.

C2 was conducted (86). Lysin resistance was not correlated with phage resistance, and was not a genetically stable trait. Knowledge of the properties of the phage lysins and the regulation of their production may provide a target through either antiphage agents or strain construction for the prevention of the destructive nature of phage against the lactic acid bacteria.

Many studies have recently focused on defining the cell-directed functions of lactic acid bacteria, which interfere with phage functioning (56). Reports on the involvement of plasmid-encoded bacteriophage-resistance mechanisms in lactic streptococci include host-dependent replication (13,115,138), prevention of phage adsorption (25,116), abortive infection (9,58), and temperature-sensitive prevention of phage development (82). Although the genetic linkage of these traits to plasmid DNA is significant, mechanisms of resistance are still unclear.

A phenomenom labeled as abortive infection was described by Klaenhammer and Sanozky (58) for lactic streptococci. Acquisition of a conjugal plasmid, pTR2030, by a strain of *S. lactis* conferred on the host the ability to inhibit the development of phage. This inhibition was characterized by an unaltered adsorption frequency, a burst size decreased from 41 to 5, and an efficiency of plating decreased from 1.0 to 0.8. This plasmid-induced effect was reversed when cells were challenged with phage at 40° versus 30°C. In contrast to more fully described abortive infections, the pTR2030-induced abortive infection results in the production of complete phage progeny, indicating that all phage genes, not just early genes, are expressed. Transfer of pTR2030 into the *S. cremoris* background did not appear to result in the expression of this same abortive infection phenotype (129,137). Instead, a much more effective blockage of phage development resulted which apparently was not accompanied by host death, as occurs with abortive infections. Differential expression of phage resistance mechanisms encoded by pTR2030 in different phage–host crosses remains to be clarified. However, the demonstration by Sing and Klaenhammer (129) that pTR2030-encoded phage resistance could be conjugally disseminated to a variety of *S. cremoris* strains set the stage for the use of this plasmid to fortify the phage resistance of commercial cultures. Sanders et al. (118) demonstrated that pTR2030 could be conjugated into wild-type *S. cremoris* and *S. lactis* strains with selection based on phage-resistance. The *S. lactis* transconjugants were not temperature-sensitive as was previously speculated (58). These strains have served as prototypes for industrial strains with phage resistance superior to their pTR2030-deficient parent strains. It is unlikely that these new strains will be totally resistant to all phage encountered in the field, because Jarvis and Klaenhammer (52) found that pTR2030-directed phage resistance was less effective against prolate-headed than isometric-headed phage. However, the utility of this approach for the construction of phage-resistant strains has been demonstrated.

Another conjugative plasmid, pNP40, originating in *S. diacetylactis* strain, when transferred to *S. lactis* rendered complete resistance to homologous phage, although phage adsorption was unaffected. Host cell survival was not affected, implying that the phage-resistance mechanism acted after adsorption but prior to phage-directed metabolism. Unlike pTR2030-directed ϕ^r, this plasmid appeared to inhibit prolate phages totally. As with pTR2030, in the LM0230 background, phage resistance was temperature-sensitive. These studies reflect significant progress in understanding the genetic control and biological functioning of phage-resistance mechanisms in industrially significant bacteria.

The genetics of non-plasmid-linked phage resistance has been studied. Resistance conferred by simple mutation has been shown to occur in some *S. cremoris* strains at a spontaneous mutation rate per generation of approximately 10^{-7} (55). Although many phage-resistant mutants of the lactic streptococci demonstrate decreased adsorption capacity (16,55,72), other investigators have isolated phage-insensitive mutants that have retained full capacity to adsorb the phage (103). Results by Oram and Reiter (103) indicated that 90% of the phage-resistant mutants isolated from secondary growth from a lysed culture were still capable of adsorbing phage. Therefore it is apparent that simple mutation can result in the prevention of phage development in more ways than the prevention of phage adsorption. Unfortunately, genetic and biochemical studies to identify mutant loci on the host genome and identify the mechanism of resistance have not, in most cases, been pursued.

Resistance due to inability of the phage to inject DNA has been suggested by Marshall and Berridge (78) for phage-resistant mutants of *S. lactis*. The conclusions were based on blender experiments designed to determine if previously adsorbed phage particles retained their DNA past the time normally required for injection. However, lysis from without was not observed, even at a phage–cell ratio of 500–1. A phage–cell ratio of 10—1 or greater resulted in lysis from without for *S. lactis* ML3 (103). Therefore the authors concluded that phage lysin was ineffective at degrading the cell wall. The gene or gene product affected by the mutation leading to this form of phage resistance was not investigated.

More extensive mechanistic work on phage development in lactic acid bacteria has been done without the benefit of genetic linkages of cell-directed functions. Watanabe et al. (163) focused on the cause of resistance in a phage-resistant mutant obtained from confluent plate lysis of *L. casei* YIT9002 (ATCC27092). A 2 log cycle difference in efficiency of plating between the mutant YIT9021 (5×10^{-5}) and its parent strain, YIT9002 (3×10^{-3}) was observed to be due to failure of the phage to inject its DNA efficiently. Conclusions were based on electron microscopic observation of the persistence of DNA-filled, absorbed phage and the lack of cell killing after phage infection, even at phage–cell ratios as high as 50. Adsorption studies showed no difference in the ability of the mutant to adsorb phage as compared to the parent strain. Characterization of the reversible and irreversible stages of phage adsorption was aided by the definition of a specific competitive inhibitor of phage adsorption, L-rhamnose (16). Earlier work by this same group demonstrated the energy-dependent nature of the injection process for *Lactobacillus* phages (165).

Alatossava et al. (4) investigated the nature of the infectivity process by *Lactobacillus lactis* phage LL-H by examining inhibition of phage development by cadmium. Cadmium at levels of 1–2 mM showed no effect on phage adsorption or injection, but specifically reduced the level of phage DNA synthesis as determined by ^{14}C-thymidine incorporation into trichloroacetic acid-insoluble DNA. The effect of cadmium was irreversible and resulted in a decrease of burst size up to fourfold.

The presence of restriction/modification systems as phage defense mechanisms has been stated by many researchers (11,13,114,115) based on host-controlled phage replication patterns. However, care must be taken in interpreting host-controlled replication data as confirmation of classical restriction/modification system (see Section 9.6.1). The first biochemical evidence for restriction enzyme activity in a lactic acid bacterium came from Fitzgerald et al. (29), who identified ScrFl, a sequence-specific endonuclease from *S. cremoris* F. These results lend credence to the operation of classical restriction/modification systems in the lactic acid bacteria as important phage defense mechanisms.

The work described in this section represents the beginning of efforts of researchers working with phage-threatened commercially significant bacteria to understand, at a molecular level, phage infection and development and the host-controlled interferences with these processes. Continued investigation of the genetics, biochemistry, and physiology of phage and their industrially significant hosts may provide directed strategies for the development of production strains exhibiting long-term resistance to phage.

9.8. CONCLUDING REMARKS

Bacterial cultures used in industrial fermentations can be negatively affected by phage. Although food fermentations are especially susceptible because of their nonaseptic nature, other commercial fermentations are phage-susceptible because of phage originating from the culture itself or through chance contamination. The modern emphasis on the production of useful end products from microorganisms developed either through the application of recombinant DNA techniques or classical microbial genetics emphasizes the potential for a heightening of the magnitude of the industrial phage problem. Techniques of detecting and coping with unwanted phage infections are available, and much can be learned from industries currently battling the problem. However, new advances in the biology, physiology, and genetics of phage interactions with industrially important bacteria have provided and will continue to provide more directed and more thorough solutions to phage infections.

REFERENCES

1. Accolas, J. -P., and H. Spillmann. 1979a. The morphology of six bacteriophages of *Streptococcus thermophilus*. J. Appl. Bacteriol. *47*:135–144.

2. Accolas, J. -P., and H. Spillmann 1979b. Morphology of bacteriophages of *Lactobacillus bulgaricus*, *L. lactis* and *L. helveticus*. J. Appl. Bacteriol. *47*:309–319.

3. Adams, M. H. 1950. Bacteriophages. Interscience, New York.

4. Alatossava, T., T. Juvonen, and R. -L. Huhtinen. 1983. Effect of cadmium on the infection of *Lactobacillus lactis* by bacteriophage LL-H. J. Gen. Virol. *64*:1623–1627.

5. Alatossava, T., and M. J. Pyhtila. 1980. Characterization of a new *Lactobacillus lactis* bacteriophage. IRCS Med. Sci. *8*:297–298.

6. Arber, W. 1974. DNA modification and restriction. Prog. Nucl. Acids Res. Mol. Biol. *14*:1–37.

7. Barksdale, L. 1959. Lysogenic conversions in bacteria. Bacteriol. Rev. *23*:202–212.

8. Barksdale, L., and S. B. Arden. 1974. Persisting bacteriophage infections, lysogeny, and phage conversions. Ann. Rev. Microbiol. *28*:265–299.

9. Baumgartner, A., M. Murphy, C. Daly, and G. F. Fitzgerald. 1986. Conjugative co-transfer of lactose and bacteriophage resistance plasmids from *Streptococcus cremoris* UC653. FEMS Microbiol. Lett. *35*:233–237.

10. Botstein, D., K. K. Lew, V. jarvik, and C. A. Swanson, Jr. 1975. Role of antirepressor in the bipartite control of repression and immunity by bacteriophage P22. J. Mol. Biol. *91*:439–462.

11. Boussemaer, J. P., P. P. Schrauwen, J. L. Sourouille, and P. Guy. 1980. Multiple modification/restriction systems in lactic streptococci and their significance in defining a phage-typing system. J. Dairy Res. *47*:401–409.

12. Cherry, W. B., and D. W. Watson. 1949. The *Streptococcus lactis* host-virus system. I. Factors influencing quantitative measurement of the virus. J. Bacteriol. *58*:601–610.

13. Chopin, A., M. -C. Chopin, A. Moillo-Batt, and P. Langella. 1984. Two plasmid determined restriction and modification systems in *Streptococcus* lactis. Plasmid *11*:260–263.

14. Collins, E. B. 1952. Action of bacteriophage on mixed strains starter cultures. I. Nature and characteristics of the "nascent phenomenon." J. Dairy Sci. *35*:371–380.

15. Collins, E. B. 1955. Action of bacteriophages on mixed strain cultures. V. Similarities among strains of lactic streptococci in commercially used cultures and use of a whey activity test for culture selection and rotation. Appl. Microbiol. *3*:145–148.

16. Collins, E. B. 1958. Changes in the bacteriophage sensitivity of lactic streptococci. J. Dairy Sci. *41*:41–48.

17. Collins, E. B., F. E. Nelson, and C. E. Parmelee. 1950. The relation of calcium and other constituents of a defined medium to proliferation of lactic streptococcus bacteriophage. J. Bacteriol. *60*:533–542.

18. Condit, R. C. 1975. F factor mediated inhibition of bacteriophage growth: increased membrane permeability and decreased ATP levels following T7 infection. J. Mol. Biol. *98*:45–56.

19. Cox, W. A. 1980. Detection and enumeration of mesophilic lactic bacteriophages, pp. 29–36. *In* Starters in the manufacture of cheese. International Dairy Federation Document 129. Brussels.

20. Daly, C. 1983. Starter culture developments in Ireland. Irish J. Food Sci. Technol. *7*:39–48.

21. Davies, F. L., and M. J. Gasson. 1984. Bacteriophages of dairy lactic acid bacteria, pp. 127–151. *In* F. L. Davies and B. A. Law (eds.), Advances in the Microbiology and Biochemistry of Cheese and Fermented Milk. Elsevier, New York.

22. Davis, C., N. F. A. Silveira, and G. H. Fleet. 1985 Occurrence and properties of bacteriophages of *Leuconostoc oenos* in Australian wines. Appl. Environ. Microbiol. *50*:872–876.

23. Deane, D. D., F. E. Nelson, F. C. Ryser, and P. H. Carr. 1953. *Streptococcus thermophilus* bacteriophage from Swiss cheese whey. J. Dairy Sci. *36*:185–191.

24. Demerec, M., and U. Fano. 1945. Bacteriophage-resistant mutants in *Escherichia coli*. Genetics *30*:119–136.

25. de Vos, W. M., H. M. Underwood, and F. L. Davies. 1984. Plasmid encoded bacteriophage resistance in *Streptococcus cremoris* SKll. FEMS Microbiol. Lett. *23*:175–178.

26. Duckworth, D. H., J. Glenn, and D. J. McCorquodale. 1981. Inhibition of bacteriophage replication by extrachromosomal genetic elements. Microbiol. Rev. *45*:52–71.

27. Erskine, J. M. 1970. Effect of spermine on host-cell lysis and reproduction by a lactic streptococcal bacteriophage. Appl. Microbiol. *19*:638–642.

28. Fields, K. L. 1969. Comparison of the action of colicins El and K on *Escherichia coli* with the effects of abortive infection by virulent bacteriophages. J. Bacteriol. *97*:78–82.

29. Fitzgerald, G. F., C. Daly, L. R. Brown, and T. R. Gingeras. 1982. ScrFl: a new sequence specific endonuclease from *Streptococcus cremoris*. Nucl. Acids Res. *10*:8171–8179.

30. Hargrove, R. E., F. E. McDonough, and R. P. Tittsler, 1961. Phosphate heat treatment of milk to prevent bacteriophage proliferation in lactic cultures. J. Dairy Sci. *44*:1799–1810.

31. Hayes, W. 1968. The Genetics of Bacteria and Their Viruses. Wiley, New York.

32. Heap, H. A., and R. C. Lawrence. 1976. The selection of starter strains for cheesemaking. N. Z. J. Dairy Sci. Technol. *11*:16–20.

33. Heap, H. A., and R. C. Lawrence. 1977. The contribution of starter strains to the level of phage infection in a commercial cheese factory. N. Z. J. Dairy Sci. Technol. *12*:213–218.

34. Heap, H. A., and R. C. Lawrence. 1981. Recent modifications to the New Zealand activity test for cheddar cheese starters. N. Z. J. Dairy Sci. Technol. *15*:91–94.

35. Heap, H. A., and R. C. Lawrence. 1984. The development of a defined starter system for casein manufacture. N. Z. J. Dairy Sci. Technol. *19*:119–123.

36. Heap, H. A., G. K. Y. Limsowtin, and R. C. Lawrence. 1978. Contribution of *Streptococcus lactis* strains in raw milk to phage infection in commercial cheese factories. N. Z. J. Diary Sci. Technol. *13*:16–22.

37. Hendrix, R. W., J. W. Roberts, F. W. Stahl, and R. A. Weisberg (eds.). 1983. Lambda II. Cold Spring Harbor Laboratory, Cold Spring Harbor, New York.

38. Hershberger, C. L., and P. R. Rosteck, Jr. 1985. Method for conferring bacteriophage resistance to bacteria. U.S. Patent 4,530,904.

39. Hirose, Y., J. Nakamura, H. Okada, K. Kinoshita, I. Kameyama, and T. Shiio. 1967. Studies on L-glutamic acid fermentation. Part II. Resistance of a streptomycin-resistant strain of *Brevibacterium lactofermentum* to bacteriophage and lysozyme. J. Agric. Chem. Soc. Japan. *41*:19–25.

40. Hongo, M., and A. Murata. 1965. Bacteriophages of *Clostridium saccharoperbutylacetonicum*. Part 1. Some characteristics of the twelve phages obtained from the abnormally fermented broths. Agric. Biol. Chem. *29*:1135–1139.

41. Hongo, M., T. Oki, and S. Ogata. 1972. Phage contamination and control, pp. 67–90. *In* K. Yamada (ed.), The Microbial Production of Amino Acids. Wiley, New York.

42. Huggins, A. R. 1984. Progress in dairy starter culture technology. Food Technol. *38*:41–50.

43. Huggins, A. R., and W. E. Sandine. 1977. Incidence and properties of temperate bacteriophages induced from lactic streptococci. Appl. Environ. Microbiol. *33*:184–191.

44. Huggins, A. R., and W. E. Sandine. 1984. Differentiation of fast and slow acid-producing strains of lactic streptococci. J. Dairy Sci. *67*:1674–1679.

45. Hull, R. R. 1977. Methods for monitoring bacteriophage in cheese factories. Aust. J. Dairy Technol. *32*:63–64.

46. Hull, R. R. 1977. Control of bacteriophages in cheese factories. Aust. J. Dairy Technol. *32*:65–66.

47. Hull, R. R. 1978. Methods of monitoring for bacteriophage in cheese manufacture, pp. 24–35. *In* R. R. Hull (ed), Factory derived cheese starters, Proceedings of the CSIRO and Australian Dairy Corp. Meeting, Oct. 4–6, Highett, Australia.

48. Hull, R. R., and A. R. Brooke. 1982. Bacteriophages more active against cheddar cheese starters in unheated milk. Aust. J. Dairy Technol. *37*:143–146.

49. Hunter, G. J. E. 1943. Bacteriophages for *Streptococcus cremoris*: Phage development at various temperatures. J. Dairy Res. *13*:136–145.

50. Ishibashi, K., T. Sasaki, S. Takesue, and K. Watanabe. 1982. *In vitro* phage-inactivating action of D-glucosamine on *Lactobacillus* phage PL-1. Agric. Biol. Chem. *46*:1961–1962.

51. Jarvis, A. W. 1984. DNA–DNA homology between lactic streptococci and their temperate and lytic phages. Appl. Environ. Microbiol. *47*:1031–1038.

52. Jarvis, A. W., and T. R. Klaenhammer. 1986. Bacteriophage resistance conferred on lactic streptococci by the conjugative plasmid pTR2030: effects on small isometric-, large isometric-, and prolate-headed phages. Appl. Environ. Microbiol. *51*:1272–1277.

53. Kaneko, T., S. Iwano, and K. Kitahara. 1955. Bacteriophage phenomena in fermentative microorganisms. Part 2. Effect of temperature on a *Leuconostoc* phage. J. Agric. Chem. *29*:788–793.

54. Keogh, B. P. 1973. Adsorption, latent period and burst size of phages of some strains of lactic streptococci. J. Dairy Res. *40*:303–309.

55. King, W. R., E. B. Collins, and E. L. Barrett. 1983. Frequencies of bacteriophage-resistant and slow acid-producing variants of *Streptococcus cremoris*. Appl. Environ. Microbiol. *45*:1481–1485.

56. Klaenhammer, T. R. 1984. Interactions of bacteriophages with lactic streptococci. Adv. Appl. Microbiol. *30*:1–29.

57. Klaenhammer, T. R., and L. L. McKay. 1976. Isolation and examination of transducing bacteriophage particles from *Streptococcus lactis* C2. J. Dairy Sci. *59*:396–404.

58. Klaenhammer, T. R., and R. B. Sanozky. 1985. Conjugal transfer of plasmids from *Streptococcus lactis* ME2 encoding phage resistance, nisin resistance, and lactose fermenting ability: evidence for a high frequency conjugal plasmid responsible for abortive infection of virulent bacteriophage. J. Gen. Microbiol. *131*:1531–1541.

59. Koibong, L. I., L. Barksdale, and L. Garmise. 1961. Phenotypic alterations associated with the bacteriophage carrier state of *Shigella dysenteriae*. J. Gen. Microbiol. *24*:355–367.

60. Kozak, W., M. Rajchert-Trzpil, J. Zajdel, and W. T. Dobrzanski. 1973. Lysogeny in lactic streptococci producing and not producing nisin. Appl. Microbiol. *25*:305–308.

61. Kruger, D. H., S. Hansen, and C. Schroeder. 1980. Host-dependent modification of bacterial virus T3 affecting its adsorption ability. Virology *102*:444–446.

62. Kruger, D. H., S. Hansen, C. Schroeder, and W. Presber. 1977a. Host-dependent modification of bacteriophage T7 and SAMase-negative T3 derivatives affecting their adsorption ability. Mol. Gen. Genet. *153*:107–110.

63. Kruger, D. H., and C. Schroeder, 1981. Bacteriophage T3 and bacteriophage T7 virus-host cell interactions. Microbiol. Rev. *45*:9–51.

64. Kruger, D. H., C. Schroeder, S. Hansen, and H. A. Rosenthal. 1977b. Active protection

by bacteriophages T3 and T7 against *E. coli* B- and K-specific restriction of their DNA. Mol. Gen. Genet. *153*:99–106.

65. Kunkee, R. E. 1967. Malo-lactic fermentation. Adv. Appl. Microbiol. *9*:235–279.

66. Lawrence, R. C., H. A. Heap, and J. Gilles. 1984. A controlled approach to cheese technology. J. Dairy Sci. *61*:1632–1645.

67. Lawrence R. C., H. A. Heap, G. Limsowtin, and A. W. Jarvis. 1978. Cheddar cheese starters: current knowledge and practices of phage characteristics and strain selection. J. Dairy Sci. *61*:1181–1191.

68. Lawrence, R. C., and T. D. Thomas. 1979. The fermentation of milk by lactic acid bacteria, pp. 182–219. *In* A. T. Bull, D. Elwood, and R. Ratledge (eds.), Microbial Technology. Current state, future prospects. Symposium 29, Soc. Gen. Microbiol. London, England.

69. Lawrence, R. C., T. D. Thomas, and B. E. Terzaghi. 1976. Reviews of the progress of dairy science: cheese starters. J. Dairy Res. *43*:141–193.

70. Ledford, R. A., and M. L. Speck. 1979. Injury of lactic streptococci by culturing in media containing high phosphates. J. Dairy Sci. *62*:781–784.

71. Lembke, J., and M. Teuber. 1979. Detection of bacteriophage in whey by an enzyme-linked immunosorbent assay (ELISA). Milchwissenschaft *34*:457–458.

72. Limsowtin, G. K. Y., and B. E. Terzaghi. 1976. Phage resistant mutants: their selection and use in cheese factories. N. Z. J. Dairy Sci. Technol. *11*:251–256.

73. Loof, M., J. Lembke, and M. Teuber. 1983. Characterization of the genome of the *Streptococcus lactis* "subsp. *discetylactis*" bacteriophage P008 wide-spread in German cheese factories. System. Appl. Microbiol. *4*:413–423.

74. Lowrie. R. J. 1974. Lysogenic strains of group N lactic streptococci. Appl. Microbiol. *27*:210–217.

75. Luria, S. E. 1953. Host-induced modifications of viruses. Cold Spring Harbor Symp. Quant. Biol. *18*:237–244.

76. Lyttle, D., and G. B. Petersen. 1984a. Method for growth and purification of bacteriophage 643 on *Streptococcus lactis* ML3. Appl. Environ. Microbiol. *48*:242–244.

77. Lyttle, D., and G. B. Petersen. 1984b. The DNA of bacteriophage 643: isolation and properties of DNA of a bacteriophage infecting lactic streptococci. Virology *133*:403–415.

78. Marshall, R. J., and N. J. Berridge. 1976. Selection and some properties of phage-resistant starters for cheese-making. J. Dairy Res. *43*:449–458.

79. Marvin, D. A., and B. Hohn. 1969. Filamentous bacterial viruses. Bacteriol. Rev. *33*:172–209.

80. Mata, M. A., Trautwetter, G. Luthaud, and P. Ritzenthaler. 1986. Thirteen virulent and temperate bacteriophages of *Lactobacillus bulgaricus* and *Lactobacillus lactis* belong to a single DNA homology group. Appl. Environ. Microbiol. *52*:812–818.

81. McKay, L. L., and K. A. Baldwin. 1973. Induction of prophage in *Streptococcus lactis* c2 by ultraviolet irradiation. Appl. Microbiol. *25*:682–684.

82. McKay, L. L., and K. A. Baldwin. 1984. Conjugative 40-megadalton plasmid in *Streptococcus lactis* subsp. *diacetylactis* DRC3 is associated with resistance to nisin and bacteriophage. Appl. Environ. Microbiol. *47*:68–74.

83. Mermelstein, N. H. 1982. Advanced bulk starter medium improves fermentation processes. Food Technol. *36*:69–76.

84. Messelson, M., R. Yaun, and J. Heywood. 1972. Restriction and modification of DNA. Ann. Rev. Biochem. *41*:447–466.

85. Mullan, M. W. A. 1979. Lactic streptococcal bacteriophage enumeration. A review of factors affecting plaque formation. Dairy Ind. Int. *44*:11–14.

86. Mullan, M. W. A. 1985. Observations on the role of phage lysin in the life cycle of phage φC2(W) for *Streptococcus lactis* C2 and some properties of lysin-insensitive streptococci. Milchwissenschaft *40*:216–220.

87. Mullan, W. M. A., and R. J. M. Crawford. 1985. Factors affecting the lysis of group N streptococci by phage lysin. Milchwissenschaft *40*:342–345.

88. Mullan, W. M. A., and R. J. M. Crawford. 1985. Lysin production by φC2(W), a prolate phage for *Streptococcus lactis* C2. J. Dairy Res. *52*:113–121.

89. Mullan, W. M. A., and R. J. M. Crawford. 1985. Partial purification and some properties of φC2(W) lysin, a lytic enzyme produced by phage-infected cells of *Streptococcus lactis* C2. J. Dairy Res. *52*:123–138.

90. Mullan, W. M. A., C. Daly, and P. F. Fox. 1981. Effect of cheese-making temperatures on the interactions of lactic streptococci and their phages. J. Dairy Res. *48*:465–471.

91. Murata, A., and K. Kitagawa. 1973. Mechanism of inactivation of bacteriophage J1 by ascorbic acid. Agric. Biol. Chem. *37*:1145–1151.

92. Murata, A., K. Kitagawa, H. Inmaru, and R. Saruno. 1972. Inactivation of bacteriophages by thiol reducing agents. Agric. Biol. Chem. *36*:1065–1067.

93. Murata, A., K. Kitagawa, and R. Saruno. 1971. Inactivation of bacteriophages by ascorbic acid. Agric. Biol. Chem. *35*:294-296.

94. Murata, A., M. Odaka, and S. Mukuno. 1974. The bacteriophage-inactivating effect of basic amino acids; arginine, histidine, and lysine. Agric. Biol. Chem. *38*:477–478.

95. Nes. I. F., and O. Sorheim. 1984. Effect of infection of a bacteriophage in a starter culture during the production of salami dry sausage; a model study. J. Food Sci. *49*:337–340.

96. Ogata, S. 1980. Bacteriophage contamination of industrial processes. Biotech. Bioengineering *22* (Suppl. 1):177–193.

97. Ogata, S., and M. Hongo. 1979. Bacteriophages of the genus *Clostridium*. Adv. Appl. Microbiol. *25*:241–273.

98. Ogata, S., H. Suenaga, and S. Hayashida. 1982. Lysogenic phages and defective phage particles of antibiotic-producing actinomycetes, p. 174. Internat. Union Microbiol Soc. 13th Intern. Congr. Microbiol., Boston, MA, Abstr. P86:5.

99. Oki, T., Y. Iijima, and A. Ozaki. 1968. Bacteriophages of L-glutamic acid-producing bacteria. XII. Inhibition of *Brevibacterium* phages by non-ionic surface-active agents. Agric. Biol. Chem. *32*:329–339.

100. Oki, T., and A. Ozaki. 1968. Bacteriophages of L-glutamic acid-producing bacteria. XI. Selection of inhibitors for phage infection and suppression of phage adsorption by phytic acid. Agric. Biol. Chem. *32*:320–328.

101. Olsen, R. H., J. S. Siak, and R. H. Gray. 1974. Characteristics of PRD1, a plasmid-dependent broad host range DNA bacteriophage. J. Virol. *14*:689–699.

102. Oram, J. D., and B. Reiter. 1965. Phage-associated lysins affecting group N and group D streptococci. J. Gen. Microbiol. *40*:57–70.

103. Oram, J. D., and B. Reiter. 1968. The adsorption of phage to group N streptococci. The specificity of adsorption and the location of phage receptor substances in cell-wall and plasma-membrane fractions. J. Gen. Virol. *3*:103–119.

104. Peake, S. E., and G. Stanley. 1978. Partial characterization of a bacteriophage of *Lactobacillus bulgaricus* isolated from yogurt. J. Appl. Bacteriol. *44*:321–323.

105. Pearce, L. E. 1978. The effect of host-controlled modification on the replication rate of a lactic streptococcal bacteriophage. N. Z. J. Dairy Sci. Technol. *13*:166–171.

106. Potter, N. N., and F. E. Nelson. 1952. Effects of calcium on proliferation of lactic streptococcus bacteriophage. I. Studies on plaque formation with modified plating technique. J. Bacteriol. *64*:105–111.

107. Reanney, D. 1976. Extrachromosomal elements as possible agents of adaptation and development. Bacteriol. Rev. *40*:552–596.

108. Reddy, M. S. 1974. Development of cultural techniques for the study of *Streptococcus thermophilus* and *Lactobacillus* bacteriophages. Ph.D. Thesis, Iowa State University, Ames, Iowa.

109. Reinbold, G. W., and M. S. Reddy. 1973. Bacteriophages for *Streptococcus thermophilus*. Dairy Ind. *38*:413–416.

110. Reiter, B. 1949. Lysogenic strains of lactic streptococci. Nature (London) *164*:667–668.

111. Roginski, H., M. C. Broome, and M. W. Hickey. 1984a. Non-phage inhibition of group N streptococci in milk. 1. The incidence of inhibition in bulk milk. Aust. J. Dairy Technol. *39*:23–27.

112. Roginski, H., M. C. Broome, D. Hungerford, and M. W. Hickey. 1984b. Non-phage inhibition of group N streptococci in milk. 2. The effects of some inhibitory compounds. Aust. J. Dairy Technol. *39*:28–32.

113. Rudolf, V. 1978. Bacteriophages in fermentation. Process Biochem. *13*:16–26.

114. Sanders, M. E., and T. R. Klaenhammer. 1980. Restriction and modification in group N streptococci: Effect of heat on development of modified lytic bacteriophage. Appl. Environ. Microbiol. *40*:500–506.

115. Sanders, M. E., and T. R. Klaenhammer. 1981. Evidence for plasmid linkage of restriction and modification in *Streptococcus cremoris* KH. Appl. Environ. Microbiol. *42*:944–950.

116. Sanders, M. E., and T. R. Klaenhammer. 1983. Characterization of phage-sensitive mutants from a phage-insensitive strain of *Streptococcus lactis*: evidence for a plasmid determinant that prevents phage adsorption. Appl. Environ. Microbiol. *46*:1125–1133.

117. Sanders, M. E., and T. R. Klaenhammer. 1984. Phage resistance in a phage-insensitive strain of *Streptococcus lactis*: temperature-dependent phage development and host-controlled phage replication. Appl. Environ. Microbiol. *47*:979–985.

118. Sanders, M. E., P. J. Leonhard, W. D. Sing, and T. R. Klaenhammer. 1986. Conjugal strategy for construction of fast acid-producing, bacteriophage-resistant lactic streptococci for use in diary fermentations. Appl. Environ. Microbiol. *52*:1001–1007.

119. Sandine, W. E. 1977. New techniques in handling lacti cultures to enhance their performance. J. Dairy Sci. *60*:822–828.

120. Sandine, W. E., and J. W. Ayres. 1981. Method and starter compositions for the growth of acid producing bacteria and bacterial compositions produced thereby. U.S. Patent 4,282,255.

121. Saxelin, M. -L., E. -L. Nurmiaho-Lassila, V. T. Merilainen, and R. I. Forsen. 1986. Ultrastructure and host specificity of bacteriophages of *Streptococcus cremoris, Streptococcus lactis* subsp. *diacetylactis* and *Leuconostoc cremoris* from Finnish fermented milk "Viili." Appl. Environ. Microbiol. *52*:771–777.

122. Seto, S., T. Osawa, and S. Yamamoto. 1968. Bacteriophages attacking an L-glutamic acid-producing strain of *Microbacterium ammoniaphilum*. IV. Inhibition of P-phages by chemical agents. Agric. Biol. Chem. *32*:261–266.

123. Shimizu-Kadota, M., M. Kiwaki, H. Kirokawa, and N. Tsuchida. 1984. A temperate *Lactobacillus* phage converts into virulent phage possibly by transposition of the host sequence. Develop. Ind. Microbiol. *25*:151–159.

124. Shimizu-Kadota, M., M. Kiwaki, H. Hirokawa, and N. Tsuchida. 1985. 1SL*1*: a new transposable element in *Lactobacillus casei*. Mol. Gen. Genet. *200*:193–198.

125. Shimizu-Kadota, M., and T. Sakurai. 1982. Prophage curing in *Lactobacillus casei* by isolation of a thermoinducible mutant. Appl. Environ. Microbiol. *43*:1284–1287.

126. Shimizu-Kadota, M., T. Sakurai, and N. Tsuchida. 1983. Prophage origin of a virulent

phage appearing on fermentations of *Lactobacillus casei* S-1. Appl. Environ. Microbiol. *45*:669–674.

127. Shimizu-Kadota, M., and N. Tsuchida. 1984. Physical mapping of the virion on the prophage DNAs of a temperate *Lactobacillus* phage φFSW. J. Gen. Microbiol. *130*:423–430.

128. Shin, C. 1983. Some characteristics of *Leuconostoc cremoris* bacteriophage isolated from blue cheese. Jap. J. Zootech. Sci. *54*:481–486.

129. Sing, W. D., and T. R. Klaenhammer. 1986. Conjugal transfer of bacteriophage resistance determinants on pTR2030 into *Streptococcus cremoris* strains. Appl. Environ. Microbiol. *51*:1264–1271.

130. Smith, J. L., and S. A. Palumbo. 1983. Use of starter cultures in meats. J. Food Protection *46*:997–1006.

131. Sozzi, T. 1972. Etude sur l'exigence en calcium des phages des ferments lactiques. Milchwissenschaft *27*:503–507.

132. Sozzi, T., H. Hose, D. Amico, and F. Gnaegi. 1982. Bacteriophage of *Leuconostoc oenos*, p. 174. Internat. Union Microbiol. Soc., 13th Intern. Congr. Microbiol., Boston, Massachusetts, Abstr. P86:2.

133. Sozzi, T., R. Maret, and J. M. Poulin. 1976. Study of plating efficiency of bacteriophages of thermophilic lactic acid bacteria on different media. Appl. Environ. Microbiol. *32*:131–137.

134. Sozzi, T., J-M. Poulin, and R. Maret. 1978. Effect of incubation temperature on the development of lactic acid bacteria and their phages. J. Dairy Res. *45*:259–265.

135. Sozzi, T., K. Watanabe, K. Stetter, and M. Smiley. 1981. Bacteriophages of the genus *Lactobacillus*. Intervirology *16*:129–135.

136. Stathouders, J. 1975. Microbes in milk and dairy products. An ecological approach. Neth. Milk Dairy J. *29*:104–126.

137. Steenson, L. R., and T. R. Klaenhammer. 1985. *Streptococcus cremoris* M12R transconjugants carrying the conjugal plasmid pTR2030 are insensitive to attack by lytic bacteriophages. Appl. Environ. Microbiol. *50*:851–858.

138. Steenson, L. R., and T. R. Klaenhammer. 1986. Plasmid heterogeneity in *Streptococcus cremoris* M12R: effect on proteolytic activity and host-dependent phage replication. J. Dairy Sci. *69*:2227–2236.

139. Stent, G. S., and R. Calendar. 1978. Molecular Genetics. An introductory narrative. Freeman, San Francisco.

140. Stetter, K. O. 1977. Evidence for frequent lysogeny in lactobacilli: temperate bacteriophages within the subgenus *Streptobacterium*. J. Virol. *24*:685–689.

143. Susskind, M. M., and D. Botstein. 1978. Molecular genetics of bacteriophage P22. Microbiol. Rev. *42*:385–413.

142. Studier, F. W. 1975. Gene 0.3 of bacteriophage T7 acts to overcome the DNA restriction system of the host. J. Mol. Biol. *94*:283–295.

143. Susskind, M. M., and D. Botstein. 1978. Molecular genetics of bacteriophage P22. Microbiol. Rev. *42*:385–413.

144. Susskind, M. M., D. Bostein, and A. Wright. 1974. Superinfection exclusion by P22 prophage in lysogens of *Salmonella typhimurium*. III. Failure of superinfecting phage DNA to enter *sieA*+ lysogens. Virology *62*:350–366.

145. Tagg, J. R., A. S. Dajani, and L. W. Wannamaker. 1976. Bacteriocins of Gram-positive bacteria. Bacteriol. Rev. *40*:722–756.

146. Terzaghi, B. E. 1976. Morphologies and host sensitivities of lactic streptococcal phages from cheese factories. N. Z. J. Dairy Sci. Technol. *11*:155–163.

147. Terzaghi, B. E., and W. E. Sandine. 1981. Bacteriophage production following exposure of lactic streptococci to ultraviolet radiation. J. Gen. Microbiol. *122*:305–311.

148. Terzaghi, E. A., and B. E. Terzaghi. 1978. Effect of lactose concentration on the efficiency of plating of bacteriophages on *Streptococcus cremoris*. Appl. Environ. Microbiol. *35*:471–478.

149. Teuber, M., and J. Lembke. 1983. The bacteriophages of lactic acid bacteria with emphasis on genetic aspects of group N lactic streptococci. Antonie van Leeuwenhoek J. Microbiol. *49*:283–295.

150. Thomas, T. D., and R. J. Lowrie. 1975a. Starters and bacteriophages in lactic acid casein manufacture. I. Mixed strain starters. J. Milk Food Technol. *38*:269–274.

151. Thomas, T. D., and R. J. Lowrie. 1975b. Starters and bacteriophages in lactic acid casein manufacture. II. Development of a controlled starter system. J. Milk Food Technol. *38*:275–278.

152. Thunell, R. K., W. E. Sandine, and F. W. Bodyfelt. 1981. Phage-insensitive, multiple-strain starter approach to cheddar cheese making. J. Dairy Sci. *64*:2270–2277.

153. Tourville, D. R., and D. B. Johnstone. 1966. Lactic streptococcal phage-associated lysin. I. Lysis of heterologous lactic streptococci by a phage-induced lysin. J. Dairy Sci. *49*:158–162.

154. Tourville, D. R., and S. Tokuda. 1967. Lactic streptococcal phage-associated lysin. II. Purification and characterization. J. Dairy Sci. *50*:1019–1024.

155. Trautwetter, A., P. Ritzenthaler, T. Alatossava, and M. Mata-Gilsinger. 1986. Physical and genetic characterization of the *Lactobacillus lactis* bacteriophage LL-H. J. Virol. *59*:551–555.

156. Trevors, K. E., R. A. Holley, and A. G. Kempton. 1983. Isolation and characterization of a *Lactobacillus plantarum* bacteriophage isolated from a meat starter culture. J. Appl. Bacteriol. *54*:281–288.

157. Trevors, K. E., R. A. Holley, and A. G. Kempton. 1984. Effect of bacteriophage on the activity of lactic acid starter cultures used in the production of fermented sausage. J. Food Sci. *49*:650–653.

158. Uetake, H., S. E. Luria, and J. W. Burrows. 1958. Conversion of somatic antigens in *Salmonella* by phage infection leading to lysis or lysogeny. Virology 5:68–91.

159. Uetake, H., S. Toyama, and S. Hagiwara. 1964. On the mechanism of host induced modification. Multiplicity activation and thermolabile factor responsible for bacteriophage growth restriction. Virology 22:202–213.

160. Uno, T., and Y. Kumamoto. 1970. Prevention of bacteriolysis by bacteriophage in L-glutamic acid fermentation. Japanese Patent 70-00,944.

161. Vedamuthu, E. R., and C. Washam. 1983. Cheese, pp. 132–313. *In* G. Reed (ed.), Biotechnology, Vol. 5. Verlag Chemie, Weinheim, Fed. Rep. Ger.

162. Waes, G. M., and R. G. Bossuyt. 1984. Impedance measurements to detect bacteriophage problems in cheddar cheesemaking. J. Food Protect. *47*:349–351.

163. Watanabe, K., K. Ishibashi, Y. Nakashima, and T. Sakurai. 1984. A phage-resistant mutant of *Lactobacillus casei* which permits phage adsorption but not genome injection. J. Gen. Virol. *65*:981–986.

164. Watanabe, K., and S. Takesue. 1975. Use of L-rhamnose to study irreversible adsorption of bacteriophage PL-1 to a strain of *Lactobacillus casei*. J. Gen. Virol. *28*:29–35.

165. Watanabe, K., S. Takasue, and K. Ishibashi. 1979. Adenosine triphosphate content in *lactobacillus casei* and the blender-resistant phage-cell complex forming ability of cells on infection with PL-1 phage. J. Gen. Virol. *42*:27–36.

166. Watanabe, K. S. Takasue, and K. Ishibashi. 1980. DNA of phage PL-1 active against *Lactobacillus casei* ATCC 27092. Agric. Biol. Chem. *44*:452–455.

167. Whitehead, H. R., and G. A. Cox. 1936. Bacteriophage phenomenon in cultures of lactic streptococci. J. Dairy Res. *7*:55–62.

168. Wigley, R. C. 1980. Advances in technology of bulk starter production and cheesemaking. J. Soc. Dairy Technol. *33*:24–30.

169. Wolf, E., A. Lembke, and R. Deininger. 1983. Protection of microorganisms against bacteriophage attacks. U.S. Patent 4,409,245.

170. Yoshimoto, A., S. Nomura, and M. Hongo. 1970. Bacteriophages of *Bacillus natto*. IV. Natto plant pollution by bacteriophages. J. Fermentation Technol. *48*:660–668.

171. Zehren, V. L., and H. R. Whitehead. 1954. Growth characteristics of streptococcal phages in relation to cheese manufacture. J. Dairy Sci. *37*:209–219.

10

CYANOPHAGE ECOLOGY

ROBERT E. CANNON

Department of Biology, University of North Carolina at Greensboro, Greensboro, North Carolina 27412-5001

10.1. INTRODUCTION

Cyanophages are viruses that attack cyanobacteria. The first cyanophage was isolated by Robert Safferman and Mary Ellen Morris in 1963 (43). They designated this isolate LPP-1 corresponding to the first letter of the names of the three genera of cyanobacteria that the virus attacked, *Lyngbya, Phormidium,* and *Plectonema.* In 1963, the host microorganisms for the virus were known as blue-green algae; therefore Safferman and Morris (43) named the virus phycovirus. It is now clear that hosts for these viruses are indeed prokaryotic microorganisms, and hence the terms cyanobacteria and

245

cyanophages will be used throughout this chapter in place of the previously used terms, blue-green algae and phycoviruses.

A number of cyanophages have been isolated and characterized over the past 20 years for a variety of cyanobacterial hosts. Some of these viruses, such as LPP-1, infect a number of different hosts. This phenomenon clearly indicates that there is some confusion and controversy about cyanobacterial taxonomy. Early investigations on the distribution of LPP-1 cyanophages discussed below demonstrated that they were ubiquitous, and there was interest in exploring possible ecological roles for cyanophages. There was also hope that cyanophages might be useful agents for biological control of nuisance cyanobacterial blooms. At one point after the initial isolation of the viruses, there was some thought given to the idea that the rapid decline of cyanobacterial blooms in the aquatic environment might be attributable to cyanophage infections. Cyanophage N-1, infecting the filamentous, nitrogen-fixing cyanobacterium *Nostoc muscorum,* was isolated from Lake Mendota in Wisconsin after a cyanobacterial bloom, but there is no evidence to support the hypothesis that death of algae (or cyanobacteria) can be attributed to the infectious activity of this phage (1).

There is a considerable literature in the field of cyanophage biology. This literature reflects the research interests of the various investigators in the field. There are a number of reviews of cyanophages by Safferman (40), Padan and Shilo (37), Brown (6), and Sherman and Brown (55). There have been fewer review articles that have dealt exclusively with cyanophage ecology, although this topic was considered in some of the reviews (10,56). The major goal of this chapter is to review aspects of the field of cyanophage ecology. We discuss the effects of various environmental effects/factors upon cyanophages, their distribution and control, continuous culture studies, analysis of studies about cyanophages in wastewater and their possible use as pollution indicators, eukaryotic algal viruses, and the prospects for future ecological research on cyanophages. The chapter begins with a brief discussion of the biology of cyanophages to provide background for the ecological discussion.

10.2. BIOLOGY OF CYANOPHAGES

Safferman et al. (41) reported on the classification and nomenclature of cyanophages for the Bacterial Virus Subcommittee of the International Committee on Taxonomy of Viruses. The report focused upon those viruses that had been well characterized biologically in the literature. Cyanophages clearly resemble bacteriophages. They have isometric heads and long or short tails of varying complexity that are either contractile or noncontractile. All the cyanophages isolated and characterized to date contain double-stranded DNA and vary some in physical size, number of structural proteins, G + C content of nucleic acid, and so on. These viruses replicate in a

TABLE 10.1 Cyanophages and Their Host Genera

Cyanophage	Family	Host range genera	Filamentous or unicellular	Refs.
AS-1	*Myoviridae*	*Anacystis* *Synechococcus*	Unicellular Unicellular	Safferman et al. (42)
N-1	*Myoviridae*	*Nostoc* reclassified *Anabaena*	Filamentous	Adolph and Haselkorn (1)
S-1	*Styloviridae*	*Synechococcus*	Unicellular	Adolph and Haselkorn (2)
SM-2	*Styloviridae*	*Synechococcus* *Microcystis*	Unicellular Unicellular	Fox et al. (22)
LPP-1	*Podoviridae*	*Lyngbya* *Phormidium* *Plectonema*	Filamentous Filamentous Filamentous	Schneider et al. (49)
SM-1	*Podoviridae*	*Synechococcus* *Microcystis*	Unicellular Unicellular	Safferman et al. (48)
LPP-2	*Podoviridae*	same as LPP-1		Safferman et al. (47)
AC-1	*Podoviridae*	*Anacystis* *Chroococcus*	Unicellular Unicellular	Sharma et al. (53)

photosynthetic, prokaryotic host that contains chlorophyll a and releases oxygen. The photosynthetic apparatus of cyanobacteria generally appears as flattened sacs called thylakoids. Some cyanophages appear to replicate within the thylakoids causing displacement of the photosynthetic lamellae, whereas others replicate in the nucleoplasm of the cells, not causing a structural disorientation of the photosynthetic apparatus. The host ranges of the various cyanophages vary considerably, reflecting the status of cyanobacterial taxonomy, but cyanophages infect hosts that are either unicellular or filamentous, not both. Table 10.1 summarizes some of the major groups of cyanophages and their host genera.

As can be seen from Table 10.1 and from a brief survey of the literature there have been relatively few cyanophages isolated when one considers the large number of species of cyanobacteria. The search for more, new cyanophages must continue and expand, especially for viruses that infect hosts of economic or environmental relevance. A number of cyanophages infect hosts that fix atmospheric nitrogen, such as *Anabaena*.

The replication of cyanophages is very similar to that of bacteriophages. During the intracellular growth phase after infection, there is a latent period while the viruses are replicating and synthesizing both structural and nonstructural proteins. The replication sites are either the nucleoplasm of the cyanobacterial cell or the photosynthetic apparatus. It should be remembered that cyanobacteria generally grow much more slowly than bacteria. Some cyanophages such as LPP-1 replicate in about 8–10 hr, whereas some others take a day or longer to complete their lytic cycles. The final burst size also varies depending upon the cyanophage and host, but generally from 100 to 1000 plaque-forming units (PFU) per cell are produced during normal cyanophage infections (see Ref. 55 for more detailed discussion of cyanophage replication).

One of the exciting aspects of the discovery of cyanophages was the fact that they infected photosynthetic, prokaryotic hosts. There is a close relationship between photosynthesis and replication of some cyanophages. Cyanobacteria had been considered obligate photoautotrophs, but recent studies have demonstrated that some species are heterotrophic under certain conditions such as growth in the dark upon various exogenous carbon sources and in the presence of the photosynthetic inhibitor 3-(3,4-dichlorophenyl)-1,1-dimethylurea (DCMU) in the light (29,59). LPP-1 cyanophage infection stops carbon dioxide photoassimilation (23). Also, cyanophage burst size is reduced considerably when the virus is grown in the dark (37). AS-1 cyanophage that infects unicellular hosts depends absolutely upon photosynthesis for replication (54). It appears that if the host is capable of heterotrophic activity while photosynthesis is impaired, certain cyanophages can replicate. Other cyanophages that replicate solely in obligate photoautotrophic cyanobacteria have an absolute requirement for light, as do their hosts. Granhall and Hofsten (25) observed that the cyanophage infecting the filamentous nitrogen fixer *Anabaena variabilis* does not infect the heterocyst, although it lyses the vegetative cells.

LPP-1 is apparently a virulent cyanophage, as are most of the cyanophages that have been isolated to date. There have been reports also about temperate cyanophages (9,38). The possible environmental implications of lysogeny versus virulence are discussed below.

10.3. ENVIRONMENTAL EFFECTS UPON CYANOPHAGES

Before we discuss the ecological studies that deal with cyanophages, it is important to consider a few laboratory studies that involve potential environmental effects upon cyanophages. Safferman and Morris (45) noted that the LPP-1 cyanophage has an absolute requirement for magnesium ions for successful infection. LPP-2 cyanophage, which is serologically distinct from LPP-1, but has the identical host range, also has this requirement for magnesium ions (47). Other cyanophages that have been isolated since LPP-1 do not show this ionic requirement. Amla (4) reported on the effects of chelating agents such as EDTA and sodium citrate upon AS-1 cyanophage. He indicated that the cyanophage was subject to chelating agent shock when virus diluted in the chelating agent was rapidly diluted in distilled water.

Intracellular cyanophages were more resistant, probably reflecting protection by the photosynthetic apparatus of the host, *Anacystis nidulans*. Shock inactivation was enhanced by increasing the temperature of the water.

Because the hosts for cyanophages are photosynthetic organisms, it has been recognized that the cyanophages, under normal growth conditions, require light for replication. Generally, in most laboratory studies cultures were incubated under constant lighting conditions at an intensity of around 2500–3000 lux, using cool white fluorescent lights and at a room temperature of around 25°C (3). The pH range tolerated by cyanophages is quite wide, from pH 4 to 11, depending upon the cyanophage. Most are stable in the range of pH 7–11, which is the optimum for cultivation of the host cyanobacteria (17,37).

Cyanobacteria are cultivated in a variety of autotrophic media of salts and trace elements. Cultures are incubated either shaking or standing, and sometimes with an air–carbon dioxide mixture bubbled into the cultures under aseptic conditions. Dhaliwal and Dhaliwal (17) have reported that addition of nucleotides and nucleotides plus amino acids to the salt medium increased LPP-1 cyanophage yield. It is unlikely, however, that these enriched environmental conditions would ever be encountered in nature. Lindmark (30) has demonstrated that LPP-1 replication could be stimulated by the addition of a carbon dioxide–air mixture to the culture medium. In the laboratory, LPP-1 cyanophage seems to replicate best when the host cyanobacteria are in the logarithmic phase of growth. Because the host, for example, *Plectonema boryanum,* is a filamentous cyanobacterium, growth curves have been difficult to quantify. Padan and Shilo (36) isolated a short trichome mutant of *Plectonema boryanum* that supported the replication of LPP-1 cyanophage equal to growth observed on the wild-type host. Other cyanophages, such as AS-1, that infect unicellular hosts lend themselves to more precise quantitation of cyanophage replication and one-step growth curve experiments. Sherman (54) and Safferman et al. (42) reported on replication of AS-1 cyanophage. Infection of *Synechococcus* by Sherman's AS-1M strain of cyanophage at a high multiplicity of infection (MOI) resulted in a burst size of about 40 PFU/mL). At low MOI, the burst size increased to around 55 PFU/mL with a lytic cycle of approximately 12 hr. Safferman's study (42) of the original AS-1 virus gave comparable burst size with slightly longer lytic cycle on the same host. The differences in these results reflect the fact that cyanophage biologists have used different viral strains as well as slightly different cultural and environmental conditions, including a variety of inorganic salts media, under a variety of similar though not identical temperature and lighting regimes. An additional example of this phenomenon comes from some research reported by Currier and Wolk (14) with N-1 cyanophage that infects *Anabaena*. They were able to alter the efficiency of plating of the cyanophage by altering the temperature at which the host cyanobacterium was incubated. They reported increased cyanophage yield when they raised the temperature to 51°C. They also presented evidence that DNA restriction might play an important role in preventing some cyanophages from replicating in closely related hosts. Whether these temperature or restriction effects play a role in the actual environment where host and cyanophage interact is still open to question.

Another example of a possible environmental effect that might influence cyanophage activity comes from studies of lysogeny of *Plectonema boryanum* by LPP cyanophages (8,9,38). An LPP-1 cyanophage strain designated LPP-1D (Delaware strain) was isolated from a lysogenic strain of *Plectonema boryanum*. The lysogenic strain had been isolated initially from a clone of the cyanobacterium that had survived a massive infection by LPP-1 virus. The lysogenic strain was inducible by mitomycin C, but not by ultraviolet light. Growth of the lysogenic strain in anticyanophage serum effectively removed free virus from the culture fluid. Free cyanophages were reisolated following the removal of antiserum.

When standing or shaking cyanobacterial cultures were infected with LPP-1, LPP-2, or LPP-1D cyanophages, all three cyanophages behaved virulently, resulting in complete lysis. Continued incubation of the cultures resulted in the development of cyanophage-resistant strains. It was not until experiments were devised to "stress" the host cyanobacterial cultures prior to cyanophage infection that lysogenization could be accomplished in the laboratory. LPP-1D and LPP-2 but not LPP-1 were able to lysogenize, which led to the conclusion that LPP-1D and LPP-2 were temperate cyanophages whereas LPP-1 had only the capacity for virulence. The stressing agents used initially were antibiotics such as chloramphenicol and streptomycin, which are known to effect prokaryotic protein synthesis. Other substances such as heavy metals (copper sulfate and mercuric chloride) and nitrogen starvation also enhanced lysogenization by LPP-1D and LPP-2. The cyanobacteria that serve as hosts for LPP cyanophages are nonblooming. Whether nonblooming is caused by the activity of cyanophages is not known, but the ability to cause lysis or lysogenization by cyanophages in the aquatic environment may influence the extent of cyanobacterial growth in the environment. When nutrients are abundant for cyanobacterial growth, the cyanophages tend to carry out the lytic cycle, but if the host population declines because of low nutrients, pollution, or antibiotics in the environment, the cyanophages can lysogenize the host cyanobacteria. Lysogenization permits a low level of both cyanophage and cyanobacterium to survive in a potentially hostile environment until conditions improve. The above material was presented as a theoretical model and has not been tested experimentally in the field.

Padan et al. (38) isolated a lysogenic strain of *Plectonema* that liberated LPP-2 cyanophage. It was immune to superinfection by LPP-2, but not LPP-1 cyanophages, and uninducible by ultraviolet light, X-rays, or mitomycin C.

10.4. CYANOPHAGE DISTRIBUTION

Enrichment assays taking advantage of the lytic activity of cyanophages have been employed to study the distribution of cyanophages. These assays allow for a large number of cyanobacterial species to be surveyed. Volumes of water or soil from a sampling site are placed into flasks of actively growing cyanobacterial host. After a period of incubation, the cultures are observed for clearing, which indicates lysis. If lytic activity is seen, aliquots are removed and filtered through small pore size filters that retain bacteria and cyanobacteria. Furthermore, samples can be treated with chloroform, which kills prokaryotes but does not affect cyanophages, which have no lipid components. The samples can then be assayed by standard plaque assay techniques or can be spotted on lawns of cyanobacteria cultured in Petri dishes. Some investigators such as Padan and Shilo (35) have concentrated samples after filtration with polyethylene glycol before doing direct plaque assay or carrying out an enrichment step. It is important to sample a large enough volume of water or soil to ensure that cyanophages, if present, will have an opportunity to replicate. Therefore, anything less than 50–100 mL is probably too small a sample to find cyanophages.

Safferman and Morris (46) surveyed a number of waste stabilization ponds around the United States for LPP cyanophages. Ponds were chosen because the original cyanophage, LPP-1, was isolated from a waste stabilization pond in Indiana (43). All ponds surveyed in 1967 received only domestic waste, not industrial waste. Out of 12 ponds sampled, 11 were positive for LPP-1 virus. SM-1 cyanophage infecting *Synechococcus* was isolated from one of the ponds (48). Using direct plaque counting techniques of chloroform-treated samples, Safferman and Morris (46) observed high direct counts of LPP-1 in some of the ponds, such as one in Redfield, South Dakota with 270 PFU/mL. They observed some seasonal fluctuations in cyanophage incidence in a pond in Indiana where no virus was detected in the winter months, but ponds in California and Arkansas gave positive samples throughout the year.

Shane (50) carried out a fairly detailed distribution analysis for LPP cyanophages using both direct count and enrichment assay techniques. Cyanophages were detected in residential ponds, farm ponds, quarries, lakes, reservoirs, industrial storage treatment tanks, and mushroom compost drainage lagoons. Highest direct counts of LPP-1 were observed in wastewater oxidation ponds. Samples taken from rivers near the University of Delaware and rivers in Pennsylvania, Ohio, New York, Maryland, and Michigan also were positive for LPP-1 cyanophages, based upon viral neutralization by specific antiserum. One part of Shane's study consisted of frequent sampling of three rivers in Delaware: Christina, Red Clay, and White Clay. Cyanophages were consistently absent from the headwaters of these rivers, but were detected frequently as the rivers flowed through more populated areas that might be considered polluted by domestic and industrial wastes. The significance of this early distribution study that was done between 1966 and 1968 is discussed below.

Padan and Shilo (34,35) investigated cyanophage distribution in fish ponds in Israel. Samples were taken during algal (cyanobacterial) blooms and lysis accompanying fish kills and also during times when cyanobacterial densities were low. They observed direct counts of cyanophages up to 13,000 PFU/L. Just as Safferman and Morris (46) were unable to detect any cyanobacteria resembling *Plectonema boryanum* in their samples, Padan and Shilo (35) were unable to isolate any host for the LPP cyanophages that they isolated, although they did observe some cyanobacteria that were morphologically similar. They discussed the possibility that cyanophage and host are in dynamic equilibrium in the aquatic environment, and it would be very difficult to isolate the host under the same sort of enrichment conditions that permit isolation of the cyanophage.

Distributional studies have also been carried out in India by Singh (57) and by Daft et al. (15) in Scotland. Singh (50) detected LPP cyanophages in freshwater ponds, sewage, and rice fields. Also reported were the genera of cyanobacteria that were observed in samples that were taken from the locations where cyanophages were found. He observed *Phormidium* and *Plectonema* in pond samples, *Phormidium* and *Oscillatoria* in sewage, and *Phormidium, Plectonema, Aphanocapsa,* and *Anabaena* in rice fields. Whether these cyanobacteria were the hosts for the LPP viruses is open to question. Safferman and Morris (45) reported that 13 different species of *Lyngbya, Phormidium,* and *Plectonema* were susceptible to LPP-1 virus. Drouet (20) reclassified the hosts for LPP-1 into one genus and species, namely, *Schizothrix calcicola*. This reflects some of the confusion about cyanobacterial taxonomy and might help to explain some of the difficulty that investigators have had in their attempts to isolate host cyanobacteria from the same environments as cyanophages.

More recently Hu et al. in 1981 (27) reported on cyanophages that attack *Anabaena* and *Nostoc* isolated from sewage settling ponds in south-central Michigan. These cyanophages attack cyanobacteria that form heterocysts and fix atmospheric nitrogen. Some of the viruses were similar morphologically to LPP cyanophages whereas others were similar to long-tailed AS cyanophages. In addition to viral morphology, growth curves, and host range studies, they employed restriction digestion and *in situ* hybridization techniques to classify the cyanophages and to group them more precisely.

Not all cyanophages have been isolated from wastewater or lakes. O-1 cyanophage that infects *Oscillatoria chlorina* was isolated from the cooling towers of a thermal power station in India (61). They observed considerable fluctuation in the population density of *O. chlorina* in the cooling towers. When the cyanobacterium was transferred *in vitro,* mass lysis of the host occurred. The O-1 virus was isolated from these lysed preparations. The lytic cycle of this cyanophage was 14–15 hr, and resulted in considerable destruction of the filamentous cyanobacterial host.

One question that has been asked many times, but has not really been answered is "What is the significance of cyanophages in the environment with respect to control of the host density?" The LPP cyanophages attack nonblooming cyanobacteria; N-1 infects blooming genera as do some others. What is the actual ecological relationship between cyanophages and their hosts? Wiggins and Alexander (64) carried out some studies with bacteriophages and bacterial host to determine minimum host density for phage replication. Their results indicated that if host cell density fell below a threshold of 10^4 colony-forming units/mL, then bacteriophages would not influence the activity or number of the host bacteria. More research is needed to determine cyanobacterial density in both the presence and absence of cyanophages to learn more about the actual environmental role of cyanophages in the distribution of their respective hosts.

10.5. CYANOBACTERIAL CONTROL BY CYANOPHAGES

Copper sulfate and chlorine have been generally used to control algal and cyanobacterial growth. After the discovery of the LPP cyanophages, Safferman and Morris (44) carried out some studies with artificial ponds to determine the effect of cyanophage infection upon "blooms" of *Plectonema* growing in 112-L tanks. The cyanophage quickly killed the host and this was followed by a 3000-fold increase in cyanophage titer. Safferman and Morris (44) were enthusiastic about the possibility of using cyanophages as aquatic algicides because the LPP-1 virus was stable over wide temperature and pH ranges and had high selective toxicity.

Jackson and Sladecek (28) carried out an extensive field study with LPP-1 cyanophage to determine its cyanobactericidal capacity under "natural" conditions. They attempted to cultivate *Plectonema* in 5000-gallon tanks using the following media: trickling filter effluent from a wastewater treatment plant that processed only domestic wastes, primary effluent from the plant, tap water from the local water system, and a cyanobacterial medium. Forty different attempts to cultivate *Plectonema* failed, though other cyanobacteria grew in the tanks. LPP-1 cyanophages were detected regularly in the wastewater, both treated and raw. They were even found in the community water supply and in the soil around the wastewater treatment facility. The authors realized that their lack of success at cultivating the host cyanobacterium was due to the pervasiveness of the cyanophage in the environment. More discussion about cyanophages in wastewater follows.

Some of the above observations were confirmed in a study of aerosol release of cyanophages around activated sludge basins (7). LPP cyanophages were detected frequently downwind from basins that were mechanically aerated, whereas tanks that employed compressed air showed little aerosol release of LPP cyanophages. Just as cyanophages were released from the large, open, mechanically aerated tanks, so were coliforms. Cyanophages were aerosolized throughout the year and appeared to be more resistant to fluctuations in temperature and to effects of solar radiation than coliforms. Samples were taken for this study in two different ways: by portable bubble gas sampler and by using Petri dishes containing phosphate buffered water that were left open to the air for varying time intervals. If cyanophages and coliforms can be aerosolized at wastewater treatment facilities, pathogenic bacteria and viruses can be as well. Long-term exposure to wastewater aerosol could pose a serious health hazard to workers at a waste treatment facility or to anyone living downwind from the plant (7). In many sewage plants open aeration basins were covered to reduce aerosol release of material and to facilitate odor control.

Desjardins et al. (16) cultivated both *Plectonema* and *Anacystis* in approximately 9000-L outdoor ponds that were enriched by cyanobacterial medium. They were successful in their host cultivation, and soon after cyanophages LPP-1 or AS-1 were added to the ponds, cyanobacteria declined. They occasionally found the lytic activity in tanks that had not been inoculated with cyanophages. As in the previously mentioned studies, they attributed this to aerosol contamination of their open tanks.

Other microorganisms may be involved in the control of cyanobacteria in the aquatic environment. For example, Fallon and Brock in 1979 (21) reported on the lytic effect of bacteria and protozoa upon *Anacystis*. The microbes were isolated from samples taken from Lake Mendota in Wisconsin. They did not test for AS-1 cyanophage in their study though it is possible that it, too, could have been affecting the cyanobacteria. They proposed that high light intensity might have a more pronounced effect upon the cyanobacteria population than the lytic organisms, though they clearly correlated an increase in the density of lytic organisms with a decline in cyanobacteria.

Goryushin and Chaplinskaya (24) reported on the isolation of a virus that lysed a number of different species of *Microcystis*. These cyanophages were isolated from artificial reservoirs in the Ukraine. They attributed the demise of cyanobacterial blooms to cyanophage lytic activity.

In studies of the collapse of *Aphanizomenon flos-aquae* blooms during an aquaculture study at pothole lakes in Canada, Coulombe and Robinson (12) reported on virus-like particles for the cyanobacterium that might have been responsible for the decline of the cyanobacterial population. The virus-like particles were observed within the cyanobacterial cells, but there was no absolute documentation that they triggered the cyanobacterial collapse. Also, the authors presented some evidence to indicate that photooxidation and death caused by oxygen toxicity might also be involved in the collapse of the *Aphanizomenon* blooms.

Blooms of *Aphanizomenon, Microcystis,* and *Anabaena* were also observed to occur in flood control lakes that were built in eastern Nebraska.

These lakes received large quantities of nutrients such as nitrate and phosphate as runoff from animal feedlots. A virus designated SM-2 infecting *Synechococcus elongatus* was isolated from these lakes as were several cyanobacterial lytic agents (32). The lytic microorganisms were found to be strains of *Alcaligenes* and *Actinomycetales*. Whether the cyanophages or bacteria had any controlling effect upon cyanobacterial blooms was not presented, though one might speculate that in a complex eutrophic environment, both viruses and bacteria could be attacking (feeding upon) the cyanobacteria.

A final interesting proposal to use cyanophages to control cyanobacterial growth comes from Peelen (39). At that time, the Netherlands was planning to build a series of dams in the Delta area of the country where three rivers flow into the North Sea. They expected the dammed areas to stagnate and become eutrophic, as a result of enrichment by domestic and industrial wastes from the rivers. The purpose of the dams was to prevent lowland flooding. Previous experience with the building of dams indicated that cyanobacterial growth could be expected. Because the lakes that were planned to form behind the dams were going to be used for drinking water as well as industrial, agricultural, horticultural, and recreational uses, there was considerable concern about potential adverse effects to the aquatic environment by cyanobacterial blooms. Peelen (39) discussed the possibility of using cyanophages to control blooms rather than resorting to application of ferrichloride or herbicides to control blooming cyanobacteria. I have seen no reports of actual field tests using cyanophages to control blooming cyanobacteria in lakes in the Netherlands during the course of the construction of the flood control dams.

10.6. CONTINUOUS CULTURE STUDIES

Over the past decade, there have been three studies on the use of continuous cultures to study the interactions of cyanophages with their hosts. The first study was carried out by Cowlishaw and Mrsa (13), using what they termed to be a "quasi-continuous" culture system where their cyanobacterial cultures were diluted 1.5-fold with fresh medium every 24 hr. They performed their experiments using *Plectonema boryanum* and LPP-1 cyanophage. During a 3½ month period, they observed four cycles of cyanophage lytic activity, with the cultures reaching a steady state after the fourth period of lysis. Isolation and testing of the cyanobacteria and viruses during the lytic cycles revealed that virus-resistant host strains had appeared in the cultures. They were unable to detect if any lysogenic strains had evolved in their system, though they did not attempt to treat their derived "resistant" strains with any inducing agents such as mitomycin C. The cyanophages that evolved during the course of cultivation showed altered plaque morphology, but were antigenically similar to input virus and still were capable of infecting the parental, cyanobacterial strain. The cyanophage isolates retained the growth characteristics of the parental virus strain; no new virus evolved. Cyanophages were never completely washed out of the culture system. This fact indicated that a small number of resistant cells were being converted into sensitive cells that were then infected to ensure the presence of a small amount of cyanophages in the culture. The extended presence of both host and virus in the culture was very similar to the results presented by Horne (26) studying interactions between T-series bacteriophage and *E. coli*. After fluctuations of both host and parasite, a persistent infection was established.

In a second continuous culture study carried out by Cannon et al. (11), the long-term cultural interactions between *Plectonema boryanum* and LPP-1, a virulent cyanophage, and LPP-2 and LPP-1D, two temperate cyanophages were investigated. The *Plectonema* placed in the chemostat initially was sensitive to infection by the three cyanophages. LPP-1 infection resulted in three cycles of oscillation of both host and cyanophage before steady-state conditions were established. Cyanophages isolated from the peaks of lytic activity were shown to be LPP-1 that was unchanged from the input virus. Cyanobacteria isolated from the chemostat were always resistant to cyanophage infection. LPP-1D and LPP-2 infection of sensitive *Plectonema* in continuous culture resulted in a single cycle of lytic activity as opposed to the three cycles observed with LPP-1 cyanophage. The experiments with LPP-1 were carried out over a period of 90 days whereas those with LPP-1D and LPP-2 were performed over 60-day periods. As with LPP-1 cultures, virus-resistant cyanobacteria repopulated the chemostat after LPP-1D and LPP-2 infections. When a strain of *Plectonema boryanum* that was resistant to infection by the three cyanophages was infected separately by each cyanophage under continuous culture conditions after the cyanobacterium had reached a steady state, the cyanophages were washed out of the chemostat completely within approximately 10 days after inoculation into the chemostat. A lysogenic strain of *Plectonema* was also cultivated in the chemostat. This strain of cyanobacterium released a low level of LPP-1D cyanophage by spontaneous induction throughout the 50-day period of cultivation. When additional LPP-1D, LPP-2, or LPP-1 cyanophages were added to the chemostat, they were apparently washed out of the chemostat within approximately 5 days. Virus isolated from the chemostat showed only the presence of LPP-1D (which is serologically indistinguishable from LPP-1). Clones of cyanobacteria isolated from the chemostat were always found to be lysogenic.

The oscillations observed when cultivating sensitive cyanobacteria with the three cyanophages might be explained by the appearance of enough sensitive host to allow a fairly large burst of cyanophages until a steady-state relationship was reached between the host and the parasite. Possibly, host range changes to the cyanophages allow them to infect a "resistant" host a defined number of times until steady state is reached. The fact that the cyanophages were unable to infect the resistant or lysogenic strains of *Plectonema* is not surprising. The resistant strain under continuous culture con-

ditions did not evolve any sensitive cells to allow the cyanophages to replicate. Possibly, if the dilution rate of the chemostat was slowed to allow the
virus and host more time to interact, there might have been some
cyanophage replication observed. Cultivation of the lysogenic strain of
cyanobacterium resembled the late stages of the interaction of the sensitive
host with cyanophages where both host and parasite coexisted in the continuous culture environment. Cannon and Shane (8) have shown that a stress
upon the host can facilitate lysogenization by LPP-2 and LPP-1D
cyanophages. Possibly, if the cyanobacteria in the chemostat had been
stressed in some way such as by limiting nutrients or light, lysogeny would
have been detected after infection of sensitive cyanobacteria by LPP-1D and
LPP-2 cyanophages.

The third continuous culture investigation was carried out by Barnet et al.
(5). They studied interactions between *Plectonema* and LPP-DUN1, an LPP-
1 type cyanophage, and between *Aphanothece stagnina* and Aph-1
cyanophage. Infections by the two cyanophages in separate continuous cultures resulted in a series of oscillations of both host and virus until an
equilibrium was reached. Two different strains of *Plectonema* that were
resistant to cyanophage infection were detected. One host strain (designated
PR1) was resistant to wild-type LPP-DUN1, but was infected by a
cyanophage that was isolated later and was considered to be some kind of
host mutant. The other resistant strain of *Plectonema* (designated PR2) was
completely resistant to both wild-type and mutant cyanophage strains. PR2
was found to grow more slowly than PR1 or the wild-type host strain of
Plectonema. In a mixed population of both PR1 and PR2 in the absence of
cyanophages, the PR2 was eliminated. The authors attributed oscillations of
the host and virus to the appearance of mutant cyanobacteria and
cyanophages with the PR1 mutant arising first, followed by development of
the PR2 strain. Later, minor fluctuations in host and virus population densities were attributed to instability in the host–parasite relationship rather than
the development of new mutants of either host or parasite. As was discussed
previously, LPP cyanophages are ubiquitous and their hosts do not usually
predominate in the aquatic environment. Clearly more research is necessary
to try to understand the actual role of the cyanobacterial host in the maintenance of the viral population in the environment. Continuous culture experiments are valuable in the study of interactions between host and virus under
laboratory conditions, but they may be far removed from what actually
occurs in nature.

10.7. CYANOPHAGES AS POLLUTION AND DISINFECTION INDICATORS

A detailed study was carried out by Shane et al (51,52) that involved extensive samplings of the Christina River that flows through Pennsylvania and
Delaware. The 1971 study focused upon plankton distribution and water

quality. The river flows through rural areas until near the mouth, where it flows through progressively more industrialized areas. Low plankton counts were made from samples taken from the more pristine headwaters of the river. Chemical analyses revealed that carbon dioxide concentrations might have been the limiting factor in keeping plankton growth low. In one experiment to test the possibility that inhibitory substances or nutrient deficiency were preventing planktonic growth, samples from each station (headwaters to mouth of river) were filtered through membrane filters and inoculated with *Plectonema boryanum* that was resistant to cyanophage infection. All samples supported luxuriant growth of the cyanobacterium, which seemed to indicate that there were sufficient nutrients to support plankton in this particular aquatic environment. The 1972 study (51) dealt with surveying the Christina River for LPP cyanophages, coliforms, and chemical parameters, and attempting to isolate host cyanobacteria for the LPP viruses. The river flow through more industrialized areas was accompanied by increased organic loading and increased coliform counts. With respect to cyanophage distribution, the upper reaches of the river that were considered to be generally unpolluted with very low coliform counts were free of cyanophages of the LPP type. As the river became more polluted, cyanophages could be detected by both direct count and enrichment culture techniques. All attempts to isolate or even observe the hosts for LPP cyanophages were unsuccessful despite the fact that the river apparently had sufficient nutrients to support growth of a laboratory-cultured, virus-resistant strain of *Plectonema boryanum.*

During a cyanophage survey, a culture of *Plectonema* (prepared to assay for LPP cyanophages) was observed to lyse more slowly than usual. From this culture, a protozoan was isolated and a study was undertaken to determine its relationship to the interaction of LPP viruses with their hosts (10). The protozoan was identified as *Hartmanella glebae,* a small, free-living amoeba commonly found in soils. It was learned that this organism fed preferentially upon filamentous cyanobacteria rather than on yeasts, molds, or green algae. The amoeba also fed upon both Gram-negative and Gram-positive bacteria. The exciting observation was that the amoeba served as a reservoir for LPP-1 cyanophages, possibly serving as an additional vector or mechanism of dispersal for cyanophages in the aquatic environment. When cyanophage-infected filaments of *Plectonema* were consumed by *Hartmanella,* whole filaments were taken into the amoeba. Cyanobacteria were digested, but complete cyanophage particles were not. They apparently remained in the vacuole and were not used as a food source by the amoeba. Yet they retained their infectivity as free virus particles within the vacuole. Later, the cyanophages were released as part of normal cellular activities or upon cell death. There was no indication that the cyanophages were capable of autonomous replication in the amoeba without host cyanobacteria. It was noted that as nutrients were depleted,*Hartmanella* had the capacity to form cysts. Some preliminary studies revealed that cyanophages could survive within them.

Plectonema is a nonblooming cyanobacterium. Could both cyanophages and protozoa serve to regulate cyanobacterial density in the aquatic environment? Even in the environments where there are apparently sufficient nutrients to support cyanobacterial growth, they are extremely difficult to isolate. A hypothetical scenario might involve the activity of the protozoa and temperate cyanophages to control cell density. When conditions are optimal for cyanobacterial growth, they are lysed by cyanophages and readily consumed by amoebas as an important part of the food chain. If nutrients become limited or if the environment is stressed by pollutants, temperate cyanophages could hypothetically be carried to other areas where there might be more host to infect. Also, cysts of the amoeba could form under environmental stress with cyanophages trapped within them. When conditions improved, the cysts could germinate, releasing cyanophages in another environment where cyanobacterial host density could support both cyanophage replication and grazing by the protozoan.

Cyanophages are distributed ubiquitously, and are commonly isolated from wastewater. Also, it has been shown that LPP cyanophages are isolated easily from environments affected by organic pollution. These observations led to two studies on the feasibility of using LPP cyanophages as pollution indicator organisms (58,60). Both studies were carried out in a secondary wastewater treatment facility in Greensboro, North Carolina that processes mostly domestic waste. Cyanophages were isolated from sampling stations throughout the treatment plant including the chlorinated effluent. Both LPP-1 and LPP-2 cyanophages were detected in the samples. The cyanophages were detected using standard direct counting methods as well as an indirect most-probable number (MPN) method. The sensitivity of the MPN test was such that it could detect cyanophages at the level of 0.001 PFU/mL. When no cyanophages were detected in the influent to the plant, cyanophages were found in samples taken after the trickling filters. These filters probably serve as a reservoir for both cyanobacteria and cyanophages in the wastewater treatment plant. Chemical analyses of the wastewater for pH, BOD, and so on, as well as the temperature of the wastewater offered conditions that were suitable for cyanophage viability. Cyanophages but not coliforms were detected in the chlorinated effluent of the plant. Preliminary experiments revealed that LPP-1 cyanophage was more resistant to chlorine disinfection than *E. coli*.

The increased chlorine resistance of LPP-1 cyanophage led Stanley and Cannon (60) to carry out a more extensive study on cyanophages isolated from the wastewater treatment plant. Cyanophages were isolated in greater numbers during warmer months of the year than during colder months. The cyanophage that predominated in the wastewater was LPP-2 by a ratio of nine to one. This was determined by doing viral neutralizations with specific antisera directed against either LPP-1 or LPP-2 cyanophages. Chlorine inactivation experiments to determine if cyanophages isolated from wastewater were more resistant to disinfection than laboratory strains of LPP-1 and LPP-2 were also done. They revealed that wastewater isolates were considerably more resistant to chlorine disinfection than stock cyanophages following exposure to chlorine residual of 2.5 mg/L for 25 min. Cyanophages are easy to assay, and the method of detection is reliable and inexpensive. Though cyanophages are not fecal organisms, they are common in environments where fecal organisms are isolated as part of water and wastewater testing. Cyanophages are more resistant to chlorine disinfection than coliforms, and their resistance approximates that of pathogenic animal viruses. If coliforms and cyanophages are removed from wastewater, it is likely that animal viruses such as poliovirus have also been removed. Also, the cyanophage assay techniques are simpler and cheaper than assays for animal viruses, and could be instituted easily at wastewater treatment plants.

Another study of the cyanophage–host interaction for biological assay purposes was done by Mallison and Cannon (31). Pesticides may enter the aquatic environment and they may affect cyanobacteria and cyanophages alone as well as the host–virus interaction. The experiments were done using concentrations of pesticides that might occur in the aquatic environment. Phototoxic agents had inhibitory effects upon *Plectonema* using disk and spot assays for testing pesticide effects. Also, Aldicarb, a cholinesterase inhibitor and systemic insecticide, had an inhibitory effect on the cyanobacterium. Malathion and Isotox are examples of two agents that inhibited the plaque-forming ability of LPP-1, LPP-2, N-1 cyanophages, and T4 bacteriophage. Phototoxic pesticides like DCMU and Atrazine had a pronounced inhibitory effect upon the one-step growth of LPP-1 cyanophage. This was not a surprising observation based upon earlier studies like those of Allen and Hutchison (3). Pesticide-resistant strains of *Plectonema* were isolated after growth of the cyanobacterium in the presence of DCMU and atrazine, but studies have not been done on cyanophage replication in these strains. Cyanobacteria and cyanophages may be affected by pesticides although they are considered to be nontarget organisms. Because cyanobacteria are like higher plants, including them among organisms to bioassay routinely when testing new pesticides might be worthy of consideration. More research on the effects of pesticides upon the host–cyanophage interaction needs to be done.

10.8. VIRUSES OF EUKARYOTIC ALGAE

Viruses and virus-like particles (VLPs) have been observed in both marine and freshwater algae (18,55), but much of the research that has been done has been descriptive. There has been little ecological research on viruses of eukaryotic algae. Just as many fungi seem to carry viruses normally, with little or no pathogenic effect, so do algae. Little is known about the life cycles of these viruses, especially the occurrence of any lytic activity. Dodds and Cole (19) indicated that VLPs might be latent because they appear only after heat shock to the algal cultures. In this section we do not review exhaustively the biology of eukaryotic algal viruses (or VLPs), but note a few interesting ecological observations about these viruses.

Mayer and Taylor (33) investigated a virus that attacked a marine phytoflagellate, *Micromonas,* which differs from other algae in that it does not have a cell wall. A VLP for this alga was observed in an enrichment culture composed of a mixture of phytoplankton. The lysate was infective for unialgal cultures of *Micromonas* only. The virus isolated in these preparations caused lysis, passed through 0.22μm filters, and was morphologically similar to other known viruses. An interesting aspect to this virus was that a fusion between algal and viral surface components occurred to bring the virus into the cell. Neither viropexis nor phagocytosis were observed to occur. No ecological studies of this virus have been done though the host occasionally reached population levels of 1000–10,000 cells per liter under natural conditions.

Some viruses have been found in cultures where algae are growing in symbiosis with hydra and paramecia (63,63). These algae were similar to *Chlorella,* a unicellular green alga. When the algae were removed from their symbiotic hosts, the virus was induced to enter the lytic cycle. Under normal circumstances, they were latent. A plaque assay has been developed for the viruses using a *Chlorella* strain to make the algal lawn. A quantitative plaque assay should facilitate more extensive study of this host–virus relationship.

Algae and cyanobacteria are considered to be ancient organisms with respect to evolution. It seems likely that eukaryotic algae and their viruses (or VLPs) have interacted for millions of years. Therefore, it is not surprising that the viruses do not appear until the algal hosts are environmentally shocked. One would expect a fairly balanced host–virus relationship to have evolved. Could it be that the viruses and VLPs that have been observed do not cause any great effect upon their hosts, but are acting as casual passengers that have known each other for a long time? Much more research needs to be done to isolate and characterize viruses that attack algae, and to investigate potential ecological implications of these viruses.

10.9. FUTURE RESEARCH

Cyanobacteria are ancient organisms evolutionarily, and this fact may account for why so few cyanophages have been discovered. It might be that cyanophages exist as integrated prophages in many species of cyanobacteria in a very stable host–virus relationship. Lytic activity might be a rare event except under the most severe conditions. Unlike lysogenic bacteria that can be induced by ultraviolet light, lysogenic cyanobacteria are instead inducible by chemicals and temperature. An associated problem is that in order to detect cyanophage lytic activity, sensitive host strains must be available for plaque assay and/or enrichment cultivation.

The search for additional cyanophages infecting environmentally important hosts must continue. The viruses must be well characterized biologically, and the effects of virus infection upon photosynthesis and nitrogen fixation should be examined. LPP-1 and LPP-2 cyanophages require magnesium ions for infectivity, whereas other cyanophages retain their viability in distilled water. Other cyanophages may require other ions or combinations of ions that are not provided in normal cyanobacterial enrichment culture media. Attempts must be made to isolate the hosts for cyanophages from the environment, and to determine if other organisms such as amoebas may be dispersing cyanophages. Any organism that feeds upon cyanobacteria could be spreading cyanophages. Further continuous culture studies with other cyanophages and their hosts with varying conditions will give additional insights into the cyanophage–host relationship that might be applicable to natural ecosystems.

Research on cyanophages has not generated as large a following of investigators as has that on bacteriophages or animal viruses. The reasons for this are numerous including funding priorities, apparent relevance of the host and virus to man's problems, and so on. Cyanophages and cyanobacteria are interesting organisms from both the perspective of basic research and the view that the viruses are attacking hosts that provide oxygen to the biosphere and are important primary producers. Cyanobacteria and algae may become more economically important in the future, and we need to understand the viruses that might infect and kill them.

ACKNOWLEDGMENTS

I would like to dedicate this chapter to Miriam Shane, my advisor in graduate school, who got me started on research with cyanophages, and to Robert S. Safferman, who discovered cyanophages and has continued to be a strong advocate of cyanophage research.

My research has been supported by the Department of Biology and the Research Council of the University of North Carolina at Greensboro, Sigma Xi, and the North Carolina Board of Science and Technology.

REFERENCES

1. Adolph, K. W., and R. H. Haselkorn. 1971. Isolation and characterization of a virus infecting the blue-green alga, *Nostoc muscorum*. Virology *46*:200–208.

2. Adolph, K. W., and R. H. Haselkorn. 1973. Isolation and characterization of a virus infecting a blue-green alga of the genus *Synechococcus*. Virology *54*:230–236.

3. Allen, M. M., and F. Hutchison. 1978. Effect of some environmental factors on cyanophage AS-1 development in *Anacystis nidulans*. Arch. Microbiol. *110*:55–60.

4. Amla, D. V. 1981. Chelating agent shock of cyanophage AS-1 infecting unicellular blue-green algae. Ind. J. Exp. Biol. *19*:209–211.

5. Barnet, Y. M., M. J. Daft, and W. D. P. Stewart. 1981. Cyanobacteria—cyanophage interactions in continuous culture. J. Appl. Bacteriol. *51*:541–552.

6. Brown, R. M., Jr. 1972. Algal viruses. Adv. Virus Res. *17*:243–277.

7. Cannon, R. E. 1983. Aerosol release of cyanophages and coliforms from activated sludge basins. J. Water Pollut. Control Fed. *55*:1070–1074.

8. Cannon, R. E., and M. S. Shane. 1972. The effect of stress on protein synthesis in the establishment of lysogeny in *Plectonema boryanum*. Virology *49*:130–133.

9. Cannon, R. E., M. S. Shane, and V. N. Bush. 1971. Lysogeny of a blue-green alga, *Plectonema boryanum*. Virology *45*:149–153.

10. Cannon, R. E., M. S. Shane, and E. DeMichele. 1974. Ecology of blue-green algal viruses. J. Env. Eng. Div., Am. Soc. Civil Eng. *100*:1205–1211.

11. Cannon, R. E., M. S. Shane, and J. M. Whitaker. 1976. Interaction of *Plectonema boryanum* (Cyanophyceae) and the LPP-cyanophages in continuous culture. J Phycol. *12*:418–421.

12. Coulombe, A., and G. G. C. Robinson. 1981. Collapsing *Aphanizomenon flos-aquae* blooms: possible contributions of photo-oxidation, O_2 toxicity, and cyanophages. Can. J. Bot. *59*:1277–1284.

13. Cowlishaw, J., and M. Mrsa. 1975. Co-evolution of a virus-alga system. Appl. Microbiol. *29*:234–239.

14. Currier, T. C., and C. P. Wolk. 1979. Characteristics of *Anabaena variabilis* influencing plaque formation by cyanophage N-1. J. Bacteriol. *139*:88–92.

15. Daft, M. J., J. Begg, and W. D. P. Stewart. 1970. A virus of blue-green algae from freshwater habitats in Scotland. New Phytol. *69*:1029–1038.

16. Desjardins, P. R., M. B. Barkley, S. A. Swiecki, and S. N. West. 1978. Viral control of nuisance blue-green algae. Report #169, California Water Resources Center, University of California, Davis, California.

17. Dhaliwal, A. S., and G. K. Dhaliwal. 1970. Physiology of algae and the *in vivo* multiplication of algal virus. Advancing Front. Plant Sci. *24*:65–74.

18. Dodds, J. A. 1979. Viruses of marine algae. Experientia *35*:440–442.

19. Dodds, J. A., and A. Cole. 1980. Microscopy and biology of *Uronema gigas*, a filamentous eucaryotic green alga, and its associated tailed virus-like particle. Virology *100*:156–165.

20. Drouet, F. 1968. Revision of the classification of the *Oscillatoriaceae*. Philadelphia Academy of Natural Science, Monograph 15.

21. Fallon, R. D., and T. D. Brock. 1979. Lytic organisms and photooxidative effects: influence on blue-green algae (cyanobacteria) in Lake Mendota, Wisconsin. Appl. Environ. Microbiol. *38*:499–505.

22. Fox, J. A., S. J. Booth, and E. L. Martin. 1976. Cyanophage SM-2:A new blue-green algal virus. Virology *73*:557–560.

23. Ginsburg, D., E. Padan, and M. Shilo. 1968. Effects of cyanophage infection on CO_2 assimilation in *Plectonema boryanum*. J. Virol. *2*:695–701.

24. Goryushin, V. A., and S. M. Chaplinskaya. 1968. The discovery of viruses lysing blue-green algae in the Dneprovsk reservoirs. Tsvetenie Vody, Dumka, Kiev.

25. Granhall, V., and A. Hofsten. 1969. The ultrastructure of a cyanophage attack on *Anabaena variabilis*. Physiol. Plantarum *22*:713–722.

26. Horne, M. T. 1970. Coevolution of *Escherichia coli* and bacteriophages in chemostat culture. Science *168*:992–993.

27. Hu, N., T. A. Thiel, T. H. Giddings, Jr., and C. P. Wolk. 1981. New *Anabaena* and *Nostoc* cyanophages from sewage settling ponds. Virology *114*:236–246.

28. Jackson, D., and V. Sladecek. 1970. Algal viruses—eutrophication control potential. Yale Scientific Magazine *44*:16–21.

29. Kenyon, C. N., R. Rippka, and R. Y. Stanier. 1972. Fatty acid composition and physiological properties of some filamentous blue-green algae. Arch. Mikrobiol. *83*:216–236.

30. Lindmark, G. 1979. Effects of environmental stresses on the relationship between *Plectonema boryanum* and cyanophage LPP-1. Institute of Limnology, University of Lund, Lund, Sweden.

31. Mallison, S. M., and R. E. Cannon. 1984. Effects of pesticides on cyanobacterium *Plectonema boryanum* and cyanophage LPP-1. Appl. Environ. Microbiol. *47*:910–914.

32. Martin, E. L., J. E. Leach, and K. J. Kuo. 1978. Biological regulation of bloom-causing blue-green algae, pp. 62–67. *In* M. W. Lovtit and J. A. R. Miles (eds.), Microbial Ecology. Springer, New York.

33. Mayer, J. A., and F. J. R. Taylor. 1979. A virus which lyses the marine nanoflagellate *Micromonas pusilla*. Nature *281*:299–301.

34. Padan, E., and M. Shilo. 1967. Isolation of "cyanophages" from freshwater ponds and their interaction with *Plectonema boryanum*. Virology *32*:234–246.

35. Padan, E., and M. Shilo. 1969. Distribution of cyanophages in natural habitats. Verhandlung Internationale Verein Limnologie *17*:747–751.

36. Padan, E., and M. Shilo. 1969. Short trichome mutant of *Plectonema boryanum*. J. Bacteriol. *97*:975–976.

37. Padan, E., and M. Shilo. 1973. Cyanophages-viruses attacking blue-green algae. Bacteriol. Rev. *37*:343–370.

38. Padan, E., M. Shilo, and A. B. Oppenheim. 1972. Lysogeny of the blue-green alga *Plectonema boryanum* by LPP-2 - SP1 cyanophage. Virology *47*:525–526.

39. Peelen, R. 1969. Possibilities to prevent blue-green algal growth in the Delta region of the Netherlands. Verhandlung Internationale Verein Limnologie *17*:763–766.

40. Safferman, R. S. 1973. Phycoviruses, pp. 214–237. *In* N. G. Carr and B. A. Whitton (eds.), The Biology of the Blue-Green Algae. University of California Press, Berkeley, California.

41. Safferman, R. S., R. E. Cannon, P. R. Desjardins, B. V. Gromov, and R. Haselkorn. 1983. Classification and nomenclature of viruses of Cyanobacteria. Intervirology *19*:61–66.

42. Safferman, R. S., T. O. Diener, P. R. Desjardins, and M. E. Morris. 1972. Isolation and characterization of AS-1, a phycovirus infecting the blue-green algae *Anacystis nidulans* and *Synechococcus cedrorum*. Virology *47*:105–113.

43. Safferman, R. S., and M. E. Morris. 1963. Algal virus isolation. Science *140*:679–680.

44. Safferman, R. S., and M. E. Morris. 1964. Control of algae with viruses. J. Am. Water Works Assoc. *56*:1217–1224.

45. Safferman, R. S., and M. E. Morris. 1964. Growth characteristics of the blue-green algal virus LPP-1. J. Bacteriol. *88*:771–775.

46. Safferman, R. S., and M. E. Morris. 1967. Observations on the occurrence, distribution, and seasonal incidence of blue-green algal viruses. Appl. Microbiol. *5*:1219–1222.

47. Safferman, R. S., M. E. Morris, L. A. Sherman, and R. Haselkorn. 1969. Serological and electron microscopic characterization of a new group of blue-green algal viruses (LPP-2). Virology *39*:775–780.

48. Safferman, R. S., I. R. Schneider, R. L. Steere, M. E. Morris, and T. O. Diener. 1969. Phycovirus SM-1: a virus infecting unicellular blue-green algae. Virology *37*:386–395.

49. Schneider, I. R., T. O. Diener, and R. S. Safferman. 1964. Blue-green algal virus LPP-1: purification and partial characterization. Science *144*:1127–1130.

50. Shane, M. S. 1971. Distribution of blue-green algal viruses in various types of natural waters. Water Res. *5*:711–716.

51. Shane, M. S., R. E. Cannon, and E. DeMichele. 1972. Pollution effects on phycoviruses and host algae ecology. J. Water Pollut. Control Fed. *44*:2294–2302.

52. Shane, M. S., E. DeMichele, and R. Cannon. 1971. Water quality and plankton ecology in the Christina River, Delaware. Environ. Pollut. *2*:81–95.

53. Sharma, C. R., G. S. Venkataraman, and N. Prakash. 1977. Cyanophage AC-1 infecting the blue-green alga *Anacystis nidulans*. Curr. Sci. *46*:496–497.

54. Sherman, L. A. 1976. Infection of *Synechococcus cedrorum* by the cyanophage AS-1M. III. Cellular metabolism and phage development. Virology *71*:199–206.

55. Sherman, L. A., and R. M. Brown, Jr. 1978. Cyanophages and viruses of eukaryotic algae, pp. 145–234. *In* H. Fraenkel-Conrat and R. R. Wagner (eds.), Comprehensive Virology, Vol. 12. Plenum, New York.

56. Shilo, M. 1972. The ecology of cyanophages. Bamidgeh *24*:76–82.

57. Singh, P. K. 1973. Occurrence and distribution of cyanophages in ponds, sewage, and rice fields. Arch. Mikrobiol. *89*:169–172.

58. Smedberg, C. T., and R. E. Cannon. 1976. Cyanophage analysis as a biological pollution indicator-bacterial and viral. J. Water Pollut. Control Fed. *48*:2416–2426.

59. Stanier, R. Y. 1973. Autotrophy and heterotrophy in unicellular blue-green algae, pp. 501–518. *In* N. G. Carr and B. A. Whitton (eds.), The Biology of Blue-Green Algae. University of California Press, Berkeley, California.

60. Stanley, J. L., and R. E. Cannon. 1977. Serological typing and chlorination resistance of wastewater cyanophages. J. Water Pollut. Control Fed. *49*:1993–1999.

61. Trivedi, J. P., and P. P. Oza. 1979. A lytic agent for blue-green alga *Oscillatoria chlorina* from cooling towers. Comp. Physiol. Ecol. *4*:207–212.

62. VanEtten, J. L., R. H. Meints, D. Kuczmarski, D. E. Burbank, and K. Lee. 1982. Viruses of symbiotic *Chlorella*-like algae isolated from *Paramecium bursaria* and *Hydra viridis*. Proc. Natl. Acad. Sci. USA *79*:3867–3871.

63. VanEtten, J. L., D. E. Burbank, D. Kuczmarski, and R. H. Meints. 1983. Virus infection of culturable *Chlorella*-like algae and development of a plaque assay. Science *219*:994–996.

64. Wiggins, B. A., and M. Alexander. 1985. Minimum bacterial density for bacteriophage replication: implications for significance of bacteriophages in natural ecosystems. Appl. Environ. Microbiol. *49*:19–23.

11

METHODS IN PHAGE ECOLOGY

SAGAR M. GOYAL

Department of Veterinary Diagnostic Investigation,
College of Veterinary Medicine,
University of Minnesota, St. Paul, Minnesota 55108

11.1. INTRODUCTION

Ever since the demonstration of bacteriophages by Twort (67) and d'Herelle (10), numerous advances have been made in understanding the replication

267

and molecular biology of bacteriophages. Phage have been used as convenient diagnostic tools for typing of bacterial strains and have even been proposed as indicators of water pollution by human viruses (25,42,60). They are useful tools for examining fundamental biological processes and molecular mechanisms of DNA replication, transcription control, and genetic recombination. The study of phage ecology, however, has not received a great deal of attention and is still in its infancy. One of the reasons for this discrepancy is believed to be the lack of suitable methods for the study of phage ecology.

Phage ecology may be defined as the interaction of phage with the environment and with their host cells and the effects of such interactions on the composition and productivity of particular habitats (53). Bacteriophages have been detected in ecosystems that support bacterial multiplication, for example, bovine and sheep rumens, natural waters, cheesemaking vats, soil, feces, sediments, and sewage (19). It has been suggested that phage may play a role in the genetic and demographic dynamics of bacterial populations in the environment. Also, cross-infection with phage may result in genetic exchange between different strains of bacteria. To study the quantitative relationship between phage and its host, it is necessary to study the distribution (seasonal and spatial), fate, and survival of phage in the environment. To accomplish this, it is important to have methods available for the isolation, enumeration, and identification of phage in food, water, soil, sewage, and air.

Waters that contain coliform bacteria probably also contain coliphages. They are indeed commonly isolated from polluted waters in the littoral zone (39). Whereas raw sewage and treated effluent can be expected to contain a large number of phage (28,56), rivers and streams may have only a few (69). In most habitats, however, phage are present in such low numbers that they can not be detected by direct assay of the samples. This is especially true for aquatic environments unless it is grossly polluted. Therefore, an enrichment or concentration step is required for phage detection in unpolluted or lightly polluted waters.

Hauduroy (33) was the first to investigate the occurrence of phage in water. Since then a number of investigators have tried to assess the distribution and implications of phage found in the water environment (6,11,16,20,41,51). Hidaka (36) observed that phage active against enterobacteria could be isolated easily from polluted estuarine and coastal waters but not from waters that were remote from terrestrial contamination. Similarly, Brown et al. (8) were able to isolate *Arthrobacter* phage from river water only when these samples were concentrated first. Earlier, Guelin (32) had used enrichment methods to detect low levels of phage active against specific host bacteria.

Several methods have been advocated for the isolation of bacteriophages from aquatic habitats. These consist of direct culture, enrichment techniques (57), dehydration or precipitation by polyethylene glycol (49,50,68), adsorption to hydroxylapatite or polyelectrolyte PE-60 (51,63), and differential centrifugation (47). An ideal method should be rapid, inexpensive, and suitable for field use, and should result in good recoveries of a wide range of bacteriophages. In addition, it should be adaptable for use with large volumes of different water sources, for example, freshwater, estuarine water, and wastewater.

In direct culture procedure, the sample (unfiltered or filtered to remove bacteria) is spotted on lawns of several bacteria. If phage is present in the sample, it will lyse sensitive strain(s) of bacteria and produce a clear spot (plaque) on the bacterial lawn. The amount of sample assayed by this procedure is necessarily small. Therefore, this method is suitable when phage are present in high numbers, which is unlikely in most situations. For detection, isolation, and enumeration of low numbers of phage, therefore, procedures that can concentrate small numbers of phage from large volumes of aquatic milieu are required. The development of such procedures would be useful for determining the effect of environmental parameters (e.g., temperature, pH, organics) on phage distribution; seasonal and spatial distribution and survival of phage in the environment; the adequacy of water treatment processes by using phage as indicators; and the extent of fecal pollution in natural water sources. Because phage are extensively used in genetic engineering studies, concern exists over the possible escape of genetically manipulated phage in the environment. The concentration techniques can be useful in monitoring the water environment for the presence of such strains in nature.

Phage are present anywhere that their hosts are present. In natural ecosystems, phage may affect the population density and/or activity of their host bacteria. In order to elucidate the ecological role of bacteriophages, Furuse et al. (22) analyzed the numbers of coliphages and of coliphage strains in the feces of healthy and ill individuals. From successive assay of coliphages and their host, they found that higher titers of phage were present in samples from patients whereas lower titers of phage were observed in healthy subjects (see also Chapter 3). Also, samples from patients contained virulent phage whereas temperate phage were found in normal individuals. They attributed this to the use of antibiotic therapy in patients, which kills bacterial hosts. The density of host is also important in determining the extent of phage replication in nature. Reanney and Marsh (55) found high titers of phage in soil only if the conditions were suitable for the proliferation of the host. Similarly, Anderson (4) found that phage of *Salmonella typhi* would replicate in nature only when the bacterial concentration was $\geq 3 \times 10^4$ CFU/mL.

In summary, to study phage ecology, it is essential to know the geographical distribution of phage and the quantitative relationship of phage–host bacteria in natural ecosystems. The effect of environmental variables on the distribution and physiology of both phage and bacterial hosts should also be known. This can be accomplished only when suitable methods are available for the study of phage distribution.

11.2. CONCENTRATION METHODS

Ideally, a method should be able to concentrate and detect small numbers of phage from large volumes of water and other environmental milieus. Such methods are useful in investigating numbers, seasonal variation, and correlation of low levels of phage with extent of pollution. Some methods make use of enrichment procedures following which, the phage are enumerated and identified.

Loehr and Schwegler (44) developed a filtration method that could be used to detect and enumerate phage in low concentrations directly from solution without the need for prior concentration. The phage-containing solution was mixed with approximately 10^{10} host bacteria, the mixture was filtered through a sterile membrane filter (Millipore, Bedford, Massachusetts), the filter was removed and incubated for 24 hr, and the resultant plaques were counted. To avoid phage reproduction, all steps prior to filtration were completed in less than 45 min. Although this method was less precise than the soft-agar technique (3), it was able to detect and enumerate lower concentrations of phage. This method, however, is limited to smaller volumes. A discussion of some of the currently used methods follows.

11.2.1. Filter Adsorption–Elution

The consumption of food or water contaminated with small numbers of human enteric viruses can produce overt infection in human beings. Such low numbers of human viruses are difficult to detect by the direct assay procedures. A number of methods have therefore been developed for the concentration and detection of small numbers of human and animal viruses from water, wastewater, and food (26). The most commonly employed procedure is called viradel (*virus ad*sorption *el*ution; see Ref. 27), which consists of virus adsorption to filters with their subsequent elution in a small volume of buffer. For optimum adsorption of virus to negatively charged filters, the water sample is adjusted to pH 3.5 and to 0.001 M aluminum chloride. This conditioned water is then passed through virus-adsorbent filters (Figure 11.1 and 11.2). Under these conditions the viruses acquire a net positive charge, enabling them to adsorb to the negatively charged membrane filters (23). Adsorbed viruses are later eluted by using high pH buffer. An alkaline pH results in reversal of charge on the virus particle so that it is rapidly eluted from the filter. Although these negatively charged filters work very well with human and animal viruses, they may not be suitable for phage concentration because pH 3.5 may inactivate some bacteriophages (57). Also, the use of high pH buffers as virus eluents may result in additional inactivation (52).

Positively charged filters such as those described by Sobsey and Jones (62) and Sobsey and Glass (61) have a definite advantage over negatively charged filters as virus adsorbents because they eliminate the need for acid

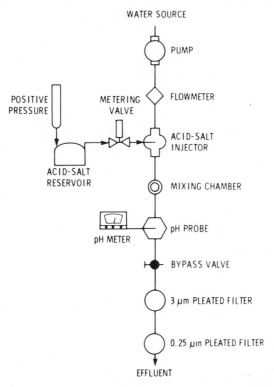

WATER SOURCE

PUMP

POSITIVE
PRESSURE

METERING
VALVE

FLOWMETER

ACID-SALT
INJECTOR

ACID-SALT
RESERVOIR

MIXING CHAMBER

pH PROBE

pH METER

BYPASS VALVE

3 μm PLEATED FILTER

0.25 μm PLEATED FILTER

EFFLUENT

Figure 11.1 Procedure for virus concentration from water. Reprinted from *Applied and Environmental Microbiology*, Vol. 35, No. 3, p. 543, 1978. Copyright by American Society for Microbiology.

or salt addition for virus adsorption. Goyal et al. (28) used positively charged Zeta Plus filters for the concentration of four different coliphages (MS-2, φX-174, T2, and T4) from tap water, sewage, and lake water. At neutral pH, these coliphages adsorbed efficiently to filters. Adsorbed viruses were eluted with a 4% beef extract solution containing 0.5 M sodium chloride at pH 9. Using this method, they were able to concentrate coliphages from 17-L volumes of tap water with recoveries ranging between 34 and 100%. Recoveries from raw and secondarily treated sewage were approximately 55%. Logan et al. (45), on the other hand, were able to recover 50–60% of coliphages from 65-L volumes of river water by using Zeta Plus filters.

The positively charged filters used in the above-mentioned studies were 2–3 mm thick and could not be pleated in cartridges for large-volume processing. Further improvements in this filter media led to the development of Virosorb filters, which can be made in thin sheets and can be pleated. Also, these filters possess a homogeneous porosity, thus eliminating the time-consuming necessity of determining which porosity filter is appropriate for a particular phage. These filters are composed of electropositive fiberglass/

I. Concentration:

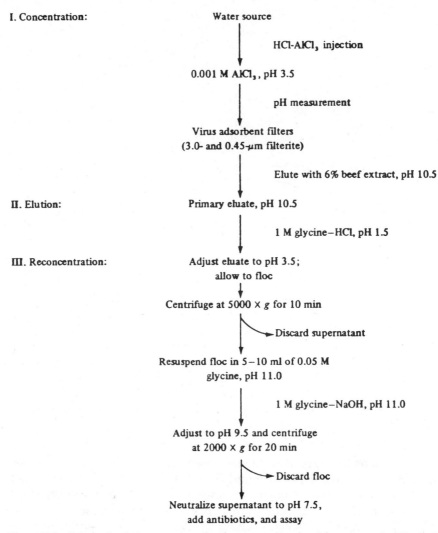

II. Elution:

III. Reconcentration:

Figure 11.2 Schematic of virus concentration from water Reprinted from *Journal of Virological Methods*, Vol. 7, p. 281, 1983. Copyright by Elsevier Science Publishers.

cellulose/surface-modified resin and may prove useful in further studies on phage concentration (59).

Singh and Gerba (60) described the use of charge-modified filter aids (AMF-Cuno, Meriden, Connecticut) to concentrate coliphages from water and wastewater. The filter aids are prepared from perlite, have large surface area per unit weight, and are available in fine, medium, and coarse grades. The filter aids are mainly used to facilitate filtration and retard clogging during filtration process. The electrokinetic adsorption by the filter aids is achieved by modifying the surface of the filter aid with resin to create a positive cationic charge on its surface. The positive charge of the filter aid assists in the electrokinetic capture of negatively charged contaminants present in suspension.

Singh and Gerba (60) evaluated the fine and coarse grades of filter aid for coliphage concentration. To achieve this, they used AP-20 prefilters as bases for overlaying the slurry of filter aid in 47-mm-diameter polypropylene filter housings. The desired quantity of filter aid was suspended in 30–50 mL of sterile distilled water, and the resultant slurry was passed through the filter housing with the aid of a syringe. Coliphages MS-2 and f2 were efficiently adsorbed from water and sewage and were subsequently eluted with 4% beef extract–0.5 M NaCl solution adjusted to pH 9.5. From 10–20 L samples of tap water, recoveries of 11–70% for T2 and 43–70% for MS-2 were reported. The authors claimed that this procedure was simple, could be conducted at ambient pH of water (7.4–7.6), and cost only a few cents per sampling. The fine grade of filter aid was found to be more efficient for virus adsorption probably because of its larger surface area per unit weight.

Seeley and Primrose (57) observed that methods used for concentrating human viruses from drinking water were not suitable for the concentration of bacteriophages from natural waters because the adsorption conditions in these methods are optimized for animal viruses, not phage. Additionally, elution conditions (high pH) resulted in phage inactivation. These authors therefore devised a procedure for phage concentration which consisted of pretreatment of water by passage through a sand filter and an anion-exchange resin (for removing dissolved organic material); adsorption of phage to fiberglass and cellulose nitrate filters by the addition of 5×10^{-4} M magnesium chloride and adjustment of pH to 3.8; elution of adsorbed phage with 3% beef extract solution (pH 8.5); and further concentration of the resulting eluate by ultrafiltration. Using this procedure, they were able to concentrate 4 L of water to 5 mL with recoveries of 18–80%. Although this method is attractive, it suffers from the disadvantages of being complicated, costly, and time consuming.

Primrose and Day (51) used hydroxylapatite (calcium phosphate) for phage concentration from 2-L volumes of distilled water. An AP-20 filter was placed in a Büchner funnel and was washed with 0.1% Tween 80 and distilled water to prevent virus adsorption to filter. A 20-mL slurry of hydroxylapatite was layered onto the filter and vacuum applied to produce a smooth bed. Phage-containing samples were then filtered through the calcium phosphate by application of a vacuum. The calcium phosphate was removed and resuspended in 10 mL of 0.8 M sodium phosphate buffer (pH 7.2). After centrifugation of this suspension, the supernatant was decanted and assayed for phage. Using this procedure, they were able to recover 33–90% selected phage from 2-L volumes of distilled water.

Logan et al. (45) evaluated positively charged Zeta Plus 60-S filters for recovery of bacteriophage from natural water. They found that a variety of phage adsorbed efficiently to these filters at water pH levels below neutrality, but that adsorption was reduced above pH 7. Based on these data, they devised a system that consisted of (*a*) prefiltration of water through a 10-inch cartridge depth filter of 5 μm porosity, (*b*) adjustment of water to pH 5.5–6.0, (*c*) adsorption of phage on to Zeta Plus 60-S filters, (*d*) elution of bound phage in small volume of arginine/1% beef extract solution (pH 9.0), and (*e*) secondary concentration of the resulting eluate by ultrafiltration. Using this procedure, phage in 65 L of river water were concentrated to 35 ml with recoveries in the range of 50–60%. Later, Logan et al. (46) scaled up this procedure to process larger volumes (up to 500 L) of water.

Recently, Shields and Farrah (59) described a procedure in which phage were adsorbed to positively charged Virosorb 1–MDS filters followed by their elution with a solution of 10% beef extract, pH 9. This eluate was further concentrated by the addition of two volumes of saturated ammonium sulfate to beef extract eluate. This resulted in the formation of an organic floc that was obtained by centrifugation and was suspended in a small volume of distilled water. Using this procedure, 2–3 L of lake water and sewage effluent was concentrated to 3–5 mL with an average phage recovery of > 70%.

In order to optimize the virus adsorption–elution technique, the following factors should be kept in mind:

1. Prefiltration of a sample may be necessary if it contains a large amount of suspended matter which may clog the virus-adsorbing filter, thus decreasing its efficiency. It should be realized, however, that most viruses have affinity with solids and may thus be removed by prefiltration of suspended matter. In such cases, both prefilter and the virus-adsorbing filter should be eluted.

2. The type of adsorbent filter used.

3. Effect of pH and ionic conditions on adsorption. Thus pH 3.8 (but not 3.5) was optimum for adsorption of MS-2 phage whereas pH 3.4 was best for P22 (57).

4. Flow rate.

5. Choice of eluent and pH. For example, phage T7 is very sensitive to pH 11.5, which is commonly used for elution.

6. Reconcentration procedure.

11.2.2. Flocculation

Bitton et al. (7) described a magnetic–organic flocculation method that is suitable for field use because it does not require centrifugation or filtration steps. This technique consists of addition of 500 mg/L of isoelectric casein and 150 mg/L of magnetite to a 2- or 4-liter sample of water or sewage. After 10 sec of stirring, the pH is adjusted to 4.5–4.6, which is the isoelectric point of casein. This results in the formation of a floc to which another 50 mg/L of magnetite is added. This additional magnetite forms a blanket on the top of the floc. After the floc is allowed to settle for 25–30 min, a magnet is held under the beaker and the supernatant fluid is poured off and discarded. The magnetic casein floc is solubilized in 0.5 M disodium hydrogen phosphate (pH 9.0). The final sample is then neutralized to pH 6.7–7.0 and assayed. Using this procedure, 68–100% of seeded coliphages were recovered from sewage effluents. It was suggested that this concentration technique could also be applied for algal viruses and perhaps animal viruses. The use of this method is limited, however, to heavily contaminated samples because of small volume (up to 4 L) concentrated.

11.2.3. Ultracentrifugation

Sensitivity of phage assays can also be increased by the application of centrifugal method of concentration and isolation of viruses. Using continuous flow centrifugation, Lammers (43) was able to isolate "aerophages" and coliphages against *Aerobacter aerogenes* and *E. coli,* respectively, from polluted stream water. A 20-L sample of water was subjected to continuous flow centrifugation at a centrifugal impulse (force × time) of 56,000 g × min. The pellet and the supernatant obtained were further processed as follows:

(*a*)   The pellet, containing organic particles >0.5 μm and clay minerals >0.2 μm, was resuspended and layered on a density gradient of 3,5-DP (methylglucamine salt of 3,5-diiodo-4-pyridone-*N*-acetic acid). This was recentrifuged at 5.6×10^4 g × min.

(*b*)   The supernatant was subjected to a second continuous flow centrifugation at 1.44×10^7 g × min. The pellet was suspended in pH 7.2 phosphate buffer containing 0.005 M magnesium sulfate.

Final samples from steps *a* and *b* were examined for phage by electron microscopy and plaquing. The procedure was complete within 30 hr of sampling. Although an efficient method, centrifugation is time-consuming and cumbersome.

11.3. ENRICHMENT PROCEDURES

In some instances, direct counts of phage for a specific host can result in phage isolation (37,65). Unless the water sample is grossly polluted, however, the direct method does not yield phage. Hilton and Stotzky (39) were unable to isolate coliphage from river water by concentrating the sample sixfold by ultracentrifugation or by direct plating. By using an enrichment procedure, however, they were able to detect phage in that sample. Guelin (32) was the first to propose enrichment methods of phage assay which may be used for detecting low levels of phage against specific host bacteria. Since then, a number of investigators have used enrichment techniques for studying the distribution of phage in the environment.

Spencer (64) described an indirect method for enrichment that was used by Hidaka (36) and Hidaka and Fujimura (38) for the isolation of phage from seawater. Sixty-eight strains of marine bacteria were first isolated from seawater. Subsequently, young cultures of bacterial isolates were mixed with 200-mL samples of filtered seawater. After overnight incubation, the cultures were centrifuged at 4500 × g for 30 min and filtered through a 0.45-μm filter. The filtrates were spotted on lawns of homologous bacteria grown on agar plates. Resultant plaques were picked and purified. The host range of these phage was determined by spotting phage lysates on overlays containing the various bacteria isolated. If the phage isolate lysed only the original host bacterium, it was considered as specific for that host.

Because the phage were isolated by an indirect method, Hidaka(36) considered the possibility that the phage isolates came not from the seawater but resulted because one of the 10 bacterial cultures added was lysogenic in nature. This possibility was, however, disproved by testing the bacterial strains for lysogenicity (36). Another problem encountered with the use of the enrichment procedure is that only one phage, usually the one that replicates the fastest or the one that predominates in the inoculum, is isolated (7,60). In addition, this technique is time-consuming, especially when studying phages of slow-growing hosts such as mycobacteriophages.

Casida and Liu (9) devised a unique soil enrichment technique for isolation of *Arthrobacter* phages. This technique does not involve the addition of potential host cells to the soil. A modification of this procedure was described by Germida and Casida (24) to study the interaction of *Arthrobacter* with its bacteriophage in nature (soil). After sieving (1.19 mm) the soil, 50 g was layered on a 0.5 cm deep glass wool plug at the base of a glass column (3 × 20 cm). Various nutrient solutions or water (50–60 mL) was then percolated over the soil. The percolate was filter-sterilized (0.3 μm-pore size) and assayed on various *Arthrobacter* species. By using a water percolate, no phage was detected. *A. globiformis* and some other phage were detected by using a nutrient broth percolate. Phage for *A. oxydans* was detected when a selective nicotine salt solution was used but not when nutrient broth was used.

If phage production in soil reflects the growth of indigenous host cells, the failure to detect a phage for a bacterium indicates their absence from that source. It is also possible that phage are present but that the nutrients percolated through the soil are not selective enough for growth (and phage production) of these bacteria. On the basis of these results, Germida and Casida (24) concluded that it was important to select percolation fluids that favor the growth of that particular strain of *Arthrobacter* for which phage are being sought.

To minimize the need for enrichment, Grant (30) used membrane filters to preselect simultaneously, yet independently, up to four mycobacteriophages

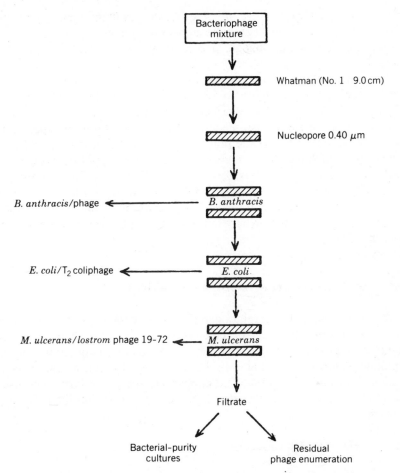

Figure 11.3 Sequential bacteriophage isolations from phage mixtures, for *Bacillus anthracis*, *Escherichia coli B*, and *Mycobacterium ulcerans*. Reprinted from *American Laboratory*, Vol. 5, No. 10, p. 18, 1973. Copyright 1973 by International Scientific Communications, Inc.

from any fluid substrate. In a later study, Grant (31) described a novel technique for sequential phage isolations from phage mixtures using a series of membrane filters. Host bacteria at the midpoint of the exponential phase were passed through a polycarbonate membrane filter (Nucleopore, Plesanton, CA) followed by the placement of another filter (Whatman) on the top of the bacterial layer with a sterile grid support to permit easier substrate flow. Up to three layers of potential host bacteria (bacterial sandwiches) can be prepared as a tier in a sleeve made of gauze and paraffin wax in a disposable container (Figure 11.3). The bacteriophage mixture was centrifuged to remove debris, filtered through a 0.4 μm filter to remove bacteria, and passed through the sandwiches while maintaining optimum pH and temperature. Viruses that adsorbed to different host bacteria were washed off and propagated in another fresh culture of bacteria. The resultant phage were filtered and characterized. This technique appears suitable for concentration of phage, isolation of new phage, and studying the kinetics of host–phage interaction.

To date all studies of heterogeneous populations of phage in natural systems suffer from the inability to determine total number of phage present. Plaque assay is no good until every host bacterium is used and conditions are optimized in each case. Also, each host should be absolutely specific for a given component of the phage population. These conditions are difficult to fulfill. The enrichment technique is useful for the rapid detection of low levels of a specific phage in water. The disadvantage is that only the phage that is active against a given host can be isolated. Also, the examination of enriched sample by electron microscopy (EM) yields qualitative data only and its efficiency lies somewhere between direct plating method and particle count by EM.

11.4. ENUMERATION AND IDENTIFICATION

To study the host–parasite relationship in the environment, it is necessary to detect the presence of phage in the environment, identify them, and make a quantitative estimate of their numbers. One difficulty with studies of phage populations, however, is that only those searched for in a specific assay can be identified, making it difficult to choose the species of host to be used in the assay. There are many methods for enumeration and identification of phage.

11.4.1. Electron Microscopy

Hidaka (36) used EM to classify bacteriophages that were isolated from seawater. Phage lysates were centrifuged at $37,000 \times g$ for 90 min and the pellet was resuspended in 1 mL of a 1% ammonium acetate solution. Equal amounts of this suspension were mixed with 2% phosphotungstic acid solution at pH 7.2 and applied as droplets directly to the 150-mesh carbon-coated collodion grids. Excess fluid was removed with filter paper, the grids were dried, and then they were examined by EM at a magnification of 50,000X.

Torrella and Morita (66) presented semiquantitative evidence of high incidence of phage in seawater of Yaquina Bay, Oregon. No specific hosts or enrichment was employed in this study. It should be realized that phage hosts remain unknown when phage counts are based exclusively on EM and that EM is of limited value in examining phage of uncharacteristic morphology, for example, actinophage. Also, positive staining with uranyl acetate may cause artifacts and make phage identification difficult (1).

Ackermann and Nguyen (2) observed that EM was suitable for the investigation of phage ecology because it is simple and rapid, and is useful for classification and ultrastructural studies on phage. As a routine, these authors enrich sewage with 35 strains of *E. coli* and study the sediments of enrichment cultures by electron microscopy. Briefly, 1-L sewage sample is centrifuged at 3000 × *g* for 15 min and 20 mL of sewage is mixed with 20 ml of double-strength trypticase soy broth. After it is seeded with 1 mL of a 3-hr old *E. coli* culture, it is agitated for 3 hr, filtered through a 0.45-μm filter, and titrated. Lysates and filtered sewage are centrifuged at 72,400 × *g* for 90 min and washed twice in 0.1 *M* ammonium acetate solution (pH 7.0). The sediment is deposited on grids, stained with phosphotungstic acid or uranyl acetate, and studied by EM.

The problem with EM is that it is a qualitative technique and does not detect all phage types present. Thus in most studies a single plaque is isolated, propagated, and examined by EM. Using the enrichment technique described above, however, Ackerman and Nguyen (2) detected up to 10 varieties of morphologically different phage in sediments of enriched cultures. All 35 cultures had phage but T-even phage predominated in 14 samples. Because most enrichment strains were polyvalent, EM was required for phage identification.

The advantages of examining enrichment cultures by EM is that it is a rapid technique and gives a more complete view of phage population as compared to isolation of single plaques. Also, phage hosts are known with a high degree of probability. The disadvantages are that it is a qualitative technique with limits of visualization; for example, if a sample contains $<10^5$ phage particles per milliliter, they are not visible by EM. In addition, the nature and number of indicator strains employed will affect the results.

Sharp (58) reviewed the methods for estimation of total numbers of plant and animal viruses employing EM for direct observation and identification of virus particles. He also developed a special centrifuge rotor that allowed sedimentation of viruses onto a collodion film on a glass cover slip. These collodion films were then mounted on specimen grids and examined by EM. Ewert and Paynter (19) described an adaptation of this procedure for counting total number of phage in complex aqueous systems, for example, sewage, bovine rumen fluid, and activated sludge mixed liquor. This procedure permitted quantitation of the total number of phage independent of bacterial host.

In this technique, 25 mL of mixed liquor is centrifuged in cold at low speed, the supernatant is diluted 10-fold with 1% (v/v) glycerol and then sedimented for 2 hr (3150 × *g*) onto a 2-mm agar block (4.5% w/v agar in water). The sedimented virus particles are removed from agar by stripping with a collodion film. The collodion film with attached virions (pseudoreplicas) is then floated off onto the surface of a 0.5% (w/v) uranyl acetate solution for 45–60 sec and a 200-mesh EM grid is placed on top of the floating membrane. The membrane-covered grid is removed from the surface of the staining solution, the pseudoreplica is carbon coated, and is examined by EM. Using this procedure, concentration of as low as 10^4 phage/mL can be assayed, as opposed to 10^6 required for sedimentation and 10^{10} for spray droplet or gel filtration assays. Also, this technique is capable of determining total phage concentration (both viable and nonviable).

11.4.2. Serology

For isolation and grouping of RNA phage, Furuse et al. (21) used the following procedure. The original phage sample is plated as such or after appropriate dilution to yield a suitable number of plaques (usually less than 20 plaques per plate) for the subsequent single-plaque isolation procedure. A phage is considered RNA phage if it lyses male indicator strains of *E. coli* (A/λ or Q13 or both) but does not lyse A/λ in the presence of 100 μg ribonuclease. The RNA phage are differentiated by serology but other phage are poorly identified, if at all. Serological testing of RNA phage has revealed two antigenic classes. Phage in one class are related to MS-2 phage and are widely distributed. The other class contains phage related to QB; these are less widely distributed. Most habitats, however, have been found to yield one or the other type of phage but not both. The method of phage identification based on serology, however, is cost prohibitive (12). In order to elucidate bacteriophage ecology Furuse et al. (20) and Osawa et al. (48) have determined the distribution of RNA coliphages in domestic sewage of several different countries because sewage is considered to constitute one of the natural habitats of RNA phage (see also Chapter 3). On the basis of their serological and physicochemical properties, Furuse et al. (20) divided RNA coliphages into four groups (I–IV), each with a unique distribution pattern. They found that sewage from mainland Japan yielded type II phage predominantly, whereas type III occurred in southern Japan and Southeast Asia. Sewage from abattoirs and feces of domestic and zoo animals yielded type I phage only but human feces contained types II and III. A similar habitat preference has been reported by Dhillon et al. (15), who isolated S13 type phages from the feces of pigs, whereas φX-174 type was isolated only from cows. Only 2 and 6% of human and animal feces, respectively, were found to contain phage (48).

11.4.3. Host Range

Most phage are differentiated by host range and plaque type. This is done by spot tests on indicator cultures by spotting the sample on overlays containing various bacterial strains.

Most phage are species-specific; some are strain-specific, but some of them may infect related strains of host bacteria (69). Because of their host specificity, a variety of hosts should be used in enumerating phage from a natural sample. Also, it is important to use hosts that are actively growing at an exponential rate because the presence of a large number of dead bacteria may lead to phage adsorption, giving a false negative or a low phage count.

The type of host bacteria used in phage assay is also important. For instance, if one looks for phage in sewage, the host should be of human fecal origin, be a consistent inhabitant of sewage, be present in large enough numbers, and be lysed by phage that do not replicate in any other host (39). It is also important to use host bacteria that are not lysogenic.

Enumeration of phage is simple, rapid, and easy. But phage detected by commonly employed *E. coli* hosts (B, C, K-12) are a heterogeneous group with different survival characteristics. It has been suggested that single-stranded RNA bacteriophages belonging to the morphological group E are a relatively homogeneous group which are capable of infecting bacteria that produce F or sex pili. Also, their shape and size are similar to those of human enteroviruses. The standard host strains for the enumeration of group E phage are F^+ or Hfr derivatives of *E. coli* K-12.

Although plaque assay is good for phage quantification, it does not give information concerning total number of phage present in a sample containing multiple types. It is important, therefore, to optimize environmental factors (pH, ionic strength, temperature, organic nutrients) for each phage–host system before plaque assay can be considered quantitative (3,4). It should be realized, however, that even if all sensitive hosts were available and were used as indicator organisms, the sum of individual phage types assayed would probably not represent the actual numbers in the system because each host would not necessarily be specific for a given phage type. Also, mutations in host and/or phage may lead to changes in host range specificities. Therefore, plaquing is not a good technique for enumeration of total number of phage in a complex system containing a mixture of virus types.

Most bacterial strains that have undergone prolonged culturing on laboratory media are excellent indicators for different types of phage present in sewage (11). This property, however, renders such bacteria unsuitable for the detection of specific types of phages. On the other hand, *E. coli* recently isolated from natural sources is very inefficient indicator for coliphages present in sewage. By genetic alteration of such bacteria it is possible to construct a suitable host.

Dhillon et al. (11) mutagenized four *E. coli* strains to obtain *lac*- mutants which were subsequently infected with an F'*lac*+ sex factor of *E. coli* K-12. Pairs of isogenic *lac*- and *lac*-/F'*lac*+ strains were used as indicators. These strains could be effectively used for a direct and almost selective enumeration of F-specific coliphages in sewage(13). Using these hosts, Dhillon et al. (11) found RNA-containing F-specific phage to be widespread in sewage (up to 5×10^3 PFU/mL of sewage). No single-stranded RNA-containing F-specific phage was detected, however.

According to Primrose et al. (53), easier and rapid estimation of the F-specific phage can be obtained by plating sample concentrates on a host mixture made up of Hfr strains of *E. coli* and *S. typhimurium*. Phage that form clear plaques on this mixture are either F-specific (80%) or polyvalent female (<20%) specific phage.

As all human enteroviruses can not be grown in a single cell line, there is no universal plating indicator for phage. Thus there is no way of enumerating all phage of a particular bacterial species. By using appropriate mixed indicators, however, it is possible to enumerate phage specific for a cell carrying a particular plasmid. Thus male-specific (F-specific) phage can not infect cells below 30°C because F pili are not synthesized below this temperature. Their isolation from water therefore indicates that they originated in the guts of warm-blooded animals. This observation can be used to (1) determine the extent of fecal pollution, and (2) quantify the rate of virus inactivation in water (53). By isopycnic centrifugation of virus concentrates coupled with the use of mixed indicators and agents such as monopalmitolein and RNAase, Primrose et al. (53) were able to enumerate all the different types of coliphage in river water.

Havelaar and Hogeboom (34) reported the construction and evaluation of an F^+ *Salmonella typhimurium* strain specifically designed for the enumeration of F-specific phages in fecally polluted waters. This strain was constructed by introduction of plasmid F′42 lac : : Tn 5 into *S. typhimurium*. These authors further stated that interference by somatic *Salmonella* phage can be ignored because in sewage, there are only a few phage that are capable of infecting F-Salmonella (usually <10 PFU/mL).

11.4.4. Phage Assay and Media

The double-agar-layer method of Adams (3) is widely used for coliphage detection. Recently, Havelaar and Hogeboom (34) described a single-agar-layer method that gave higher recoveries than obtained by the double-agar-layer method. Both these procedures are based on assaying 1-mL amounts. Kott (41) devised a most-probable number assay with a theoretical detection limit of 2 PFU/100 mL. In this procedure, 65-mL samples are tested and the results are prorated to determine coliphage counts in 100 mL. Grabow and Coubrough (29) recently described a direct single-agar-layer plaque assay for the direct detection of coliphages in 100-mL samples of water. In this procedure, a 100-mL sample of water, an agar medium containing divalent cations, and the host *E. coli* are mixed and then plated on 10 14-cm diameter Petri dishes. The theoretical minimum detection limit for this method is 1 PFU/100 mL.

The addition of certain organic and inorganic cofactors in phage isolation media is beneficial for phage–host cell adsorption/interaction. Anderson (5) found that coliphage T4 required 2×10^{-4} M free L-tryptophan for optimum attachment to cells. Puck et al. (54) found that the addition of magnesium sulfate accelerated phage adsorption. Similarly, Ca^{2+} has also been found to increase phage adsorption to their hosts (3).

All environmental samples can be expected to contain indigenous bacterial flora that may interfere with growth of bacterial host, development of phage plaques, and resolution of plaques on host bacterial lawns (40). It is essential, therefore, to eliminate or inactivate this bacterial flora in sample concentrates. Several procedures that have been tried for decontamination include:

(a)   Chloroform extraction: The sample is treated with chloroform to inactivate bacteria. However, all bacteria may not be sensitive to chloroform. Also, some phage may be inactivated by chloroform.

(b)   Filtration of sample through bacteria-retaining filters has been used. The disadvantage is that a certain portion of phage population may also be retained by these filters by the process of adsorption.

(c)   Addition of antibiotics in assay media: This may kill certain interfering organisms but may not be able to kill all flora. Also, the host bacteria used in the assay must be resistant to the antibiotic applied.

(d)   Selective media: The use of certain selective media may result in suppression of bacterial interference on direct assay plates and may obviate the need for bacterial decontamination procedures. Because phage are resistant to detergents (saponin, sodium dodecyl sulfate, sodium deoxycholate, Emulsol-607, Zephiran, and cetyl pyridinium chloride), the use of selective media may not have any deleterious effect on phage. Kennedy et al. (40) compared nonselective nutrient and lactose broths with selective media (EC broth, GN broth, nutrient broth with 0.05% sodium dodecyl sulfate, and lactose broth with 0.05% sodium dodecyl sulfate) and found that, with the exception of GN broth, direct assay of all samples with selective media generally resulted in significantly higher ($p < 0.05$) recoveries than did assays with nutrient broth only.

11.4.5. Phage Characterization

Criteria that can be used for characterization of phage isolates are described below (11.4.5.1–11.4.5.5).

11.4.5.1. Plaque Morphology.   Different phage produce different types of plaques. Virulent phage are known to produce plaques that are clear in the center, whereas temperate phage produce plaques that are turbid in center but more or less clear along the periphery. Some phage, such as coliphage HK243, produce uniformally opaque, white areas against amber growth of indicator lawn (17).

11.4.5.2. Lysotype.   Lysogenic characteristics of the host can also be used to characterize phage. Thus a prophage is classified as ultraviolet light inducible if the irradiated culture shows a >10-fold higher titer than the control (14). Prophage immunity, superinfection immunity, and whether a bacterium is a single or double lysogen can also be used for phage characterization (14,15).

11.4.5.3. Phage Morphology. Structure of head, tail, and end plate of a phage can be used for phage classification (see Chapter 2). The head may be hexagonal and the tail may be contractile or noncontractile. If contractile, the tail may be stretched or contracted. It may be long, thin, short, conical, or slightly curved. The end plate may have spikes or short lobate projections.

11.4.5.4. Determination of Host Range. To determine the host range of a phage isolate, it is spotted on lawns of several bacteria. Formation of plaques indicates the sensitivity of that particular host to phage.

11.4.5.5. Other Criteria. Several other properties that can be utilized for phage characterization are sensitivity to chloroform, pH, and temperature; storage time; type of one-step growth curve; latent period; burst size; nucleic acid content; phage virulence; specificity for original host; and the effect of environmental variables (e.g., salinity) on phage distribution.

11.5. CONCLUSIONS

To study phage ecology, it is essential to determine the distribution, fate, and survival of phage in the environment. This can be accomplished only if sensitive methods are available for isolation, enumeration, and identification of phage from environment. Some of the methods that have been used for this purpose include flocculation, ultracentrifugation, filter adsorption–elution, and enrichment techniques. To concentrate small numbers of phage from large volumes of water, the viradel procedure seems to be preferred by most investigators. Subsequent to concentration and/or enrichment, the phage can be enumerated and identified by several procedures, for example, electron microscopy, phage serology, host range, and other phage characteristics. The advantages and disadvantages of these procedures must be balanced.

REFERENCES

1. Ackermann, H.-W., P. Jolicoeur, and L. Berthiaume. 1974. Advantages et inconvenients de l'acetate d'uranyle en virologie comparee: etude de quarte bacteriophages caudes. Can. J. Microbiol. *20*:1093–1099.

2. Ackermann, H.-W., and T. M. Nguyen. 1983. Sewage coliphages studied by electron microscopy. Appl. Environ. Microbiol. *45*:1049–1059.

3. Adams, M. H. 1959. Bacteriophages. Interscience, New York.

4. Anderson, E. S. 1957. The relations of bacteriophages to bacterial ecology, pp. 189–217. *In* R. E. O. Williams and C. C. Spicer (eds.), Microbial Ecology. Cambridge University Press, Cambridge.

5. Anderson, T. F. 1948. The activation of the bacterial virus T4 by L-tryptophan. J. Bacteriol. *55*:637–649.

6. Bell, R. G. 1976. The limitations of the ratio of fecal coliforms to coliphages as a water pollution index. Water Res. *10*:745–748.

7. Bitton, G., L. T. Chang, S. R. Farrah, and K. Clifford. 1981. Recovery of coliphages from wastewater and polluted lake water by the magnetite-organic flocculation. Appl. Environ. Microbiol. *41*:93–97.

8. Brown, D. R., J. G. Holt, and P. A. Patter. 1978. Isolation and characterization of *Arthrobacter* bacteriophages and their application to typing of soil *Arthrobacter*. Appl. Environ. Microbiol. *35*:185–191.

9. Casida, L. E., Jr., and K.-C. Liu. 1974. *Arthrobacter globiformis* and its bacteriophage in soil. Appl. Microbiol. *28*:951–959.

10. d'Herelle, F. 1917. Sur un microbe invisible antagonistic des bacilles dysenterique. C. R. Acad. Sci. Paris. *165*:373–375.

11. Dhillon, T. S., Y. S. Chan, S. M. Sun, and W. S. Chau. 1970. Distribution of coliphages in Hong Kong sewage. Appl. Microbiol. *20*:187–191.

12. Dhillon, T. S., and E. K. S. Dhillon. 1972. Studies on bacteriophage distribution. II. Isolation and host range based classification of phages active on three species of *Enterobacteriaceae*. Jap. J. Microbiol. *16*:296–307.

13. Dhillon, E. K. S., and T. S. Dhillon. 1974. Synthesis of indicator strains and density of ribonucleic acid containing coliphages in sewage. Appl. Microbiol. *27*:640–647.

14. Dhillon, T. S., and E. K. S. Dhillon. 1976. Temperate coliphage HK022: Clear plaque mutants and preliminary vegetative map. Jap. J. Microbiol. *20*:385–396.

15. Dhillon, T. S., E. K. S. Dhillon, H. C. Chan, W. K. Li, and A. H. C. Tsang. 1976. Studies on bacteriophage distribution: Virulent and temperate bacteriophage content of mammalian feces. Appl. Environ. Microbiol. *32*:68–74.

16. Dhillon, E. K. S., T. S. Dhillon, Y. Y. Lam, and A. H. C. Tsang. 1980a. Temperate coliphages: classification and correlation with habitats. Appl. Environ. Microbiol. *39*:1046–1053.

17. Dhillon, T. S., E. K. S. Dhillon, S. Toyama, and S. Linn. 1980b. Coliphage HK243: Biological and physicochemical characteristics. Microbiol. Immunol. *24*:515–524.

18. Ewert, D., and M. J. B. Paynter. 1980a. Enumeration of bacteriophage and host bacteria in sewage and activated sludge treatment process. Appl. Environ. Microbiol. *39*:576–583.

19. Ewert, D. L., and M. J. B. Paynter. 1980b. Technique for determining total bacterial virus counts in complex aqueous systems. Appl. Environ. Microbiol. *39*:253–260.

20. Furuse, K., A. Ando, S. Osawa, and I. Watanabe. 1981. Distribution of ribonucleic acid coliphages in raw sewage from treatment plants in Japan. Appl. Environ. Microbiol. *41*:1139–1143.

21. Furuse, K., T. Aoi, T. Shiba, T. Sakurai, T. Miyake, and I. Watanabe. 1973. Isolation and grouping of RNA phages. IV. A survey in Japan. J. Keio Med. Soc. *50*:363–376.

22. Furuse, K., S. Osawa, J. Kawashiro, R. Tanaka, A. Ozawa, S. Sawamura, Y. Yanagawa, T. Nagao, and I. Watanabe. 1983. Bacteriophage distribution in human feces: continuous survey of healthy subjects and patients with internal and leukemic diseases. J. Gen. Virol. *64*:2039–2043.

23. Gerba, C. P., S. R. Farrah, S. M. Goyal, C. Wallis, and J. L. Melnick. 1978. Concentration of enteroviruses from large volumes of tapwater, treated sewage, and seawater. Appl. Environ. Microbiol. *35*:540–548.

24. Germida, J. J., and L. E. Casida. 1981. Isolation of *Arthrobacter* bacteriophages from soil. Appl. Environ. Microbiol. *41*:1389–1393.

25. Goyal, S. M. 1983. Indicators of viruses, pp. 211–230. *In* G. Berg (ed.), Viral Pollution of the Environment, Vol. 1. CRC Press, Boca Raton, Florida.

26. Goyal, S. M., and C. P. Gerba. 1982. Concentration of viruses from water by membrane filters, pp. 59–116. *In*: C. P. Gerba and S. M. Goyal (eds.), Methods in Environmental Virology. Dekker, New York.

27. Goyal, S. M., and C. P. Gerba. 1983. Viradel method for detection of rotavirus from seawater. J. Virol. Methods *7*:279–285.

28. Goyal, S. M., K. S. Zerda, and C. P. Gerba. 1980. Concentration of coliphages from large volumes of water and wastewater. Appl. Environ. Microbiol. *39*:85–91.

29. Grabow, W. O. K., and P. Coubrough. 1986. Practical direct plaque assay for coliphages in 100 ml samples of drinking water. Appl. Environ. Microbiol. *52*:430–433.

30. Grant, J. 1971. Selective mycobacteriophage isolation with membrane filters. Appl. Microbiol. *21*:1091–1096.

31. Grant, J. 1973. Membrane filter techniques for the study of bacteriophage isolation and replication kinetics. Am. Lab. *5*:17–23.

32. Guelin, A. 1952. Application de la recherche des bacteriophages a l'etude des eaux polluees. I. La survie des enterobacteriacees dans les eaux. II. Bacteriophages des eaux a grandes et petites plages. Ann. Inst. Pasteur, Paris. *82*:78–89.

33. Hauduroy, P. 1923. Recherches du bacteriophage d'Herelle dans differents milieux. C. R. Seances Soc. Biol. Filiales *88*:1084–1092.

34. Havelaar, A. H., and W. M. Hogeboom. 1983. Factors affecting the enumeration of coliphages in sewage and sewage-polluted waters. Antonie van Leeuwenhoek J. Microbiol. *49*:387–397.

35. Havelaar, A. H., and W. M. Hogeboom. 1984. A method for enumeration of male-specific bacteriophages in sewage. J. Appl. Bacteriol. *56*:439–447.

36. Hidaka, T. 1971. Isolation of marine bacteriophages from sea water. Bull. Jap. Soc. Sci. Fish. *37*:1199–1206.

37. Hidaka, T. 1977. Detection and isolation of marine bacteriophage systems in the southwestern part of the Pacific Ocean. Mem. Fac. Fish. Kagoshima Univ. *26*:55–62.

38. Hidaka, T., and T. Fujimura. 1971. A morphological study of marine bacteriophages. Mem. Fac. Fish. Kagoshima Univ. *20*:141–154.

39. Hilton, M. C., and Stotzky, G. 1973. Use of coliphages as indicators of water pollution. Can. J. Microbiol. *19*:747–751.

40. Kennedy, J. E., Jr., G. Bitton, and J. L. Oblinger. 1985. Comparison of selective media for assay of coliphages in sewage effluent and lake water. Appl. Environ. Microbiol. *49*:33–36.

41. Kott, Y. 1966. Estimation of low numbers of *Escherichia coli* bacteriophage by use of the most probable number method. Appl. Microbiol. *14*:141–144.

42. Kott, Y., N. Roze, S. Sperber, and N. Betzer. 1974. Bacteriophages as viral pollution indicators. Water Res. *8*:165–171.

43. Lammers, W. T. 1968. Bacteriophage populations in natural water. Union Carbide Corp. Nuclear Div., Oak Ridge Gaseous Diffusion Plant, Rep. No. K-1760.

44. Loehr, R. C., and D. T. Schwegler. 1965. Filtration method for bacteriophage detection. Appl. Microbiol. *13*:1005–1009.

45. Logan, K. B., G. E. Rees, N. D. Seeley, and S. B. Primrose, 1980. Rapid concentration of bacteriophages from large volumes of freshwater: Evaluation of positively charged microporous filters. J. Virol. Methods *1*:87–97.

46. Logan, K. B., G. E. Scott, and S. B. Primrose. 1981. A portable device for the rapid concentration of viruses from large volumes of natural freshwater. J. Virol. Methods *3*:241–249.

47. Metcalf, T. G. 1975. Evaluation of shellfish sanitary quality by indicators of sewage pollution. *In* A. L. H. Gameson (ed), Discharge of Sewage from Sea Outfalls. Pergamon, London.

48. Osawa, S., K. Furuse, M. S. Choi, A. Audo, T. Sakurai, and I. Watanabe. 1981. Distribution of ribonucleic acid coliphages in Korea. Appl. Environ. Microbiol. *41*:909–911.

49. Padan, E., and M. Shilo. 1969. Distribution of cyanophages in natural habitats. Verh. int. Ver. Limnol. *32*:745–751.

50. Padan, E., M. Shilo, and N. Kisley. 1967. Isolation of cyanophages from freshwater ponds and their interaction with *Plectonema boryanum*. Virology *32*:234–246.

51. Primrose, S. B., and M. Day. 1977. Rapid concentration of bacteriophages from aquatic habitats. J. Appl. Bacteriol. *42*:417–421.

52. Primrose, S. B., N. D. Seeley, and K. Logan. 1981. The recovery of viruses from water: methods and applications, pp. 211–231. *In* M. Goddard and M. Butler (eds.), Viruses and Wastewater Treatment. Pergamon, Oxford.

53. Primrose, S. B., N. D. Seeley, K. B. Logan, and J. W. Nicolson. 1982. Methods for studying aquatic bacteriphage ecology. Appl. Environ. Microbiol. *43*:694–701.

54. Puck, T. T., A. Garen, and J. Cline. 1951. Mechanism of virus attachment to host cells: role of ions in primary reaction. J. Exp. Med. *93*:65–88.

55. Reanney, D. C., and S. C. N. Marsh. 1973. The ecology of viruses attacking *Bacillus stearothermophilus* in soil. Soil Biol. biochem. *5*:399–408.

56. Scarpino, P. V. 1978. Bacteriophage indicators, pp. 201–227. *In* G. Berg (ed.), Indicators of Viruses in Water and Food. Ann Arbor Science, Ann Arbor, Michigan.

57. Seeley, N. D., and S. B. Primrose, 1979. Concentration of bacteriophages from natural waters. J. Appl. Bacteriol. *46*:103–116.

58. Sharp, D. G. 1960. Sedimentation counting of particles via electron microscopy, pp. 542–548. *In* Proceedings of the Fourth International Congress on Electron Microscopy, Berlin, 1958. Springer, Berlin.

59. Shields, P. A., and S. R. Farrah. 1986. Concentration of viruses in beef extract by flocculation with ammonium sulfate. Appl. Environ. Microbiol. *51*:211–213.

60. Singh, S. N., and C. P. Gerba. 1983. Concentration of coliphage from water and sewage with charge-modified filter aid. Appl. Environ. Microbiol. *45*:232–237.

61. Sobsey, M. D., and J. S. Glass. 1980. Poliovirus concentration from tapwater with electropositive adsorbent filters. Appl. Environ. Microbiol. *40*:201–210.

62. Sobsey, M. D., and B. L. Jones. 1979. Concentration of poliovirus from tapwater using positively charged microporous filters. Appl. Environ. Microbiol. *37*:588–595.

63. Sorber, C. A., B. A. Sagik, and J. F. Malina. 1971. Monitoring of low-level virus in natural waters. Appl. Microbiol. *22*:334–338.

64. Spencer, R. 1955. A marine bacteriophage. Nature *175*:690–691.

65. Spencer, R. 1963. Bacterial viruses in the sea, pp. 350–365. *In* C. H. Oppenheimer (ed.), Symposium on Marine Microbiology. Charles C Thomas, Springfield, Illinois.

66. Torrella, F., and R. Y. Morita. 1979. Evidence by electron micrographs for a high incidence of bacteriophage particles in the waters of Yaquina Bay, Oregon: ecological and taxonomical implications. Appl. Environ. Microbiol. *37*:774–778.

67. Twort, F. W. 1915. An investigation on the nature of ultramicroscopic viruses. Lancet *2*:1241–1243.

68. Yamamoto, K. R., B. M. Alberts, R. Benzinger, L. Lawthorne, and G. Treiber. 1970. Rapid bacteriophage sedimentation in the presence of polyethylene glycol and its application to large scale virus purification. Virology *40*:734–744.

69. Zaiss, U. 1981. Dispersal and fate of coliphages in the River Saar. Zbl. Bakt. Hyg., I. Abt. Org. B. *197*:160–173.

12

BACTERIOPHAGES IN FOODS

JAMES E. KENNEDY, JR.

American Home Food Products, Central Research Laboratory, Milton, Pennsylvania 17847

GABRIEL BITTON

Department of Environmental Engineering Sciences, University of Florida, Gainesville, Florida 32611

12.1. INTRODUCTION

The microbial flora found in foods can be quite diverse and reflect the unique history of that food—for example, its origin, nutritional and chemical composition, human or environmental contact during handling and processing, and physicochemical parameters associated with processing, packaging, and/or storage conditions. The initial microbial flora of fresh and/or perishable foods is generally heterogeneous and is derived from plant and animal sources. Other sources include soil, water, dust, food handlers and their tools or equipment, as well as the intestinal contents, hides, hooves, feathers, and feed of animals. Because bacteriophages are associated with most bacterial species and are found ubiquitously in the same environments as their bacterial hosts, bacteriophages associated with the bacterial flora of foods or environments related to food production may also be found in foods.

A number of bacteriophages and phage–host systems have been isolated from various foods. The consequences of phage–host interactions in the production of fermented foods have been recognized and studied for more than 50 years, but only recently have phage–host relationships among food spoilage bacteria and bacteria of public health or sanitary significance in foods been investigated. In this chapter we discuss the incidence of bacteriophage or phage–host systems in foods, the behavior or stability of phage with regard to food processing or storage conditions, and the significance of phage in relation to the bacterial ecology and the sanitary quality of foods.

12.2. BACTERIOPHAGE–HOST INTERACTIONS IN FOODS

Bacteriophages infecting numerous bacterial species that constitute the microflora of various foods as well as phage–host systems have been recovered from foods (Table 12.1). Bacteriophages can influence bacterial populations in foods or other environments directly by selective inhibition of susceptible bacterial populations by virulent phage and possible replacement by phage-resistant populations, or indirectly by lysogenization of bacteria and conference of phage-resistance or new phenotypic characteristics or transduction of new genetic properties. Valuable applications of phage–host interactions include potential inhibition of spoilage bacteria in refrigerated or perishable foods and development of phage-typing schemes for precise identification of food spoilage or pathogenic bacteria. Detrimental implications include inhibition of bacteria used for fermentation of foods and conferment of virulence factors among related bacteria via lysogenic conversion.

12.2.1. Bacteriophages Associated with Food Spoilage Bacteria

Members of the genus *Pseudomonas* are the most important and predominant bacterial group implicated in the low-temperature, aerobic spoilage of

TABLE 12.1 Summary of Bacteriophage Types Recovered from Foods[a]

Original Bacterial Host	No. of Host Strains	Source of Phage Isolation	Ref.
Pseudomonas aeruginosa	1	Fresh oysters, mussels	25
Pseudomonas fluorescens	1	Fresh meat	8
	16	Spoiled beef rib steaks	48
Pseudomonas fragi	12	Fresh pork sausage, ground beef, chicken, oysters, raw milk	145
	2	Spoiled beef rib steaks	48
Pseudomonas putida	5	Spoiled beef rib steaks	48
Pseudomonas putrefaciens	2	Spoiled fish filets	23,24
Pseudomonas syringae	1	Fresh meat	8,9
Pseudomonas spp.	7	Spoiled beef rib steaks	48
	27	Fresh pork sausage, ground beef, chicken, oysters, raw milk	145
Leuconostoc spp.	8	Fresh pork sausage, ground beef, chicken, oysters, raw milk	145
Brochothrix thermosphacta	14	Spoiled beef rib steaks	49
Staphylococcus aureus[b]	84	Poultry	116
	229	Poultry	44
Escherichia coli	3	Fresh oysters	135
	1	Fresh oysters	74
	1	Fresh chicken and pork, sausage, cooked turkey	68
	3	Fresh chicken, pork sausage, ground beef, oysters, lettuce and mushrooms; cooked luncheon meats; refrigerated biscuit dough, frozen chicken pot pie	70
Enterobacteriaceae[c]	3	Fresh pork sausage, chicken, ground beef, oysters, raw milk	145
Coliform bacteria[d]	12	Raw milk	78
Salmonella spp.[e]	10	Raw milk	78

[a] Does not include bacteriophages associated with fermented foods and/or bacteria used in food fermentations.

[b] Bacteriophages recovered from lysogenic *S. aureus* isolates derived from poultry.

[c] Differentiation between psychrotrophic isolates of *Enterobacteriaceae* used as hosts were not reported.

[d] Differentiation was not reported between the following bacterial hosts: *Bacterium coli commune* (*E. coli*) and *Bacterium lactis aerogenes* (*Enterobacter aerogenes*).

[e] Differentiation was not reported between the following bacterial hosts: "typhique and paratyphique group" (*Salmonella typhi* and *S. paratyphi*).

fresh meats, seafood, and eggs (6,39,61,114). Bacteriophages specific for various species and strains of *Pseudomonas* as well as phage–host systems involving *Pseudomonas* spp. have been isolated from various foods (see Table 12.1). Additionally, the psychrophilic nature of many phage of corresponding psychrophilic *Pseudomonas* hosts has been documented (23,24,48,50,91,92,144–146).

Because the identification of pseudomonads using conventional biochemical tests has often been difficult, there has been much interest in the development of phage sensitivity tests for characterization of this bacterial group. Billing (8,9) isolated pseudomonad phage from various sources, including fresh meat, diseased plants, soil, and manure, which were infective for pseudomonads isolated from fresh chicken, water, and soil, as well as for various phytopathogenic pseudomonads. Phage sensitivity patterns were developed for a large number of pseudomonads and found useful for identification in conjunction with other tests (8,9). Delisle and Levin (23,24) isolated 180 phage from marine water, sewage, and spoiled fish fillets which were infective for many psychrophilic pseudomonads associated with spoilage of refrigerated fish. Fish fillets were tested only for phage active against strains of *Pseudomonas putrefaciens*. These authors concluded that phage sensitivity patterns generally could not differentiate between marine and terrestrial pseudomonads and that they were not useful for identifying unknown pseudomonad isolates. However, some ecological specificity was noted between phage and hosts for *P. putrefaciens* (24), and distinct lysotypes were noted for most pseudomonad strains.

More definitive investigations of phage–host systems, that is, both the bacterial host and corresponding phage isolated from the same sample, in refrigerated foods undergoing spoilage have been conducted by Whitman (144), Whitman and Marshall (145,146), and Greer (48–51). Whitman and Marshall (145) isolated various phage–host systems from 22 of 45 refrigerated food samples including ground beef, pork sausage, chicken, oysters, and raw milk, but found no phage–host sytems in fresh egg whites or processed luncheon meats. Bacteriophages were recovered from 38 of 216 bacterial hosts that were obtained from food samples; these included isolates of *Pseudomonas, Enterobacteriaceae,* and *Leuconostoc.* A relationship was observed between phage recovery and psychrotrophic bacterial counts, with most phage being isolated in samples with high bacterial counts. Host-range studies showed that (*a*) most phage were specific for the bacteria upon which they were isolated, (*b*) cross-reactions between phage and related bacteria from different samples were infrequent, and (*c*) no cross-reactions occurred between *Pseudomonas* group I and group II (115), different genera, or different families. Phage–host systems involving pseudomonads and *Brochothrix thermosphacta* in spoiled beef have recently been investigated by Greer (48–51). A total of 38 homologous phages were recovered from 25 of 40 isolates of *Pseudomonas,* including strains of *P. fluorescens* and *P. putida* obtained from spoiled beef rib steaks (48). Each pseudomonad isolate tested from spoiled beef represented a distinctive lysotype based upon host–range studies. Similar strain distinctions were also reported for pseudomonads isolated from spoiled fish (23,24) and other refrigerated foods (145), as previously discussed. Greer (48) suggested that phage typing could provide a means of identifying specific pseudomonad strains and determining strain distribution during beef spoilage because the collective population of pseudomonads in refrigerated foods appears to consist of a variety of distinct lysotypes.

Greer (49) has also reported recovery of bacteriophages of *Brochothrix thermosphacta* from spoiled beefsteaks. *B. thermosphacta* has been recognized a predominant psychrotrophic spoilage bacteria in recent years, second only to *Pseudomonas* in frequency of isolation in meats (32,47). Strains of *B. thermosphacta* obtained from spoiled beef rib steaks were used as hosts for isolation of 21 psychrotrophic bacteriophages, with 13 of 15 bacterial hosts recovering phage (49). Although biochemical and cultural characteristics of all *B. thermosphacta* strains subjected to subsequent phage typing were identical, 14 distinct lysotypes were observed. *B. thermosphacta* phage were also species-specific; that is, no cross-reactions occurred with related genera such as *Microbacterium, Lactobacillus,* or *Listeria.* As with pseudomonads in spoiled beef, Greer (49) noted that a large number of distinctive lysotypes constituted the population of *B. thermosphacta* in spoiled beef and suggested that phage typing would be used to detect sources of *B. thermosphacta* contamination and determine strain distribution during spoilage or processing of foods.

12.2.2. Characteristics of Psychrotrophic Phage Isolated from Foods

Before we address the effects of phage–host interactions in refrigerated foods, it would be useful to discuss the characteristics of some psychrophilic phage that have been isolated from these foods. Phage capable of lysing numerous strains of *Pseudomonas* and *B. thermosphacta* at temperatures below 7°C have been isolated from various foods (23,24,48,49,144–146). Phage of *P. putrefaciens* (24), *Pseudomonas fragi* (146), and *B. thermosphacta* (49) produced plaques on respective host lawns at 1–2°C. The size of plaques were inversely related to incubation temperature between 1 and 25°C for all three phage–host systems. Phage of *P. fragi* would not form plaques at 32°C despite the growth of their respective hosts at the temperature (146).

Two psychrotrophic phages of *P. fragi* isolated from ground beef were further characterized and found to differ with regard to morphology, physiology, and stability (146). One *P. fragi* phage was classified morphologically as Bradley's (10) group A and the other as group C. Phage of *B. thermosphacta* could not be precisely categorized but had characteristics of both Bradley's groups A and B (49). Burst size and latent period of *P. fragi* phage at 7°C ranged from 100 to 500 PFU per host cell and from 2 to 4 hr, respectively (146). Inactivation rates of *P. fragi* phage in broth at 60°C varied from greater than 99% reduction in 1 min to 39% reduction in 30 min, whereas both phage were resistant to freezing in broth for 24 hr at −20°C. Exposure to pH 4 buffer for 1 hr resulted in 30 and 99% inactivation of respective *P. fragi* phages. The resistance of bacteriophages to inactivation by various physicochemical factors relevant to food systems are discussed in a broader context below.

12.2.3. Effect of Bacteriophage on Bacterial Flora in Foods

The isolation of bacteriophages infective for numerous strains of food spoilage bacteria has generated much interest with regard to the potential influence of these phage on host populations during food spoilage (48,50,51,144). By inhibiting the growth of certain strains of spoilage bacteria during refrigerated storage of foods, it has been suggested that psychrotropic phage could alter the spoilage characteristics and/or shelf life of refrigerated foods as well as provide a unique method for biological control of spoilage of certain foods such as fresh beef (50,51).

In general, the influence a virulent phage exerts on a susceptible bacterial population in a given environment depends on (*a*) suitability of the environment, for example, temperature, nutrients, pH, and osmolarity, for growth or appropriate physiological activity of host bacteria as well as for phage adsorption and multiplication; and (*b*) reasonably high probability of phage–host contact, for example, sufficiently high concentration of both phage and host as well as the appropriate medium for movement or contact of phage and hosts. The net effects of phage–host interaction may also be influenced by the rate of mutation of bacteria to phage-resistant strains as well as the mutation rate of phage to phenotypes virulent for bacteria resistant to the parent phage (3).

The ability of phage to inhibit the growth of susceptible spoilage or starter culture bacteria has been readily demonstrated in various liquid growth media (9,48,86,132,144), whereas corresponding phage–host interactions have rarely been observed or investigated in food systems such as fresh meat (51,84,132,144). Likewise, no apparent multiplication of coliphages or effect of coliphages on *Escherichia coli* or coliform populations in sewage or wastewater has been reported despite their ability to inhibit growth of specific hosts in laboratory cultures (27,98,140). Net proliferation of total phage in the activated sludge process has been demonstrated by direct electroscopic counts in conjunction with plaque assays, although the predominant bacterial populations affected could not be determined (35). Homologous phage significantly inhibited the growth of *P. fragi* in skim milk but not in sterile ground beef when inoculated with 10^3 host cells/g at phage–host ratios of 1–1 or 0.01–1 during 72 hr incubation at 7°C (144). Greer (48) investigated the effect of five different phage on homologous *Pseudomonas* hosts, for example, four strains of *P. fluorescens* and one strain of *P. putida,* inoculated into tryptic soy broth and aqueous extract of beef muscle at initial levels of 10^{10} and 10^7 phage and bacteria per milliliter, respectively, and incubated at 7°C. The growth of *Pseudomonas* hosts was significantly inhibited, with bacterial lag phases extended by 2.5–5.6 days. After 8 days of incubation, bacteria isolated from phage–host mixtures were generally phage resistant, thus indicating the selection of phage-resistant mutants. A more recent study (51) demonstrated that homologous bacteriophages could proliferate on the surface of beefsteaks and inhibit the growth of a spoilage pseudomonad inoculated onto the steaks. The shelf life of the inoculated steaks was significantly increased using a high titer phage lysate (10^8 PFU/ mL). Recent studies on phage–host systems involving bacteriophages of *Lactobacillus plantarum,* a bacterium used as a starter culture in the production of various fermented sausages, may also be relevant to phage–host interactions in fresh meats (86,132). The growth and lactic acid production of *L. plantarum* in liquid medium and a fermenting sausage product were significantly delayed by phage B2 (86). At initial levels of 1.5×10^6 and 10^6 phage and host bacteria per gram, respectively, growth of *L. plantarum* in sausages was delayed for 3–4 days with subsequent growth of phage-resistant bacteria following a 95% initial inactivation of parent bacteria. Other workers reported that phage fri of *L. plantarum* did not affect lactic acid production of *L. plantarum* in a fermenting sausage product inoculated with approximately 10^9 phage and 10^8 bacteria per gram (132). However, phage fri did inhibit growth and lactic acid production of *L. plantarum* in liquid cultures. It has been reported that both phage and host concentrations must exceed 10^6 organisms per milliliter in natural environments to ensure necessary physical contact between phage and host for phage to exert a selective effect on host populations (3). This observation may help explain the apparent failure of phage to affect susceptible *Pseudomonas* in ground beef inoculated at levels of 10^3 phage and host per gram (144). The extent of phage–host interactions among indigenous phage and bacterial populations in foods undergoing spoilage is not clear based on available information, although conditions appear suitable for such interactions considering phage and total bacterial levels of 6×10^6 PFU and 10^9 bacteria per gram, respectively, reported in some refrigerated foods undergoing spoilage (145). Based upon current information, the potential of phage for controlling the development of spoilage bacteria and/or extending the shelf life of certain refrigerated foods appears limited owing to the wide spectrum of spoilage bacteria involved and the proliferation of phage-resistant bacteria. However, further investigation into applications of phage in this context would be of great interest to the food microbiologist.

Bacteriophage may also influence bacteria with regard to food safety and quality in a more indirect manner, for example, lysogenization or transduction. Lysogenic bacteria may also serve as sources of phage contamination in foods. Lysogenization has been reported in strains of lactic streptococci used as starter cultures in cheesemaking (77), in strains of *P. putrefaciens* isolated from spoiled fish (4), and in virtually all strains of *Staphylococcus aureus,* including those isolated from poultry (43,44,116). In laboratory cultures, lysogeny of specific phage types of *S. aureus* isolated from poultry skin conferred a competitive advantage to that strain of *S. aureus* over other *S. aureus* isolates and resulted in its predominance in a mixed culture (43). Lysogenic conversion of bacteria of food safety concern is exemplified by the conferment of toxigenicity in types C and D *Clostridium botulinum* by temperate phage (33,119) and by changes in the O-antigens of type E *Salmonella* by a temperate phage (133).

12.2.4. Bacteriophage–bacteria Interactions in Fermented Foods

The effects and potential risks of bacteriophages infective for bacterial starter cultures in various food fermentation processes are addressed in Chapter 9 of this book as well as in various reviews (77,90,103). Therefore, this section presents only an overview of the impact of bacteriophages in the fermented food industry. In general, the effects of phage on a given fermentation process are influenced by a number of factors including the type of fermentation, the type of phage, the level of phage contamination, time of infection, and the multiplicity of phage infection, as well as by nutritional, physical, and chemical conditions in the fermenting product (90). The serious problems caused by bacteriophages infecting starter cultures used in the production of cheese and other fermented dairy products have been recognized and extensively investigated for some time (77,143). These problems are exacerbated by the relative thermostability of many bacteriophages associated with dairy starter cultures and their resulting resistance to pasteurization procedures as well as conditions favorable for phage contamination and proliferation during the fermentation process (22,73,147,150). Among bacteriophages infecting lactic acid bacteria, phage infective for strains of *Streptococcus* and *Lactobacillus* have been isolated and characterized (77,86,132,141). There is no indication that phage infection of starter cultures poses a problem in fermented meat products as with dairy products. However, phage B2, infective for *L. plantarum* starter culture, has been reported to delay growth and lactic acid production in a fermented sausage product under laboratory conditions (86).

12.3. STABILITY OF BACTERIOPHAGES IN FOODS

Bacteriophages and other microorganisms in foods are subject to inactivation by a number of physicochemical factors, for example, temperature extremes, pH, osmotic pressure, water activity, and various inorganic and organic compounds added to or naturally occurring in foods. In addition, the rates of inactivation are influenced by interactions between these factors and other food components such as fats, sugars, and salts. Certain bacteria and bacteriophages are also capable of proliferation in various foods under suitable conditions as previously discussed. The relative stability of bacterio-phages in various food systems is important when considering the interac-tions between phage and surviving bacterial populations in foods as well as the use of certain phage as indicator organisms for viruses or bacteria of public health concern. Very little information is available concerning the stability of bacteriophages in food environments as a function of processing, food composition, or storage conditions. Studies of viral stability in some food systems as well as basic studies on the relative effects of various physi-cochemical factors on bacteriophages and viruses can provide some per-spective. However, it is often difficult to predict the stability of an organism in foods based upon studies of related groups or species because of the variation in stability reported with even closely related organisms and the complexity of the food environment. Because of the paucity of information concerning the physicochemical stability of bacteriophages in food systems, relevant studies of human viruses in this regard are also discussed.

12.3.1. Effects of High Temperature

The thermal inactivation of viruses, bacteriophages, and bacteria generally follows first-order kinetics, although biphasic inactivation plots are often observed with increasing time. The reader is referred to Hiatt (58) for an in-depth discussion of thermal inactivation kinetics of viruses. The rate of inactivation of microorganisms is influenced by a number of factors, includ-ing the nature and number of organisms and various characteristics of the suspending medium such as water activity and pH as well as the presence of solutes, fats, and proteins (1,2,4,5,38,63,64,76,123,127). The thermostability of viruses and bacteriophages generally increases with increasing salt concentrations and/or decreasing water activity but is salt specific (1,12,31,76,137,139). Proteinaceous and/or more complex media such as nu-trient broth enhance the thermal stability of coliphages as compared to water or simple buffers (1,12,76). Bovine serum albumin, nutrient broth, skim milk, and cystine have a similar protective effect for polioviruses exposed to heat (136). Reducing substances such as cystine, cysteine, dithiothreitol, and glutathione have also been reported to increase the thermostability of en-teroviruses (52,96). Various components of food have also been shown to contribute to increased heat resistance of viruses and phage. A bacterio-phage of *Streptococcus cremoris* was more heat resistant in milk than cul-ture broth at 65°C (73). The heat resistance of polioviruses was reported to be enhanced by increasing levels of fat in ground beef (38) and by collagen in phosphate-buffered saline (83). Enteroviruses (63,64), adenoviruses (127), and reoviruses (127) were less heat resistant in milk than in ice cream mix, which contains higher levels of fat, sugar, and milk solids than milk. The heat resistance of phage and viruses is also influenced by the pH of the suspending medium with the effect of pH varying according to the type or strain of virus (22,40,62,100,138,147).

Extreme variation in relative heat sensitivity has been demonstrated in viruses, particularly bacteriophages, and considerable diversity has been noted among strains of the same group or serological type of virus. A comparative summary of thermal resistance reported for various bacteriophages, animal viruses, and representative bacteria or bacterial spores, presented in Table 12.2, illustrates this point. The thermoduric nature of various phage has been demonstrated in broth media and suggests that some phage of various bacterial types are capable of surviving various heating and/or pasteurization processes, depending on initial numbers of phage in the food and the type of food. Some enteroviruses (29,38,63,64,125) and thermoduric bacteria, for example, *Streptococcus faecalis* (21,94,113), *S. faecium* (113), *Micrococcus* sp. (130), *Microbacterium* sp. (130), and *Lactobacillus* sp. (130), have been reported to survive various thermal processes but may not be as heat resistant as certain phage, based upon comparative studies (see Table 12.2).

Very few data are available concerning the heat stability of bacteriophages in food systems. Using coliphage T4 as a model for viral inactivation, DiGirolamo et al. (30) reported high uptake of T4 by crabs in water and subsequent survival rates of 21 and 2.5% in crabs after 5 and 20 min of boiling, respectively. The corresponding internal temperatures of crabs after 5 and 20 min of boiling were 70 and 84°C, respectively. Coliphages have been recovered from cooked meat products, although the concomitant recovery of coliform bacteria indicated that postprocessing contamination did occur (68,70). Thus coliphages recovered from cooked meats could reflect their thermostability and/or postprocessing contamination (68,70).

Thermostability studies of enteroviruses in various foods may be of some interest in relation to analogous stability of some phage because enteroviruses and cubic RNA phage are very similar in size, structure, and morphology. Survival rates of poliovirus in oysters ranged from approximately 7 to 13% after stewing, steaming, frying, or baking for 8, 30, 8, and 20 min, respectively (29). The internal temperatures of oysters following these processes were 75, 94, 100, and 90°C, respectively. The D values at 70°C of poliovirus in ground beef subjected to thermal processing were 2.7 and 4.4. min in the presence of 3 and 27% fat, respectively (38). Sullivan et al. (125) reported percentage survival of 1.4 and 1.7% for poliovirus 1 and coxsackie virus B-2, respectively, in rare, broiled hamburgers heated to an internal temperature of 60°C over 7.9 min and quickly cooled. However, no virus was detected if the hamburgers were allowed to maintain 60°C for 3 min after cooking or if they were broiled to 71 or 76.7°C, corresponding to medium and well done, respectively.

12.3.2. Effects of Low Temperature and Other Physicochemical Factors

The inactivation of microorganisms including viruses and bacteriophages as a result of freezing, storage at subfreezing temperatures, and thawing is influenced by physicochemical factors similar to those affecting heat stability. Substances such as egg white, sucrose, corn syrup, glycerol, and meat extracts protected various food-borne bacteria from freeze inactivation, whereas lowering pH tended to increase inactivation (41). Likewise, T1 and T7 coliphages were not freeze-inactivated when suspended in nutrient broth, casein, peptone, or beef extract but were significantly inactivated in Tris buffer (109). However, the presence of 0.1% egg albumin protected T1 but not T7 from freeze inactivation. Differences in freeze sensitivity between strains of phage and between suspending media were also noted for *P. fragi* phage, where phage ps1 was 70% inactivated by freezing and storage at −20°C for 24 hr in skim milk but phage wy underwent no corresponding inactivation (146). On the other hand, neither *P. fragi* phage was affected by freezing and storage at −20°C for 24 hr in trypticase soy broth or ground beef extract.

Because bacteriophages, animal viruses, and bacteria as well as related strains within these groups may vary greatly in their resistance to freeze inactivation, it is difficult to generalize about the comparative freeze resistance of these groups. Among bacteria, the cocci are generally more freeze resistant than Gram-negative rods, with endospores being virtually unaffected by freezing. For example, survival of fecal streptococci and coliform organisms in chicken gravy at −21°C for 91 days was approximately 100 and 2%, respectively (72).

Several studies suggest that many viruses including bacteriophages can survive frozen and refrigerated storage in foods for extended periods. The percentage survival of coliphage T4 in cooked and uncooked crabs was 35 and 17%, respectively, after 30 days of storage at −20°C, whereas the survival after 5 days at 8°C was 40 and 29% for cooked and uncooked crabs, respectively (28). *S. aureus* phage 80 in oyster plasma, held at 5°C for 5 days, was not detectably inactivated (37). Coliphages have been recovered from 50% of commercially frozen chicken pot pies, although their levels prior to frozen storage were not known (70). The survival rate of poliovirus in frozen oysters held at −17.5°C was 91, 40, and 10% after 14, 42, and 84 days, respectively, with 40% of poliovirus surviving storage at 5°C for 15 days (29). In another study, no reduction of poliovirus viability occurred in six of seven oyster lots after 28 days of storage at 5°C (131). Titers of coxsackie virus A9 in ground beef were reduced by only 16% during 8 days of storage at 4°C (57). Poliovirus 1 and coxsackie viruses B1 and B6, added to various commercially frozen or convenience foods, maintained 5–100% of viability after 4–5 months of storage at −20°C or 2 weeks at 10°C, depending upon the food type (81). Inactivation of polioviruses in various foods stored at 4 or 20°C was affected by the relative acidity of the foods with survival rates at 4°C ranging from approximately 27 to 37% and 0.5 to 2% for nonacid and acidic foods, respectively (55). Cliver et al. (18) also noted that acidity significantly enhanced the inactivation of enteroviruses in low moisture foods held at 5 or 22–24°C, depending upon the protein and salt content of

TABLE 12.2 Comparative Heat Resistance of Various Bacteriophages, Viruses, and Food-borne Bacteria

Organism/Strain	Medium[a]	Temperature(s) (°C)	Respective D Value(s)[b]	Ref.
Bacteriophage				
Escherichia coli				
T7	N. broth	60, 65	1, 0.2 min	97
T1	N. broth	65, 75	40, 5 min	97
T5	N. broth	63.4, 69.5	106, 10 min	1
f2	TYE broth	60, 65	47, 10 min	11
M13	H broth	85	30 min	106
Pseudomonas aeruginosa				
12 strains	TS broth	60	6.8–100 min	89
12 strains	TS broth	70	4.6–100 min	89
5 strains	TGE broth	60	83 min–>50 hr	92
P. putida	TGE broth	60	<10 min	92
P. fluorescens (3 strains)	TGE broth	60	<10–15 min	92
P. fragi (2 strains)	TS broth	60	<5–140 min	146
Streptococcus cremoris				
c1	milk	60, 65	14, 1.1 min	73
r1	milk	65	16–64 min[c]	73
S. lactis (2 strains)	milk	62.8	20–46 min	150
Bacillus megaterium (3 strains)	TYE broth	60	5–200 min	40
Viruses				
Poliovirus 1	Tris buffer, 1 *M* MgCl₂	55, 60	60, 2 min	31

Poliovirus				
1 (8 strains)	Medium 199	50	80–234 min	148
2 (6 strains)	Medium 199	50	160–340 min	148
3 (5 strains)	Medium 199	50	87–200 min	148
Poliovirus 1	Ground beef	70, 80	4.4, 1.7 min	38
Adenovirus 12	Ice cream mix	55, 60	2.1, 0.2 min	127
Reovirus 1	Ice cream mix	55, 60	0.8, 0.1 min	127
Bacteria				
E. coli	1% yeast extract	60	0.2 min	104
	Raw milk	60	0.8 min	101
	Whole milk	62.5	0.4 min	21
P. fluorescens	Whole milk	62.5	0.1 min	21
Salmonella enteriditis	Whole milk	62.5	0.3 min	20
S. seftenberg	Whole milk	62.5	2.8 min	20
Streptococcus faecalis	Skim milk	62.8	7.5 min	113
	Whole milk	62.5	3.3 min	21
	Chicken a la king	65.5	1.9 min	94
S. faecium (S. durans)	Skim milk	62.8	36 min	113
Bacterial spores				
Clostridium botulinum	Canned foods	100	1–12.5 hr	123
Bacillus stearother-mophilus	Canned foods	100	125–317 hr	123

[a] Abbreviations of common media are as follows: N = nutrient, TYE = tryptone yeast extract, TS = trypticase soy, TGE = tryptone glucose yeast extract. See references for further explanation of media.

[b] D value is the time required for 90% inactivation at a given temperature. For comparative purposes, some D values were estimated from data presented in the reference using calculations of Stumbo (123).

[c] D values represent two distinct portions of a diphasic inactivation curve.

the food. It should be noted that bacteriophages are generally less resistant to extremes in acidity or alkalinity than are enteroviruses (105).

The stability of enteroviruses in various foods held at room temperature has also been investigated and generally found to be less than the corresponding stability at refrigeration or subfreezing temperatures (18,55,57,81). The influence of spoilage bacteria or their metabolites on the inactivation of enteroviruses in foods has also been investigated (17,18,55,57). Although proteolytic enzymes derived from bacteria in growing cultures have been reported to inactivate enteroviruses (17), the presence of high levels of spoilage bacteria does not seem to affect enteroviruses in foods underoing spoilage at refrigeration or room temperatures significantly (18,55,57). The inactivation of phage or viruses by various food additives has not been extensively investigated, although poliovirus was apparently inactivated by sodium bisulfite added to commercially prepared cole slaw (81).

Little information is available concerning the stability of viruses and phage with regard to other food processes such as fermentation, freeze-drying, or irradiation. Approximately 10% of *L. plantarum* phage (86) and 14% of coxsackie viruses (57) survived 12 and 1 days, respectively, in fermenting meat sausage products. Freeze-drying or freeze dehydration of various foods reduced the titer of polioviruses by 3–4 log cycles in one study (55) and 2 log cycles in another (18). However, it was noted that polioviruses were quite stable in the freeze-dried foods for 15 weeks at 5°C following an initial reduction of 2 log cycles during the freeze-drying process (18). With regard to preservation of foods by irradiation, viruses are generally much more resistant to gamma radiation than bacteria or endospores, with the possible exception of *Micrococcus radiodurans* (102). The use of radappertization, a radiation process designed to inactivate spores of *C. botulinum* by 12 log cycles, would result in survival of viruses and resistant bacteria in meat products (102,124,126). However, the inactivation of these organisms is thought to be accomplished by a pre-irradiation heat process used to inactivate autolytic enzymes (102).

12.3.3. Phage Resistance to Disinfection

Viruses and phage are generally more resistant to inactivation by various environmental stresses (7,36,56,60,65) and disinfection (13,15,36,75,79,107) than are vegetative bacteria. A wide range of sensitivity to chlorine action has been reported among different types of coliphages (75,107,112) and viruses (34,79). Some studies indicate that coliphages f2 and MS-2 are more resistant to chlorine inactivation than polioviruses and other types of coliphage (75,112). Coxsackie virus A2 was found to be 7–46 times more resistant to chlorine inactivation than *E. coli* (15), and most of the 20 strains of enteroviruses tested by Liu et al. (79) were approximately 10 times more resistant to chlorine inactivation than enteric bacteria. On the other hand, adenovirus type 3 has been found equivalent to *E. coli* in chlorine susceptibility (16,79). Many bacteriophages and viruses are also more resistant than enteric bacteria to inactivation by various environmental factors. Coliphages have been found more resistant than coliform bacteria and/or *E. coli* to inactivation in aerosol emissions of wastewater treatment facilities (36), in raw, primary, and activated-sludge sewage samples (7,60), in river water (56), and in seawater (65), as indicated by decreasing ratios of coliform to coliphage counts during the process or over time. It appears probable that bacteriophages and viral contaminants could be quite persistent in food processing environments subjected to inadequate disinfection and/or cleaning. Additionally, unsanitary conditions suitable for proliferation of bacterial contaminants in food processing or handling areas and equipment could result in replication of phage if susceptible bacteria are among those contaminants.

12.4. BACTERIOPHAGES AS INDICATOR ORGANISMS IN FOODS

Certain bacteriophages, particularly coliphages, have received increased attention in recent years as expedient and inexpensive indicators of fecal pollution, for example, bacterial (45,59,66,74,107,108,142,149) and viral (36,46,75,107,108,118,122,135) pathogens of fecal origin in water and wastewater. Significant correlations ($p < 0.001$) have been reported between levels of coliphages and bacterial indicators such as coliforms and fecal coliforms in various water systems (66, 142) and between levels of coliphages and enteroviruses during water treatment (122). Various studies also indicate that coliphages are superior to coliform bacteria as indicators of viral inactivation or removal during water and wastewater treatment (13,36,46,75,80,122). Phage of *Bacteroides* spp., a predominant intestinal bacteria, have recently been recovered from human and animal feces as well as from water, wastewater, and sediments (128). Levels of *Bacteroides* phage were found to correlate with the degree of fecal pollution in water and sediments. The usefulness of bacteriophages, that is, coliphages, as indicators of fecal pollution in natural waters and shellfish has also been questioned by workers with regard to fluctuations in host specificity and/or lack of appropriate host bacteria (59,111,135), as well as the origin of certain coliphages (95,110,111,135). The reader is referred to recent reviews (107,108,111) and Chapter 8 of this book for a more detailed discussion of phage as indicator organisms in water and wastewater.

As with water, the use of various bacterial indicator organisms such as *E. coli,* fecal coliforms, coliforms, and fecal streptococci has become well established for assessing the sanitary quality of many foods (14,85,117,129). Bacteriophages of *Enterobacter aerogenes, Escherichia coli, Salmonella typhi,* and *S. paratyphi,* as well as coliform bacteria, were recovered from the same samples of raw milk in 1937 by Lipska (78), although the relationship between phage and bacteria was not determined. Coliphages have since been recovered from shellfish (74,135) and various other foods (68,70) and their incidence and numbers compared to those of bacterial indicators (68,70,74) and enteroviruses (135).

12.4.1. Distribution of Coliphages and Bacterial Indicators in Various Foods

Kott and Gloyna (74) reported a high correlation between MPN counts of coliphage and coliforms in oysters from water stations at various distances from a sewage outfall. The ratio of coliforms to coliphage in oysters from water stations was approximately 10–1 over a range of 20–2000 coliphage per 100 g of oyster meat. Coliphages were also recovered in market oysters at levels up to 17/100 g. Vaughn and Metcalf (135) found that coliphages were widely distributed in oyster- and shellfish-growing waters but found no correlation between their incidence and that of enteroviruses in oysters or water samples. Coliphages were also isolated from 80% of retail oyster samples by Kennedy et al. (70). Significant correlations ($p < 0.01$) were noted between *E. coli* C or *E. coli* C-3000 coliphage counts and coliform MPN counts in oysters but not between coliphage counts and *E. coli* or fecal coliform MPN counts. In another study, coliphages of *E. coli* C were recovered from 100% of fresh chicken and pork sausage samples and from 33% of cooked delicatessen meat samples in numbers up to 25,000, 3500, and 540 PFU/100 g in chicken, pork sausage, and cooked meats, respectively (68). Ratios of coliphage to fecal coliforms were quite variable among food samples although a positive correlation was noted between coliphage and fecal coliform counts in fresh meat samples.

A more comprehensive study of the distribution of coliphages and bacterial indicator organisms in a wider variety of foods has recently been conducted (70). A summary of coliphage and bacterial indicator recoveries from this study is presented in Table 12.3. Coliphages of either *E. coli* C or a male strain, *E. coli* C-3000, were recovered from 63% of food samples tested, whereas *E. coli* and fecal coliforms were recovered from 52 and 81% of samples, respectively. Recovery of coliphages in foods using *E. coli* C or *E. coli* C-3000 were similar, and recoveries of coliphage with these host strains were consistently much higher than those with *E. coli* B. Similar observations have been noted for recovery of coliphages in water and wastewater (42,53,60,67). Correlation analysis for 100 food samples indicated that coliphage counts were more strongly correlated with *E. coli* and fecal coliform counts than with coliform or total bacterial counts(70). Although the incidence and numbers of coliphages and *E. coli* or fecal coliforms in many foods were similar, a quantitative relationship between levels of coliphage and bacterial indicators was not found for most food samples (70). Among individual types of foods, significant correlations ($p < 0.01$) between coliphages and all bacterial indicators were noted only for pork sausage samples. Significant correlations ($p < 0.01$) were noted between coliphages and

TABLE 12.3 Distribution of Coliphages and Bacterial Indicator Organisms in Various Foods[a,b]

Food product	Coliphage Recovery ($\log_{10}$ PFU/100 g) on host			Bacterial Indicators ($\log_{10}$MPN/100 g)		
	C	*E. coli* C-3000	B	*E. coli*	Fecal coliforms	Total coliforms
Raw chicken						
Range	2.7–4.0	1.8–4.4	ND[c]–2.5	2.5–4.6	3.4–4.6	4.6–6.4
% incidence	100	100	100	100	100	100
Pork sausage						
Range	1.0–4.8	1.3–4.4	ND–3.4	ND–5.0	1.6–5.0	1.6–5.0
% incidence	100	100	90	90	100	100
Ground beef						
Range	1.0–3.1	ND–2.5	ND–1.3	2.4–5.4	2.4–5.4	5.0–6.4
% incidence	100	80	10	100	100	100
Raw oysters						
Range	ND–4.9	ND–4.9	ND–4.7	ND–2.9	ND–2.9	2.6–6.4
% incidence	80	70	20	70	90	100
Roast turkey[d]						
Range	ND–1.6	ND–2.9	ND	ND–4.9	ND–4.0	2.4–5.0
% incidence	30	30	ND	30	90	100
Luncheon meat[d]						
Range	ND–1.9	ND–1.5	ND	ND–3.4	ND–3.4	1.6–3.4
% incidence	10	10	ND	20	70	100
Chicken pot pie						
Range	ND–1.8	ND–2.7	ND	ND–2.0	ND–2.9	ND–4.2
% incidence	40	50	ND	20	80	90
Biscuit dough						
Range	ND	ND–2.6	ND	ND–3.6	ND–3.6	ND–3.6
% incidence	ND	30	ND	50	80	90
Fresh lettuce						
Range	ND–2.0	ND–3.8	ND	ND–2.4	ND–5.0	3.6–6.4
% incidence	20	30	ND	30	40	100
Fresh mushrooms						
Range	ND–2.3	ND–2.4	ND–1.0	ND–2.4	ND–2.6	5.2–6.4
% incidence	80	40	10	10	60	100

Source: Adapted from Kennedy et al. (70).

[a] Ten samples of each food type were analyzed.

[b] Approximate sensitivity levels for coliphage and indicator analyses were 10 PFU/100 g and 30 MPN/100 g, respectively.

[c] ND = none detected at sensitivity level of analysis.

[d] Roast turkey and luncheon meat (pickle and pimento loaf) obtained at delicatessens.

E. coli for refrigerated biscuits and between coliphages and coliforms for oysters, fresh chicken, and frozen pot pie. The relationship between coliphages and bacterial indicators such as fecal coliforms and coliforms may be difficult to assess in many refrigerated foods because numbers of these bacteria may reflect initial contamination and/or proliferation of psychrotrophic species of these groups depending upon the duration of storage (87,88). Other factors influencing the observed relationships between coliphages and indicator bacteria in foods include the uneven distribution of microorganisms in most foods and food processing environments, the failure to recover all coliphages with the host strain(s) used, the failure to recover all *E. coli* using the MPN procedure, and the inherent lack of precision associated with the MPN procedure.

12.4.2. Sanitary Significance of Coliphages in Foods

The recovery of coliphages in numbers and frequency similar to that of *E. coli* in a wide variety of foods (70) as well as the ecological relationship between *E. coli* and coliphages suggest consideration of coliphages as indicators of enteric organisms in many foods. The sanitary significance of indicator organisms such as *E. coli* or coliforms in foods is generally unique to that food and the degree of processing or preparation it has received. For example, the presence of some level of *E. coli* or coliforms in fresh meats or vegetables may be an unavoidable and acceptable component of the initial microbial contamination whereas their presence in heat-processed or pasteurized foods would not be expected or acceptable. Owing to the uneven distribution of enteric pathogens in foods, bacterial indicators have not been shown to be quantitative or precise indicators of food safety with regard to specific pathogens (84,121). However, their presence in inordinate numbers may indicate a lack of hygiene in handling of raw materials and in distribution and in storage of foods, as well as inadequate processing in cooked foods (14,85). General criteria to be considered for suitability of an organism as an indicator of fecal contamination of foods include the following: (*a*) they must be easily quantifiable, (*b*) they must originate in the feces of humans or animals, (*c*) they must occur in high enough numbers in feces or fecally contaminated foods to be easily detected and (*d*) they must possess similar or greater resistance to the extraenteral environment, for example, food and food processing environments, as enteric pathogens.

Methodology for expedient and efficient recovery of coliphages from various foods has recently been developed with recovery rates of T2 and MS-2 ranging from 48 to 68% and 58 to 100%, respectively, depending upon the food type (69). Results of coliphage analysis are obtained within 16–18 hr (69) as compared to 8–10 days required for confirmed *E. coli* or 2–4 days required for fecal coliform analyses of foods by standard AOAC procedures (134). Because no universal plating host for coliphage has been found, the number of coliphages enumerated in a sample depends upon the host strain of *E. coli* used. However, the enumeration of coliphages in foods or the environment can be optimized with regard to maximal recovery of coliphages by using efficient elution and/or recovery procedures as well as appropriate broad-range hosts and plate-assay techniques (53,60,67,69,111).

Both *E. coli* (120) and coliphages (26,93) occur in varying concentrations in the feces of many warm-blooded animals including humans, cows, pigs, poultry, and horses. Both *E. coli* and coliphages are generally present wherever fecal pollution occurs, and are capable of persistence and proliferation in the extraenteral environment depending upon suitable nutritional, physical, and chemical factors (3,14,95,99,107,110,135). Thus the age of fecal pollution may be difficult to establish in environments capable of supporting the growth of *E. coli* and/or coliphages. Replication of coliphages in food processing areas or perishable foods during storage implies the concomitant presence and growth of appropriate *E. coli* hosts and, subsequently, undesirable or unsanitary conditions. Coliphages have recently been characterized according to temperature of infectivity as high-temperature (HT), middle-temperature (MT), or low-temperature (LT) phage corresponding to plaque production between the temperatures of 30–45°, 15–42°, and 15–30°C, respectively (110). Seeley and Primrose (110) found that the feces of humans and animals contained only HT and MT phage, whereas sewage contained primarily MT phage. Ratios of HT and MT to LT phage in river water decreased with distance from sewage outfalls. Similar findings were reported by Parry et al. (95). When coliphages isolated from foods were characterized according to temperature of infectivity (71), ratios of HT and MT to LT phage in most food types resembled temperature profiles reported for phage isolated from feces or sewage (110). A large proportion of LT phage was noted only for oyster samples (71), as might be expected in relation to their source. The possibility of some coliphages having alternate bacterial hosts other than *E. coli* which may be less specifically associated with the intestinal environment does exist (146) and would diminish the value of coliphages as fecal indicators in food or water. However, recent studies suggest that the incidence of somatic, polyvalent coliphages in water and sewage is relatively low (54,99,149). Although male-specific coliphages may have alternate bacterial hosts carrying an F plasmid, the F pili required for adsorption of male-specific coliphages are not produced by bacteria below 30°C (99,111). Thus replication of these phage may be generally limited to the intestinal environment (99-111).

As previously discussed, coliphages and viruses are generally more resistant to inactivation by temperature extremes and chlorination than are enteric indicator or pathogenic bacteria. As a result, coliphages may be better indicators of food environment sanitation or of fecal contamination in frozen foods than are *E. coli*. On the other hand, certain heat-resistant coliphages appear capable of withstanding many thermal processes designed to destroy enteric bacteria in various foods, and as a result, their presence in low numbers may have little value as an indicator of processing adequacy or postprocessing contamination with regard to enteric bacteria. For example, a survival rate of 2.5% has been reported for coliphage T4 in crabs boiled for 20 min (30). However, coliphages were generally not recovered in the absence of coliforms in cooked meat samples despite their consistent recovery in fresh meats (70). As with heat-resistant enterococci (14), the presence of inordinately high numbers of coliphages in heat-processed foods nevertheless could indicate highly contaminated raw materials, inadequate heat processing, postprocessing contamination, or improper storage temperatures. Cliver and Salo (19) reviewed the potential of various organisms including coliphages as indicators of viral contamination in processed foods. These authors concluded that the heat stability of coliphages as a group was too heterogeneous and that too few data were available concerning the heat inactivation of coliphages in foods for their consideration as valid indicators of viruses in processed foods. Noting the similarities between some coliphages, that is, male-specific, RNA phages such as f2 and MS-2, and enteroviruses with regard to size, structure, and resistance to environmental stress, Metcalf (82) suggested their use as indicators of enteroviruses in shellfish. Additionally, coliphage f2 has been proposed as an indicator organism for achieving acceptable heat inactivation of all pathogens including viruses, bacteria, and parasites during composting based upon comparative thermostability data (11). Comparative studies of the heat stability of specific coliphages and enteroviruses in various foods are indicated to demonstrate the feasibility of using coliphage models for evaluating viral inactivation in thermally processed foods.

12.5. SUMMARY

Foods can harbor an often complex microbial ecosystem that uniquely reflects their source, composition, handling, processing, and storage conditions. Bacteriophages of numerous bacterial strains comprising the microflora of various foods have been isolated and must be considered concomitant components of the contaminating and perhaps proliferating microflora. The recovery of indigenous psychrotrophic phage–host systems from refrigerated foods illustrates this point. Although the influence of indigenous phages on susceptible bacterial populations in foods is not clear, the high levels of psychrotrophic phage found in some refrigerated foods suggest the potential for selective phage–host activity in bacterial populations developing in such foods. Phage of spoilage bacteria isolated from foods have been useful in developing phage-typing schemes that could provide a rapid means of identifying spoilage bacteria, detecting sources of contamination of specific strains of bacteria, and monitoring strain distribution during food spoilage or processing. In addition, the use of pools of bacteriophages to inhibit the development of spoilage bacteria and extend the shelf life of certain foods has been suggested, based upon laboratory studies of phage isolated from refrigerated foods. More research is required to substantiate this potential application of phage–host interactions to actual food systems.

Bacteriophages and enteroviruses appear to be generally more resistant to various physical and chemical stresses to which foods can be subjected than are most vegetative bacteria. The reported thermostability of some phage also exceeds that of enteroviruses. The relative thermoresistance of a coliphage and various enteroviruses in some heat-processed foods has been demonstrated. Likewise, both coliphage and enteroviruses have been reported to survive freezing and extended periods of subfreezing and refrigerated storage in many foods.

The potential of coliphages as indicators of fecal contamination and viral inactivation in water and wastewater has been demonstrated in numerous studies. Coliphages have recently received attention as potential indicators of fecal contamination and/or sanitary quality of foods. Coliphages have been recovered from a wide variety of foods using expedient techniques. The distribution of coliphages in foods is similar to that of *E. coli* or fecal coliforms, although a quantitative relationship between coliphage and bacterial indicator levels has not been observed for most food types. The reported correlations between coliphage and bacterial indicator distributions in a variety of foods as well as the ecological specificity of coliphages recovered from foods indicate that coliphages do have potential as fecal indicators in certain foods.

REFERENCES

1. Adams, M. H. 1949. The stability of bacterial viruses in solutions of salts. J. Gen. Physiol. *32*:579–594.

2. Adams, M. H. 1959. Bacteriophages. Wiley-Interscience, New York, N.Y.

3. Anderson, E. S. 1957. The relationship of bacteriophages to bacterial ecology, pp. 189–217. *In* R. E. D. Williams and C. C. Spicer (eds.), Microbial Ecology: 7th Symposium of the Society for General Microbiology. University Press, Cambridge.

4. Anellis, A., J. Lubas, and M. M. Rayman. 1954. Heat resistance in liquid eggs of some strains of the genus *Salmonella*. Food Res. *19*:377–395.

5. Baird-Parker, A. C., M. Boothroyd, and E. Jones. 1970. The effect of water activity on the heat resistance of heat sensitive and heat resistant strains of salmonellae. Appl. Bacteriol. *33*:515–522.

6. Barnes, E. M., and C. S. Impey. 1968. Psychrophilic spoilage bacteria of poultry. J. Appl. Bacteriol. *31*:99–107.

7. Bell, R. G. 1976. The limitations of the ratio of fecal coliforms to coliphage as a water pollution index. Water Res. *10*:745–748.

8. Billing, E. 1963. The value of phage sensitivity tests for the identification of phytopathogenic *Pseudomonas* spp. J. Appl. Bacteriol. *26*:193–210.

9. Billing, E. 1970. Further studies on the phage sensitivity and the determination of phytopathogenic *Pseudomonas* spp. J. Appl. Bacteriol. *33*:478–491.

10. Bradley, D. E. 1967. Ultrastructure of bacteriophages and bacteriocins. Bacteriol. Rev. *31*:230–314.

11. Burge, W. D., D. Colacicco, and W. N. Cramer. 1981. Criteria for achieving pathogen destruction during composting. J. Water Pollut. Control Fed. *53*:1683–1690.

12. Burnet, F. M., and M. Mckie. 1930. Balanced salt action as manifested in bacteriophage phenomena. Aust. J. Exp. Biol. Med. Sci. *7*:183–198.

13. Burns, R. W., and O. J. Sproul. 1967. Virucidal effects of chlorine in waste water. J. Water Poll. Control. Fed. *39*:1834–1849.

14. Buttiaux, R., and D. A. A. Mossel. 1961. The significance of various organisms of fecal origin in foods and drinking water. J. Appl. Bacteriol. *24*:353–364.

15. Clark, N. A., and P. W. Kabler. 1954. The inactivation of purified coxsackie virus in water by chlorine. Am. J. Hyg. *59*:119–127.

16. Clark, N. A., R. E. Stevenson, and P. W. Kabler. 1956. The inactivation of purified type 3 adeonovirus in water by chlorine. Am. J. Hyg. *64*:314–319.

17. Cliver, D. O., and J. E. Herrmann. 1972. Proteolytic and microbial inactivation of enteroviruses. Water Res. *6*:797–805.

18. Cliver, D. O., K. D. Kostenbader, Jr., and M. R. Vallenas. 1970. Stability of viruses in low moisture foods. J. Milk Food Technol. *33*:484–491.

19. Cliver, D. O. and R. J. Salo. 1978. Indicators of viruses in foods preserved by heat, pp. 329–354. In G. Berg (ed.), Indicators of Viruses in Water and Food. Ann Arbor Science, Ann Arbor, MI.

20. Dabbah, R., W. A. Moat, and V. M. Edwards. 1971. Heat survivor curves of food-borne bacteria suspended in commercially sterilized whole milk, I. Salmonellae. J. Dairy Sci. *54*:1583–1588.

21. Dabbah, R., W. A. Moats, and V. M. Edwards. 1971. Heat survivor curves of food-borne bacteria suspended in commercially sterilized whole milk, II. Bacteria other than salmonellae. J. Dairy Sci. *54*:1172–1179.

22. Daoust, D. R., H. M. El-Bisi, and W. Litsky. 1965. Thermal destruction kinetics of a lactic streptococcal bacteriophage. Appl. Microbiol. *13*:478–485.

23. Delisle, A. L., and R. E. Levin. 1969. Bacteriophages of psychrophilic pseudomonads. I. Host range of phage pools active against fish spoilage and fish-pathogenic pseudomonads. Antonie van Leeuwenhoek *J. Microbiol. 35*:307–317.

24. Delisle, A. L., and R. E. Levin. 1969. Bacteriophages of psychrophilic pseudomonads. II. Host range of phage active against *Pseudomonas putrefaciens*. Antonie van Leeuwenhoek *J. Microbiol. 35*:318–324.

25. Denis, F. A. 1975. Contamination of shellfish with strains of *Pseudomonas aeruginosa* and specific bacteriophages. Can. J. Microbiol. *21*:1055–1057.

26. Dhillon, T. S., E. K. S. Dhillon, H. C. Chau, W. K. Li, and A. H. C. Tsang. 1976. Studies on bacteriophage distribution: Virulent and temperate bacteriophage content of mammalian feces. Appl. Environ. Microbiol. *32*:68–74.

27. Dias, F. F., and J. V. Bhat. 1965. Microbial ecology of activated sludge II. Bacteriophages, *Bdellovibrio,* coliforms, and other organisms. Appl. Microbiol. *13*:257–261.

28. DiGirolamo, R., and M. Daley. 1973. Recovery of bacteriophage from contaminated chilled and frozen samples of edible west coast crabs. Appl. Microbiol. *25*:1020–1022.

29. DiGirolamo, R., J. Liston, and J. R. Matches. 1970. Survival of virus in chilled, frozen and processed oysters. Appl. Microbiol. *20*:58–63.

30. DiGirolamo, R., L. Wiczynski, M. Daley, F. Miranda, and C. Viehweger. 1972. Uptake of bacteriophage and their subsequent survival in edible west coast crabs after processing. Appl. Microbiol. *23*:1073–1076.

31. Dimmock, N. J. 1967. Differences between the thermal inactivation of picornaviruses at "high" and "low" temperatures. Virology *31*:338–353.

32. Egan, A. F., and F. H. Grau. 1981. Environmental conditions and the role of *Brochothrix thermosphacta* in the spoilage of fresh and processed meat, pp. 211–221. In T. A. Roberts, G. Hobbs, J. H. B. Christian, and N. Skovgaard (eds.), Psychrotrophic Microorganisms in Spoilage and Pathogenicity. Academic, New York.

33. Eklund, M. W., and F. T. Poysky. 1973. Bacteriophages and toxigenicity of *Clostridium botulinum*, pp. 31–39. In B. C. Hobbs and J. H. B. Christian (eds.), The Microbiological Safety of Food. Academic, New York.

34. Englebrecht, A. S., M. J. Weber, B. L. Salter, and C. A. Schmidt. 1980. Comparative inactivation of viruses by chlorine. Appl. Environ. Microbiol. *40*:249–256.

35. Ewert, D., and M. J. B. Paynter. 1980. Enumeration of bacteriophage and host bacteria in sewage and activated sludge treatment process. Appl. Environ. Microbiol. *39*:576–583.

36. Fannin, K. F., J. J. Gannon, K. W. Cochran, and J. C. Spendlove. 1977. Field studies on coliphages and coliforms as indicators of airborne animal viral contamination from wastewater treatment facilities. Water Res. *11*:181–188.

37. Feng, J. S. 1966. The fate of a virus, *Staphylococcus aureus* phage 80, injected into the oyster, *Crassostrea virginica*. J. Invertr. Path. *8*:496–504.

38. Filppi, J. A., and G. J. Banwart. 1974. Effect of the fat content of ground beef on the heat inactivation of poliovirus. J. Food Sci. *39*:865–868.

39. Florian, M. L. E., and P. C. Trussell. 1957. Bacterial spoilage of shell eggs. IV. Identification of spoilage organisms. Food Technol. *11*:56–60.

40. Friedman, M., and P. B. Cowles. 1953. The bacteriophages of *Bacillus megaterium*. I. Serological, physical, and biological properties. J. Bacteriol. *66*:379–385.

41. Georgala, D. L., and A. Hurst. 1963. The survival of food poisoning bacteria in frozen foods. J. Appl. Bacteriol. *26*:346–358.

42. Gerba, C. D., C. H. Stagg, and M. G. Abadie. 1978. Characterization of sewage associated viruses and behavior in natural waters. Water Res. *12*:805–812.

43. Gibbs, P. A., J. T. Patterson, and J. Harvey. 1980. Interactive growth of *Staphylococcus aureus* strains with a poultry skin microflora in a diffusion apparatus. J. Appl. Bacteriol. *48*:191–205.

44. Gibbs, P. A., J. T. Patterson, and J. K. Thompson. 1978. Characterization of poultry isolates of *Staphylococcus aureus* by a new set of poultry phages. J. Appl. Bacteriol. *44*:387–400.

45. Giraldi, V., and P. Donati. 1976. Determination of bacteriophages in water as an index of bacterial contamination. Rivista della Societa Italiana de Scienza dell'Alimentazione *5*:349.

46. Grabow, W. O. K., B. W. Bateman, and J. S. Burger. 1978. Microbiological quality indicator for routine monitoring of wastewater reclamation systems. Prog. Water Technol. *10*:317–327.

47. Greer, G. G. 1981. Rapid detection of psychrotrophic bacteria in relation to retail beef quality. J. Food Sci. *46*:1669–1672.

48. Greer, G. G. 1982. Psychrotrophic bacteriophages for beef spoilage pseudomonads. J. Food Prot. *45*:1318–1325.

49. Greer, G. G. 1983. Psychrotrophic *Brochothrix thermosphacta* bacteriophages isolated from beef. Appl. Environ. Microbiol. *46*:245–251.

50. Greer, G. G. 1984. Psychrotrophic bacteriophages for beef spoilage bacteria. J. Food Prot. *47*:822.

51. Greer, G. G. 1986. Homologous bacteriophage control of *Pseudomonas* growth and beef spoilage. J. Food Prot. *49*:104–109.

52. Halsted, C. C., D. S. Y. Seto, J. Simkins, and D. H. Carver. 1970. Protection of en-

teroviruses against heat inactivation by sulfhydryl-reducing substances. Virology *40*:751–754.

53. Havelaar, A. H., and W. M. Hogeboom. 1983. Factors affecting the enumeration of coliphages in sewage and sewage-polluted waters. Antonie van Leeuwenhoek *49*:387–397.

54. Havelaar, A. H., and W. M. Hogeboom. 1984. A method for enumeration of male-specific bacteriophages in sewage. J. Appl. Bacteriol. *56*:439–447.

55. Heidelbaugh, N. D., and D. J. Giron. 1969. Effect of processing on recovery of poliovirus from inoculated foods. J. Food Sci. *34*:239–241.

56. Hejkal, T. W. 1985. Effect of environmental factors on ratios of coliphages to indicator bacteria. Abstr. Ann. Meet. Am. Soc. Microbiol., p. 225.

57. Herrmann, J. E., and D. O. Cliver. 1973. Enterovirus persistance in sausage and ground beef. J. Milk Food Technol. *36*:426–428.

58. Hiatt, C. W. 1964. Kinetics of the inactivation of viruses. Bacteriol. Rev. *28*:150–163.

59. Hilton, M. C., and G. Stotzky. 1973. Use of coliphages as indicators of water pollution. Can. J. Microbiol. *19*:747–751.

60. Ignazzitto, G., L. Volterra, F. A. Aulicina, and A. M. D'Angelo. 1980. Coliphages as indicators in a treatment plant. Water Air Soil Pollut. *13*:391–398.

61. Ingram, M., and Dainty, R. H. 1971. Changes caused by microbes in spoilage of meat. J. Appl. Bacteriol. *34*:21–39.

62. Janda, A., and V. Vonka. 1963. Thermoresistance testing for identification of LSc 2ab virus progeny. Arch. Ges. Virusforsch. *14*:227–237.

63. Kaplan, A. S., and J. L. Melnick. 1952. Effect of milk and cream on the thermal inactivation of human poliomyelitis virus. Am. J. Pub. Health *42*:525–534.

64. Kaplan, A. S., and J. L. Melnick. 1954. Effect of milk and other dairy products on the thermal inactivation of coxsackie viruses. Am. J. Pub. Health *44*:1174–1184.

65. Kapuscinski, R. B., and R. Mitchell. 1982. Sunlight induced mortality of viruses and *Escherichia coli* in coastal seawater. Environ. Sci. Technol. *17*:1–6.

66. Kenard, R. P., and R. S. Valentine. 1974. Rapid determination of the presence of enteric bacteria in water. Appl. Microbiol. *27*:484–487.

67. Kennedy, J. E., Jr., G. Bitton, and J. L. Oblinger. 1985. Comparison of selective media for assay of coliphages in sewage effluent and lake water. Appl. Environ. Microbiol. *49*:33–36.

68. Kennedy, J. E., Jr., J. L. Oblinger, and G. Bitton. 1984. Recovery of coliphages from chicken, pork sausage and delicatessen meats. J. Food Prot. *47*:623–626.

69. Kennedy, J. E., Jr., C. I. Wei, and J. L. Oblinger. 1986. Methodology for enumeration of coliphages in foods. Appl. Environ. Microbiol. *51*:956–962.

70. Kennedy, J. E., Jr., C. I. Wei, and J. L. Oblinger. 1986. Distribution of coliphages in various foods. J. Food Prot. *49*:944–951.

71. Kennedy, J. E., Jr., C. I. Wei, and J. L. Oblinger. 1986. Characterization of coliphages recovered from foods according to temperature of infectivity. J. Food Prot. *49*:952–954.

72. Kereluk, K., and M. F. Gunderson. 1959. Studies on the bacteriological quality of frozen meats. IV. Longevity studies on the coliform bacteria and enterococci at low temperatures. Appl. Microbiol. *7*:327–328.

73. Koka, M., and E. M. Mikolajcik. 1967. Kinetics of thermal destruction of bacteriophages active gainst *Streptococcus cremoris*. J. Dairy Sci. *50*:1025–1031.

74. Kott, Y., and E. F. Gloyna. 1965. Correlating coliform bacteria with *E. coli* bacteriophages in shellfish. Water Sewage Works *112*:424–426.

75. Kott, Y., N. Rose, S. Sperber, and N. Betzer. 1974. Bacteriophages as viral pollution indicators. Water Res. *8*:165–171.

76. Lark, K. G., and M. H. Adams. 1953. The stability of phages as a function of the ionic environment. Cold Harbor Symp. Quant. Biol. *18*:171–183.

77. Lawrence, R. C. 1978. Action of bacteriophages on lactic acid bacteria: Consequences and protection. N.Z. J. Dairy Sci. Technol. *13*:129–136.

78. Lipska, I. 1937. Les colibacilles et les bacteriophages du lait de consommation en nature a Varsovie. Le Lait. *17*:236–241.

79. Liu, O. C., H. R. Seraichekas, E. W. Akin, D. A. Brashear, E. L. Katz, and W. J. Hill, Jr. 1971. Relative resistance of twenty human enteric viruses to free chlorine in Potomac water, pp. 171–195. In V. Snoeyink (ed.), Virus and Water Quality: Occurrence and Control. Proceedings of the 13th Water Quality Conference, Department of Civil Engineering. University of Illinois, Urbana-Champaign. University of Illinois Bulletin, Vol. 69, No. 1.

80. Lothrop, T. L., and O. J. Sproul. 1969. High-level inactivation of viruses in waste water by chlorination. J. Water Pollut. Control. Fed. *41*:567–575.

81. Lynt, R. K., Jr. 1966. Survival and recovery of enteroviruses from foods. Appl. Microbiol. *14*:218–222.

82. Metcalf, T. G. 1978. Indicators of viruses in shellfish, pp. 383–415. In G. Berg (ed.), Indicators of Viruses in Water and Food. Ann Arbor Science, Ann Arbor, Michigan.

83. Milo, S. E., Jr. 1971. Thermal inactivation of poliovirus in the presence of selective organic molecules (cholesterol, lecithin, collagen, and β-carotene). Appl. Microbiol. *21*:198–202.

84. Miskimin, D. K., K. A. Berkowitz, M. Solberg, W. E. Riha, Jr., W. C. Franke, R. L. Buchanan, and V. O'Leary. 1976. Relationships between indicator organisms and specific pathogens in potentially hazardous foods. J. Food Sci. *41*:1001–1006.

85. Mossel, D. A. A. 1982. Microbiology of foods: The ecological essentials of assurance and assessment of safety and quality, 3rd ed. The University of Utrecht, Utrecht, The Netherlands.

86. Nes, I. F., and O. Sorheim. 1984. Effect of infection of a bacteriophage in a starter culture during the production of salami dry sausage, a model study. J. Food Sci. *49*:337–340.

87. Newton, K. G. 1979. Value of coliform tests for assessing meat quality. J. Appl. Bacteriol. *47*:303–307.

88. Oblinger, J. L., J. E. Kennedy, Jr., and D. M. Langston. 1982. Microflora recovered from foods on violet red bile agar with and without glucose and incubated at different temperatures. J. Food Prot. *45*:948–952.

89. O'Callaghan, R. J., W. O'Mara, and J. B. Grogan. 1969. Physical stability and biological and physicochemical properties of twelve *Pseudomonas aeruginosa* bacteriophages. Virology *37*:642–648.

90. Ogata, S. 1980. Bacteriophage contamination in industrial processes. Biotechnology and Bioengineering *22*(Suppl. 1):177–193.

91. Olsen, R. H. 1967. Isolation and growth of psychrophilic bacteriophage. Appl. Microbiol. *15*:198.

92. Olsen, R. H., E. S. Metcalf, and J. K. Todd. 1968. Characteristics of bacteriophages attacking psychrophilic and mesophilic pseudomonads. J. Virol. *2*:357–364.

93. Osawa, S., K. Furuse, and I. Watanabe. 1981. Distribution of ribonucleic acid coliphages in animals. Appl. Environ. Microbiol. *41*:164–168.

94. Ott, T. M., H. M. El-Bisi, and W. B. Esselen. 1961. Thermal destruction of *Streptococcus faecalis* in prepared frozen foods. J. Food Sci. *26*:1–10.

95. Parry, O. T., J. A. Whitehead, and L. T. Dowling. 1981. Temperature sensitive coliphage in the aquatic environment, pp. 277–279. In M. Goddard and M. Butler (eds.), Viruses and Wastewater Treatment. Pergamon, Oxford.

96. Pohjanpelto, P. 1958. Stabilization of poliovirus by cystine. Virology 6:472–487.

97. Pollard, E., and M. Reaume. 1951. Thermal inactivation of bacterial viruses. Archiv. Biochem. Biophys. 32:278–287.

98. Pretorius, W. A. 1962. Some observations on the role of coliphages in the number of *Escherichia coli* in oxidation ponds. J. Hyg. 60:279–281.

99. Primrose, S. B., N. D. Seeley, K. B. Logan, and J. W. Nicolson. 1982. Methods for studying aquatic bacteriophage ecology. Appl. Environ. Microbiol. 43:694–701.

100. Rafajka, R. R., and J. C. Young. 1964. Thermal and pH stability of adenovirus types 12, 14, and 18. Proc. Soc. Exp. Biol. Med. 116:683–685.

101. Read, R. B., Jr., C. Schwartz, and W. Litsky. 1961. Studies on thermal destruction of *Escherichia coli* in milk and milk products. Appl. Microbiol. 9:415–418.

102. Rowley, D. B., R. Sullivan, and E. S. Josephson. 1978. Indicators of viruses in foods preserved by ionizing radiation, pp. 355–381. In G. Berg (ed.), Indicators of Viruses in Water and Food. Ann Arbor Science, Ann Arbor, Michigan.

103. Rudolf, V. 1978. Bacteriophages in fermentation. Process Biochem. 13:16–26.

104. Russell, A. D., and D. Harries. 1968. Factors influencing the survival and revival of heat-treated *Escherichia coli*. Appl. Microbiol. 16:335–339.

105. Sabatino, C. M., and S. Maier. 1980. Differential inactivation of three bacteriophages by acid and alkaline pH used in the membrane adsorption-elution method of virus recovery. Can. J. Microbiol. 26:1403–1407.

106. Salivar, W. O., H. Tzagoloff, and D. Pratt. 1964. Some physical-chemical and biological properties of the rod-shaped coliphage M13. Virology 24:359–371.

107. Scarpino, P. V. 1975. Human enteric viruses and bacteriophages as indicators of sewage pollution, pp. 49–61. In A. L. H. Gamegon (ed.), Discharge of Sewage from Sea Outfalls. Pergamon, Oxford.

108. Scarpino, P. V. 1978. Bacteriophage indicators, pp. 201–227. In G. Berg (ed.), Indicators of Viruses in Water and Food. Ann Arbor Science, Ann Arbor, Michigan.

109. Schiffenbauer, M., and M. Calderon. 1985. Inactivation of coliphage T1 and T7 by subrefrigeration temperature. Abstr. Ann. Meet. Am. Soc. Microbiol., p. 268.

110. Seeley, N. D., and S. B. Primrose. 1980. The effect of temperature on the ecology of aquatic bacteriophages. J. Gen. Virol. 46:87–95.

111. Seeley, N. D., and S. B. Primrose. 1982. The isolation of bacteriophages from the environment. J. Appl. Bacteriol. 53:1–17.

112. Shah, P. C., and J. McCamish. 1972. Relative resistance of poliovirus I and coliphages f2 and T2 in water. Appl. Microbiol. 24:658–659.

113. Shannon, E. L., G. W. Reinbold, and W. S. Clark, Jr. 1970. Heat resistance of enterococci. J. Milk Food Technol. 33:192–196.

114. Shaw, B. G., and J. M. Shannon. 1968. Psychrophilic spoilage bacteria of fish. J. Appl. Bacteriol. 31:89–96.

115. Shewan, J. M., G. Hobbs, and W. Hodgkiss. 1960. A determination scheme for the identification of certain genera of gram-negative bacteria, with special reference to the *Pseudomonadaceae*. J. Appl. Bacteriol. 23:379–390.

116. Shimizu, A. 1977. Isolation and characterization of bacteriophages from staphylococci of chicken origin. Am J. Vet. Res. 38:1389–1392.

117. Silliker, J. H. 1982. Selecting methodology to meet industry's microbiological goals for the 1980's. Food Technol. 35:65–70.

118. Simkova, A., and J. Cervenka. 1981. Coliphages as ecological indicators of enteroviruses in various water systems. Bull. World Hlth. Org. 59:611–618.

119. Smith, L. D. S. 1977. Botulism, the Organism, its Toxins, the Disease. Charles C Thomas, Springfield, IL.

120. Smith, W. H., and W. E. Crabb. 1961. The faecal bacterial flora of animals and man: its development in the young. J. Path. Bacteriol. *82*:53–66.

121. Solberg, M., D. K. Miskimin, B. A. Martin, G. Page, S. Goldner, and M. Libfeld. 1977. Indicator organisms, foodborne pathogens and food safety. Assoc. Food Drug Off. Quart. Bull. *41*(1):9–21.

122. Stetler, R. E. 1984. Coliphages as indicators of enteroviruses. Appl. Environ. Microbiol. *48*:668–670.

123. Stumbo, C. R. 1973. Thermobacteriology in Food Processing, 2nd ed. Academic, New York.

124. Sullivan, R., A. C. Fassolitis, E. P. Larkin, R. B. Reed, Jr., and J. T. Peeler. 1971. Inactivation of thirty viruses by gamma radiation. Appl. Microbiol. *22*:61–65.

125. Sullivan, R., R. M. Marnell, E. P. Larkin, and R. B. Reed, Jr. 1975. Inactivation of poliovirus I and coxsackievirus B-2 in broiled hamburgers. J. Milk Food Technol. *38*:473–475.

126. Sullivan, R., P. V. Scarpino, A. C. Fassolitis, E. P. Larkin, and J. T. Peeler. 1971. Gamma radiation inactivation of coxsackie virus B-2. Appl. Microbiol. *26*:14–17.

127. Sullivan, R., J. T. Tierney, E. P. Larkin, R. B. Read, Jr., and J. T. Peeler. 1971. Thermal resistance of certain oncogenic viruses suspended in milk and milk products. Appl. Microbiol. *22*:315–320.

128. Tartera, C., and J. Jofre. 1985. *Bacteriodes* bacteriophages as potential indicators of human viruses in the environment. Abstr. Ann. Meet. Am. Soc. Microbiol., p. 270.

129. Thatcher, F. S., and D. S. Clark. 1968. Microorganisms in Foods: Their Significance and Methods of Enumeration. University of Toronto Press, Toronto, Canada.

130. Thomas, S. B., R. G. Durce, G. J. Peters, and D. G. Griffith. 1967. Incidence and significance of thermoduric bacteria in farm milk supplies: A reappraisal and review. J. Appl. Bacteriol. *30*:265–298.

131. Tierney, J. T., R. Sullivan, J. T. Peeler, and E. P. Larkin. 1982. Persistence of polioviruses in shellstock and shucked oysters stored at refrigeration temperature. J. Food Prot. *45*:1135–1137.

132. Trevors, K. E., R. A. Holley, and A. G. Kempton. 1984. Effect of bacteriophage on the activity of lactic acid starter cultures used in the production of fermented sausage. J. Food Sci. *49*:650–653.

133. Uetake, H., T. Nakagawa, and T. Akiba. 1955. The relationship of bacteriophage to antigenic changes in Group E salmonellas. J. Bacteriol. *31*:571–579.

134. U.S. Food and Drug Administration. 1978. Bacteriological Analytical Manual for foods, 5th ed. Association of Official Analytical Chemists. Washington, D.C.

135. Vaughn, J. M., and T. G. Metcalf. 1975. Coliphages as indicators of enteric viruses in shellfish and shellfish raising estuarine waters. Water Res. *9*:613–616.

136. Wallis, C., and J. L. Melnick. 1961. Stabilization of poliovirus by cations. Tex. Rep. Biol. Med. *19*:683–700.

137. Wallis, C., and J. L. Melnick. 1962. Cationic stabilization—a new property of enteroviruses. Virology *16*:504–506.

138. Wallis, C., and J. L. Melnick. 1962. Effect of organic and inorganic acids on poliovirus at 50°C. Proc. Soc. Exp. Biol. Med. *111*:305–308.

139. Wallis, C., J. L. Melnick, and F. Rapp. 1965. Different effects of $MgCl_2$ and $MgSO_4$ on the thermostability of viruses. Virology *26*:694–699.

140. Ware, G. C., and M. A. Mellon. 1956. Some observations on the coli/coliphage relationship in sewage. J. Hyg. *54*:99–101.

141. Watanabe, K., S. Takesue, K. Jin-Nai, and T. Yoshikawa. 1970. Bacteriophage active against the lactic acid beverage producing bacterium *Lactobacillus casei*. Appl. Microbiol. *20*:409–415.

142. Wentzel, R. S., P. E. O'Neil, and J. F. Kitchens. 1982. Evaluation of coliphage detection as a rapid indicator of water quality. Appl. Environ. Microbiol. *43*:430–434.

143. Whitehead, H. R., and G. J. E. Hunter. 1939. Bacteriophage–organism relationships in the group of lactic streptococci. J. Dairy Res. *10*:403–409.

144. Whitman, P. A. 1971. Bacteriophages of psychrophilic bacteria associated with refrigerated food products. Ph.D. Dissertation, University of Missouri, Columbia, Missouri.

145. Whitman, P. A., and R. T. Marshall. 1971. Isolation of psychrophilic bacteriophage–host systems from refrigerated food products. Appl. Microbiol. *22*:220–223.

146. Whitman, P. A., and R. T. Marshall. 1971. Characterization of two psychrophilic *Pseudomonas* bacteriophages isolated from ground beef. Appl. Microbiol. *22*:463–468.

147. Wilkowske, H. H., F. E. Nelson, and C. E. Parmelee. 1954. Heat inactivation of bacteriophage strains active against lactic streptococci. Appl. Microbiol. *2*:250–253.

148. Younger, J. S. 1957. Thermal inactivation studies with different strains of poliovirus. J. Immunol. *78*:282–290.

149. Zaiss, U. 1981. Dispersal and fate of coliphages in the River Saar. Zbl. Bakt. Hyg., I. Abt. Org. B. *1974*:160–173.

150. Zottola, E. A., and E. H. Marth. 1966. Thermal inactivation of bacteriophages active against lactic streptococci. J. Dairy Sci. *49*:1338–1342.

INDEX

317